Entomology in the Doon Valley (Garhwal Himalaya)

A stronghold for insect research

(Addressing questions in systematics, ecology, behaviour, conservation and economic importance of the magnificent insect fauna)

Prof. Dr B.K. Tyagi

Visiting Professor, Bharathidasan University, Tiruchirappalli, TN;
Core Faculty, SERB-DST School in Insect Biology, Hyderabad University, Telangana
[Scientist "G" (Director) & Director in-Charge *(Emerited)*
Centre for Research in Medical Entomology (ICMR), Madurai (TN), India]

&

Dr Vijay Veer

Ex- Scientist "G" & Director
Defense Research Laboratory (DRDO)
Tezpur (Assam), India

SCIENTIFIC PUBLISHERS

Published by
SCIENTIFIC PUBLISHERS (INDIA)

Jodhpur –
5 A, New Pali Road
P.O. Box 91
Jodhpur - 342 001 INDIA

Delhi –
4806/24, Ansari Road
Daryaganj
New Delhi - 110 002 INDIA

ISBN: 978-93-86102-75-1

Visit the Scientific Publishers (India) website at
http://www.scientificpub.com

Printed in India

Printed at Replika Press Pvt. Ltd.

CONTENTS

CONTENTS

Foreword

Prof. B. Kiauta,
photographed on the occasion of being
honoured in 2002 by Her Majesty -
the Queen of The Netherlands,
with the highest civilian award,
Knighthood In The Order Oranje-Nassau)

The great mountain ranges and rivers of India are creating some of the most marvelous valleys of the world. Among these, the Doon Valley stands out as one of the most beautiful valleys of the Himalayan foothills. Had the King Jehangir first come to this area, there is hardly any doubt that he would have extoled its beauty by resonating what he said of the elegant Kashmir Valley: '*Agar Firdaws ba roy-i zamin ast, hamin ast-u hamin ast-u hamin ast*', meaning '*If there is a Paradise on Earth, it is this, it is this, it is this*'. The Doon Valley, with its undulating landscapes, scores of rivers and streams, sparkling cascades, torrents, waterfalls, lakes and, above all, the lush greenery and fragrant forests dotted with villages, represents one of the most attractive sets of the Indian ecosystems and harbours an exceptionally rich fauna and flora.

The history of forest and insect explorations in Dehra Dun covers a century, includes the names of dozens of prominent Indian workers and research institutions, among these, for example, the legendary Professor M.S. Mani, and it resulted in a large number of papers and monographs, containing also the descriptions of numerous new species, some of these endemic. Now certainly the time has arrived to present a well-balanced review of the fauna, filling up the gaps and pointing out the lacunae in our knowledge that need further research in the future.

This is done in the present volume, written with much skill and a profound knowledge of the subject. The Doon Valley research experience and the related personal bibliography of the two authors are appreciable, hence they are exceptionally qualified to conclude this project. Dr Brij Kishore Tyagi published a score of papers pertaining to the odonate and termite fauna of the Doon Valley, including a widely acknowledged PhD dissertation on chromosome cytology of over 40 of its odonate species. Likewise, Dr Vijay Veer's thysanopterological work is highly notable.

The book will be certainly much appreciated by professionals and non-professionals alike and it will serve as an indispensable tool to regional nature conservationists for a long time to come.

Dr B. Kiauta, D.Sci.
Professor Emeritus of Invertebrate Cytogenetics and Cytotaxonomy of the University of Utrecht;
Past-Head of the Netherlands Centre for Alpine Biological Research, Utrecht;
Past Executive Editor of *Odonatologica*;
etc., etc.

डॉ० सविता, भा.व.से.
निदेशक व.अ.सं.
एवं
कुलपति व.अ.सं. सम विश्वविद्यालय

Dr. SAVITA, IFS
Director FRI
and
Vice-Chancellor FRI Deemed University

दूरभाष / Phones :
कार्यालय / Offi. : 0135-2755277
0135-2224444
निवास / Resi. : 035-2751679
0135-2224545
फैक्स / FAX : 91-0135-2756865
E-mail : dir_fri@icrfe.org

वन अनुसंधान संस्थान
(भारतीय वानिकी अनुसंधान एवं शिक्षा परिषद्)
(पर्यावरण, वन एवं जलवायु परिवर्तन मंत्रालय, भारत सरकार की एक स्वायत्त परिषद्)
डाकघर न्यू फॉरेस्ट, देहरादून–248006
FOREST RESEARCH INSTITUTE
(Indian Council of Forestry Research and Education)
(An autonomous bocy of Ministry of Enviornment, Forests & Climate Change Govt. of India)
P.O. New Forest, Dehra Dun - 248006

अ०शा०सं०
D.O.No.

दिनांक
Dated. the 15th April, 2016

FOREWORD

The Doon landscape is one of the most affluent ecosystems in the country, full of life with its highly rich fauna and flora, undulating landscapes, beautiful glens and vales, scores of rivers and rivulets, sparkling cascades, magnificent lakes and above all lush green and fragrant forests *inter alia* dotted with villages known worldwide for their hospitality. It is also a home to some of country's best educational, scientific, research and defense institutions.

Due to its valuable assets, Dehradun was considered appropriate location for establishing the country's first ever Forest Research Institute. With the century old history of forestry research, including appreciable quantum of entomological investigations, the Doon Valley really boasts of unparalleled annals of information on a variety of disciplines pertaining to insect/pest ecology, biology, chorogeography and its control. The erstwhile FRI & Colleges, Dehradun now popularly known as Forest Research Institute under the aegis of Indian Council of Forestry Research & Education has to its credit some of the greatest entomologists and their legacies in forest research on timber-insect interaction. Outstanding research explorations on agriculture-pest interrelationship and environment bioremediation using certain insect groups as bioindicators have been carried out in association with equally significant institutions of ZSI and the DAV (PG) College, at Dehradun. Though, huge entomological data on various research activities in different time periods is recorded in assorted documents and treatises, but a comprehensive monograph detailing not only the achievements in different insect orders but also the gaps in research for future has been lacking.

This book entitled, "Entomology in the Doon Valley (Garhwal Himalaya)", authored by Dr. B.K. Tyagi and Dr. Vijay Veer - the Doon Valley's very own researchers, is indeed a welcome addition to the knowledge of the Doonites on this very important subject of entomology. The layout with the aim to provide insect order-wise information; the lucid language and the comprehensible format of the book have made the book a highly readable material not only for the professionals but also for the amateur biologists, nature lovers and those involved in conservation of natural resources.

15/04/2016
(Dr. Savita)

Preface

"It is singularly appropriate that the first comprehensive discussion of the ecology of high altitude insects should come from India, for the most extensive mountainous areas of the world are in and adjacent to India."

– Gordon Alexander, 1962

The Doon Valley in the foothills of Himalaya is an unparalleled conservative natural habitat inhabited by thousands of insect species, many of them endemic to the valley, and representing virtually all the orders. With the establishment of the world famous the then Forest Research Institute & Colleges more than a century ago the research in timber-interacting insect species, mostly as pests, the science of forest entomology got hugely strengthened and diversified. Enormous amount of research was carried out on the insect pest-host plant species interrelationship under the phenomenal guidance of a streak of dedicated foresters and forest entomologists such as E.P. Stebbing, C.F.C. Beeson, A.D. Imms, M.L. Roonwal, R.N. Mathur, P.K. Sen-Sarma, Pratap Singh, just to exemplify. The classics penned by each of Stebbing (*Insect intruders in Indian homes*, 1909), Beeson (*Ecology and control of forest insects in India and neighbouring countries*, 1941) and Imms (*A General Textbook of Entomology*, 1925) remain the most exhibitive cornerstones in the evolution of both the forest and general entomology. Particularly Imms's *magnum opus* attracted entomologists most and it became, as it still is, the premier entomological textbook for next nine decades. The value of this masterpiece was so high that it was revised repeatedly under the wise care of Owain Richards and Richard Gareth Davies who were able to bring out their final edition, the tenth, in 1977. Currently this is revised in a two-volume set, and continues to remain an inevitably indispensable reference work of general entomology both for the seasoned and amateaur entomologists alike. With the arrival of Dr. M.L. Roonwal, as the Forest Entomologist in-Charge of the Forest Entomology Division at the FRI & Colleges, another golden era began through diverse research programmes on the bioecology and control of serious pest species of forest timber. This legacy was discernibly upheld by his successors, e.g., Dr. P.K. Sen-Sarma, an authority on forest entomology especially with reference to the insect order Isoptera (Termites). Dr. Roonwal and Dr. Sen-Sarma made significant contributions to the knowledge on morphology, taxonomy, biology, ecology and control of termites. Besides discovering several new taxa including a new family (Indotermitidae), Dr. Roonwal co-authored two books on termites, viz., *Contributions to the Systematics of*

Oriental Termites and *Fauna of India (Isoptera)*. One of the outstanding contributions of Dr. Roonwal and Sen-Sarma had been the elucidation of evolutionary and taxonomic significance of wing microsculpturing.

Close on the heels of this outstanding Roonwal/Sen-Sarma era of forest entomology, another wonderful school of odonatological and thrips research was slowly emerging in the historical Department of Zoology of the D.A.V. College, Dehradun mentored by Dr. S.K. Sangal and Dr. S.K. Kulshrestha, respective Head of the Department of Zoology between 1980 and 2000. Professor Dr. Bastian Kiauta of Institute of Genetics, Utrecht University, The Netherlands and himself a great odonatologist, had once written to Dr. Sangal about his influence in promoting odonatological research in the majestic Doon Valley as "developing into a stronghold for dragonfly research" – subsequently amply justified by the masterly works of Dr. Arun Kumar and Dr. M. Prasad of Zoological Survey of India, Dehradun and Dr. B.K. Tyagi at the DAV College. Thysanoptera or thrips got shot in arm with the extraordinary research investigations by Dr. Vijay Veer, Dr. Neena Chauhan, Dr. A.K. Shrivastava. Together they have described over a dozen new species of dragonflies and thrips. The Doon Valley earned such a high reputation for its wealth of insect fauna that even the legendary Prof. M.S. Mani, accompanied by Dr. Santokh Singh, a coleopterologist, and Dr. H.N. Baijal, a collembologist, paid it a visit *en route* to his epochial three high alpine entomological expeditions in 1950s and 1960s (Mani, 1968).

So much and more had been entomologically explored on the Doon Valley's fauna, yet it is surprising that there was not a single treatise or monograph which could summarise or at least address to the entire range of insect orders with respect to questions pertaining to systematics, ecology, behaviour, conservation and economic importance of these magnificent creatures of the Valley. The present book is therefore written with a clear view to characterize all the insect orders as prevailing in the Doon Valley, hoping that both the undergraduates and the postgraduates, as well as research scholars, will be benefitted in their quest for knowledge on insects of this grandiose ecosystem, the Doon Valley.

Dr. B.K. Tyagi
Dr. Vijay Veer

18th April, 2016, Dehradun

Dedicated

to our
beloved teachers, guides and mentors

Dr. S.K. Sangal,
Dr. P.K. Sen-Sarma, Dr. T.N. Ananthakrishnan,
Dr. S.K. Kulshrestha and Dr. B. Kiauta

Acknowledgement

In completing this book we have received enormous help from several friends, colleagues and well-wishers. To all of them we duly acknowledge their help and guidance, and offer them our deep sense of gratitude.

As always in past our wives, *Ajita* (Dr. B.K. Tyagi) and *Neena* (Dr. Vijay Veer) have silently endured our long absences for months from extending a helping hand in their daily chores but, at the same time, obliged by their unfailing support for our book to be timely completed in all its details. We remain thankful to them.

Our thanks also go to Shri Pawan Kumar Sharma, Scientific Publishers (India), Jodhpur for encouraging throughout the course of preparation of this book, and ensuring to bring about timely the publication. He has been indeed very courteous, motivative and committed towards this literary achievement.

1

INTRODUCTION TO THE DOON VALLEY

> *"Down to Dehra for the day. In the valley, spring is giving way to summer...the scent of mango blossom fills the air. ....wild patches of lantana and ageratum, a bluish–purple flowering weed that brings colour to hedges and empty plots."*
>
> *– Ruskin Bond, Dec. 4, 2014*

India is a land of great scenic beauty: Lush green forest, beautiful valleys, undulating hills, sparkling mountain streams, cascades and torrents, magnificent lakes and several major rivers and rivulets. The great mountain ranges and rivers of India create few of most beautiful valleys in the world. The Doon or Dehradun Valley, blessed with astonishing glens and vales, landscapes, waterfalls, tributaries, natural forests etc., is one of the most serene valleys in India known for its diverse flora and fauna (Fig. 1).

The beautiful Doon Valley, a historic entity which associated with the times of Lord Rama and the epic battle of Mahabharat, is situated in the western ranges of the Garhwal Himalayan region at an elevation of 3,556 m. The pristine valley lies between the two sacred rivers of India – Ganga and Yamuna. The valley is known for its amazing flora and fauna. It is surrounded by the glittering peaks of mountain ranges, on one side, and the dense forests, on the other. The scintillating beauty of the valley attracts the tourists from all over the world. The valley provides the exotic view of the Jundhar Glacier, which lies at the height of 4,300 m. above the valley. The valley is the home to the trees of Bhojpatra flower '*Brahmakamal*'. The cradle shaped valley is a paradise for the trekkers as well. Tourists find divine pleasure in visiting the valley – the site of meditation by the epic Mahabharatha-famed preceptor, *Guru* Dronacharya. There is also a temple dedicated to the Kaurava King, Duryodhana. The Doon Valley, dotted with a large number of famous temples, is considered as the holy place among the Hindus in India. It is a complete paradise for the nature lovers, bird and butterfly watchers, and trekkers.

The lush green grassy land and the dense forests of Chestnuts, walnuts, willows and chinars makes it an ideal location for the nature lovers and photographers. The scented pine trees and profusion of flowers make it a complete fairyland. The area is also quite rich with the medicinal herbs and the aquatic shrubs. These wonderful creations of the nature add dash of splendour to nature's abundance that exists in the environs of the Doon Valley.

Fig. 1 & 2. The Doon Valley as seen from Mussoorie during (1) the daytime and (2) the night (Source: Wikipedia, the free encyclopedia).

The valley undulates seemingly endlessly into many glens and vales intersected by many fast-flowing rivulets and tributaries surrounded by forests harbouring different kinds of wild animals such as the Himalayan black bear, tahr, serow, monal, Indian leopard, musk deer, and *chir* which can be easily spotted. The Doon Valley has rich tradition of agriculture, horticulture and forestry. The famous '*Basmati*' rice is abundantly produced among many other agriculture produce like cereals, *Rajma* etc. which are among the main crops grown in this area.

Doon Valley's main city is Dehradun which occupies an indelible spot in the country's academic and research front. The name Dehradun is made up of two words: '*dehra*' is derived from the word "dera", deriving from griha and meaning home, while '*dun or doon*' is a term used for the valley that lies between the Himalaya and the Shivalik mountain ranges. At times, Dehradun is also remembered as "Dronanagari" **(द्रोणनगरी)**, a name given to the city after '*guru*' Dronacharya, the great teacher of Kaurava and Pandava who lived here.

Dehradun is prolifically renowned for its natural resources, publishing services and particularly for its educational and research institutions. It is one of the highest rain receiving areas of North India. It hosts training institutions of national importance such as the Indian Military Academy and Indian Forest Service. It is home to national foundations such as the Ordnance Factory Dehradun, Opto Electronics Factory, and other defence establishments, in addition to the DIT University, IMS Unison University, Uttaranchal University, Indian Institute of Petroleum, Oil and Natural Gas Corporation, University of Petroleum and Energy Studies, Law College Dehradun, Survey of India, Wadia Institute of Himalayan Geology, Indian Institute of Remote Sensing, Wildlife Institute of India, Forest Research Institute & Colleges, DAV College, Doon School, Welhem School, etc.

The scenic beauty of the Doon Valley is infinite; it needs to be appreciated, cultivated, and conserved with utmost care and sensitivity. If only the lyrics (by Shakeel Badayuni) were to be seen and the melodious recital by Mohammed Rafi in "Choudvi ka Chand" enjoyed in the quaintly quiet climes of the Doon Valley, where … we have recently begun to miss Doon's sylvan surroundings and its cool breeze that would tenderly ruffle our hair and the teasing summer showers that caressed our faces like drops of mist …, then the everlasting memorable song would appear to have been enlivened certainly absolutely completely in her (Doon Valley's) praise:

चौदहवीं का चाँद हो या आफ़ताब हो
जो भी हो तुम खुदा की कसम लाजवाब हो।

• • •

2

TOPOGRAPHY AND PHYSIOGRAPHY OF THE DOON VALLEY

"Though losses and crosses
Be lessons right severe,
There's wit there, you'll get there,
You'll find no other where."

The Doon Valley is an asymmetrical synclinal valley (80 km in length, 20 km in average width) along the NW-SE regional Himalayan strike in Garhwal sub-Himalaya, bound by young topographic relief of frontal Siwalik range with 800 m. average elevation. 'Doon' is a local word for valley. The Doon valley is an open (longitudinal) valley between two intermittent ranges of the Himalayas (the Sivaliks and higher Himalayan foothills). By virtue of valley it is a low-lying land, relatively flatter, bounded on all sides by mountains, with one range running from the west to the east (with Mussorie at the centre) in a semi-circular arc; and one running at the south from Poanta Sahib to Haridwar ("http://en.wikipedia.org/wiki/Himalaya"). Doon Valley is fragile, mainly composed of river alluvium (sand, silt and clay). Doon gravels comprised of boulder, pebble and gravel and older Doon gravels mainly consists of silt and clay of recent to Pleistocene age. The valley forms a watershed between the Yamuna and Ganges river systems, with Yamuna forming western boundary and the Ganges east. The Yamuna, Asan, Saung, Suswa, Tawa and many other seasonal waterbodies drain the Doon valley.

The city of Dehradun, located between latitudes 29° 58' N and 31° 2' N and longitudes 77° 34' E and 78° 18' E, is centrally pivoted in the Doon Valley at a varying height from 410 m in Clement Town to above 600 m at Jakhan which is 4 km south of the city. However, general elevation is 450 m above sea level. Jakhan is the starting point of Upper Shivalik Range that extends to Mussoorie and beyond. Tiuni hills in Vikasnagar District rises to 3,700 m above sea level. The hilly region of Mussoorie goes up to a height of 1870-2170 m above sea level. The Doon Valley contains the settlements including Dehradun, Doiwala, Herbertpur, Vikasnagar, Sahaspur, Rishikesh, Raiwala and Subash Nagar and Clement Town (Fig. 3).

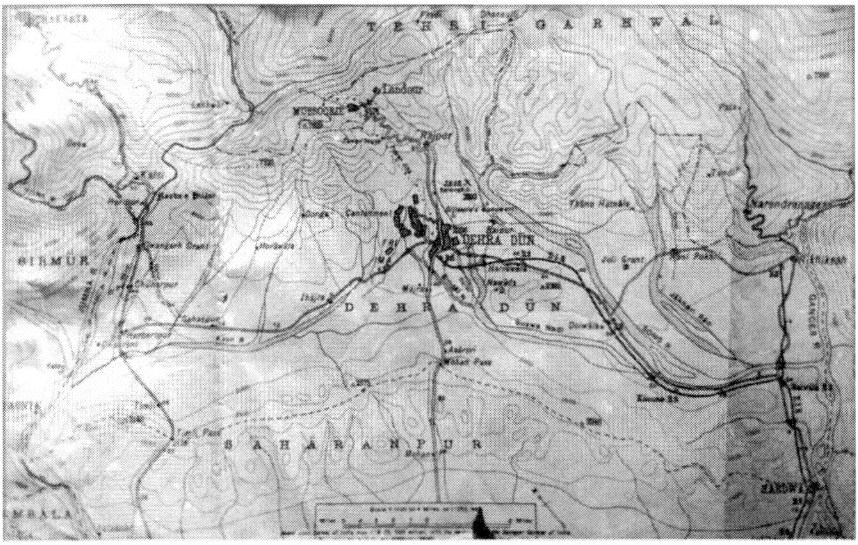

Fig. 3. The contoured map of Dehradun and the Doon Valley (*Source:* B.K. Tyagi's unpublished PhD thesis, 1978).

The district contains Rajaji National Park which is home to several elephants. The Doon Valley has the Terai and Bhabar forests within it as well as the Shiwalik hills and Lesser Himalayan Range containing hill stations such as Mussoorie and Chakrata. The district is bordered by the Himalayas in the north, the Sivalik Hills to the south, the river Ganges to the east, and the Yamuna river to the west. Towns in the foothills of the mountain ranges include Mussoorie, Sahastradhara, Chakrata, Lahkhamondal, Gautam Kund Chandrabani and Dakpathar. This district is divided into two major parts: the main city Dehradun surrounded by Shivalik and the Jaunsar Bavar, which is in the foothills of Himalaya. In the north and northwest it borders on the district of Uttarkashi and in the east by Pauri. In the west, it is bordered by Yamunanagar district of Haryana and the Tons and Yamuna rivers. To the south of Dehradun are the sacred towns of Haridwar and Rishikesh, and Uttar Pradesh's agriculture town of Saharanpur district. This district consists of six tehsils, viz., Dehradun, Chakrata, Vikasnagar, Kalsi, Tiuni and Rishikesh; six community development blocks, i.e., Chakrata, Kalsi, Vikasnagar, Sahaspur, Rajpur and Doiwala; 17 towns and 764 villages.

•••

3

DEMOGRAPHY OF DEHRADUN

In 2011, Dehradun had a population of 1,696,694 of which male and female were 892,199 and 804,495 respectively which was a staggering 25% of more to that of 2001 census when Dehradun had a population of 1,282,143 of which males were 679,583 and remaining 602,560 were females. In 2011, Dehradun District population constituted 16.82 per cent of total Maharashtra population. In 2001 census, this figure for Dehradun District was at 15.10 percent of Maharashtra population.

There is currently a change of 32.33 percent in the population compared to population as per 2001. In the previous census of India 2011, Dehradun District recorded an increase of 25 per cent to its population compared to 2001 (Table 1).

Table 1. Census of Dehra Dun

Description	2011	2001
Actual Population	1,696,694	1,282,143
Male	892,199	679,583
Female	804,495	602,560
Population Growth	32.33%	25.00%
Area Sq. Km	3,088	3,088
Density/km2	549	415
Proportion to Uttarakhand Population	16.82%	15.10%
Sex Ratio (Per 1000)	902	887
Child Sex Ratio (0-6 Age)	889	894
Average Literacy (%)	84.25	78.98
Male Literacy (%)	89.40	85.87
Female Literacy (%)	78.54	71.20

Till about 1950s the Doon Valley was aparrently a paradise rife with life of innumerable fauna and flora thriving amongst scores of rivers and rivulets, cascades, torrents, fountains, mountain well, Sulphur springs, landscapes, forests, vales and glens. The demography of Dehradun took a steep escalation in the late 20th century which, along with the mostly unplanned expansion of the urban areas, had drastically altered the environment of the Doon Valley by way of rapid industrialization and environmental pollution of of air, water and land. The growth rate of the Dehradun urban area from 1971 to 1981 was 44%. The forests of the Doon Valley decreased by 25% from 1972 to 1982. The level of groundwater in the city of Dehradun reportedly dropped by 30 m. in the same period. Poosibly the biggest devastation was brought about by incessant quary minig of the mountain-sides resulting in habitat destruction and a non-ending streak of lime kilns-generated pollution driving away the insects to other ecosystems on the periphery of the Doon Valley! *Palpopleura sexmaculata sexmaculata,* a beautiful multi-hued small dragonfly that used to be once playfully discerned in the environs of Raipur or the slow running stream breeder, *Copera marginipes,* near Clement Town suburb, are no more seen anywhere near! The odonate *Zyxomma petiolatum,* the evening visitor to the algae-filled fresh watered cement tanks is hardly ever noticed, while another cement tank breeding libellulid, *Bradinopyga geminata* is more often than not seen laden with ectoparasites bestowed courtesy polluted aquatic ecosystems.

Possibly the most devastating vicissitude in topo-physiography is noted at *Shahastradhara* or the Sulphur Springs. The piquantly serene climes of the fragrant Sulphur Springs Valley once bubbling with numerous mountain wells as source of recharge to the underlying rivers, and supporting breeding of as arcane and elegant endemic damselfly, *Calicnemis doonensis,* but also the commoner *Neurobasis chinensis chinensis,* and many species of Trithemis, Orthetrum etc., is now seemingly converted to a fast-growing concrete jungle with much of the interior breeding habitats of many elegant and rare insect species completely destructed. One wonders where would have gone the endangered gomphid, *Burmagomphus siwalikensis* (Tyagi, 1982)?

Conservation of animals' habitats ensures unchallenged sustenance of human being itself. Many of the insects, such as mayflies and dragonflies, are direct indicators for purity of aquatic ecosystems. Any amount of pollution to their habitats will drive these species away and the resultant source of water could turn out to be one's Japanese Minamata river or Maharashtra's Kali river tragidy (Tyagi, 1985).

• • •

4

CLIMATE OF THE DOON VALLEY

The latitudinal extent of the region defines its climate. As the city is located between 30-31°N the climate is exclusively classified under Humid Subtropical Climate and is often Continental type. The climate is similar to Haldwani, Panchkula, Jammu or Siliguri. The city is also referred as the 'Rainy City of India' due to continuous and incessant rainfalls especially in monsoon. Dehradun is India's wettest State capital; recording 2,865 mm of rainfall between 1 June and 30 September, 2015. The climate of Dehradun is humid continental, although it varies from tropical hot in summers to severely cold, depending upon the season and the altitude. The nearby hilly regions often get snowfall during winter (Table 2).

Table 2. Climate data for Dehradun

Month	Jan	Feb	Mar	Apr	May	Jun	Jul	Aug	Sep	Oct	Nov	Dec	Year
Record high °C (°F)	24.6 (76.3)	31.2 (88.2)	37.2 (99)	40.8 (105.4)	42.8 (109)	44.6 (112.3)	40.6 (105.1)	37.2 (99)	36.6 (97.9)	36.1 (97)	30.6 (87.1)	27.4 (81.3)	44.6 (112.3)
Average high °C (°F)	19.3 (66.7)	21.5 (70.7)	26.4 (79.5)	32.1 (89.8)	35.6 (96.1)	34.8 (94.6)	30.5 (86.9)	29.4 (84.9)	29.7 (85.5)	28.5 (83.3)	25.0 (77)	21.1 (70)	27.8 (82)
Average low °C (°F)	6.0 (42.8)	7.8 (46)	12.0 (53.6)	16.7 (62.1)	20.7 (69.3)	23.0 (73.4)	22.8 (73)	22.4 (72.3)	20.8 (69.4)	15.7 (60.3)	10.4 (50.7)	6.8 (44.2)	15.4 (59.7)
Record low °C (°F)	−1.1 (30)	−1.1 (30)	2.2 (36)	7.2 (45)	11.3 (52.3)	13.1 (55.6)	13.2 (55.8)	18.0 (64.4)	14.3 (57.7)	8.4 (47.1)	2.8 (37)	0.0 (32)	−1.1 (30)
Average precipitation mm (inches)	55.0 (2.165)	58.8 (2.315)	49.0 (1.929)	22.5 (0.886)	41.7 (1.642)	201.8 (7.945)	672.6 (26.48)	728.2 (28.669)	296.5 (11.673)	49.8 (1.961)	8.6 (0.339)	24.4 (0.961)	2,208.9 (86.965)
Average relative humidity (%)	72	66	57	46	48	66	85	86	81	69	68	71	67.9

Source #1: IMD "Dehradun Climatological Table 1901–2000" (PDF). Indian Meteorological Department. Retrieved 24 January 2015;
Source #2: "Climate profile for Dehradun"· MyWeather.com. Retrieved 24 January 2015.

Although the temperature in Dehradun can reach below freezing during severe cold, this is not common. Summer temperatures can reach up to 44°C for a few days and hot wind called *Loo* blows over North India, whereas winter temperatures are usually between 1 and 20°C and fog is quite common in winters like plains. During the monsoon season, there is often heavy and protracted rainfall. Agriculture benefits from fertile alluvial soil, adequate drainage and plentiful rain. Mountain areas are also used for agriculture. Dehradun is known internationally for its variety of *Basmati* rice and *Lychees.*

●●●

5

HISTORY OF ENTOMOLOGICAL RESEARCH IN THE DOON VALLEY

*"Leaving no memorial but a world
Made better by their lives."*

The history of entomological research in the Doon Valley, vastly appreciated for its rich natural forests and fauna, has its indelible traces in the annals of forestry research in the British India, being at that time well known for its high quality of timbers. Roots of Indian forest entomology can be traced back to the times of Carl von Linne (Carolus Linnaeus), the originator of 'binomial nomenclature' in taxonomy, who mentioned in his *magnum opus*, 'Systema Naturae', a large number of Indian species, though he had never had a chance to travel to India. Soon thereafter J. G. Koenig published the life history of *Odontotermes redemanni*, the Indian termite species that destructs the construction wood etc., in 1779. Till then and much thereafter most of the entomological studies were carried out by non-professionals in their tour reports and publications. Nearly a century later Morsefield and Moore published a catalogue on food plants and described larvae of forest insects, which occurred as a major important work on the inurious timber borers, introducing in a true sense the science of forest entomology in India. Subsequent classics like "*Observations on timber borers of South India*" by H. Cleghorn and S.N. Ward (1861), and "*Report on insects destructive to woods and forests*", with remarkable and informative illustrations and photographs for accurate identification of insect species, by Thompson in 1868 had clearly re-inforced the necessity of forest entomology in India. Their efforts soon culminated in starting a pioneer journal "*The Indian Forester*", in June 1875, which became a medium for publication of notes on damage to the forest trees and timber by insects. In the first issue of the journal a note on damage done by toon shoot borer was published; interesting to note that at that time this species of insect was not yet named, Moore in 1886 described this species as *Hypsipyla robusta* Moore (Pyralidae). In the same year, i.e., 1875 the Indian Museum, Calcutta was established and it became the center of information on injurious insects. In 1893 E.C. Cotes, Deputy Superintendant at the Museum prepared "*An Elementary Manual of Zoology*". In this manual the author could quote only few known Indian insects and had to rely on foreign insect examples. From this point onwards interest in Indian forest insects increased and specimens sampled from all faunal regions and ecosystems, including the Doon Valley, were sent for identification to the Indian Museum (Singh & Singh, 2012).

By late 1900, Stebbing was appointed as the Forest Zoologist. Credit of this appointment goes to Mr. B. Ribbentrop, Inspector General of Forests, although this post was sanctioned for only two years. On his posting Stebbing commented "It was the foundationstone on which a great edifice was to arise in the course of next two decades." (Stebbing, 1923).

Forest Entomology Division was one of the six founder branches, which were initially started with the establishment of the Forest Research Institute in 1906. Mr. E. P. Stebbing was appointed as first Imperial Zoologist in April 1906 till 1909. This post was renamed in 1922 as Forest Entomologist. Before 1904 forest zoology was located in the Forest College and some portions in a house in the Dehradun city, but thereafter a separate building at Chandbagh (present Doon School) was used, where the zoological exhibits were displayed in the College Museum. From the institute a series of publications like Memoires, Records, Bulletin etc. were published to disseminate knowledge generated by the researchers. Stebbing contributed abundantly to each series and in 1907 completed "A Manual of Elementary Forest Zoology for India". He left India in December, 1909 and after his departure published in 1914 "Indian Forest Insects of Economic Importance- Coleoptera", a book with 648 pages, 401 text-figures and 64 plates; the most sumptuous work ever produced on Indian forest beetles. In December, 1909 Subramaniya Iyer (PFS, 1899-1913) was appointed Temporary Forest Zoologist and held the post till April, 1911. S. Maulik was his assistant in 1910 and 1911 (later the author of volumes in the Fauna of British India on Chrysomelidae). In October, 1911, A.D. Imms, Professor of Biology at Allahabad University was appointed with one assistant, N.C. Chatterjee. During his sixteen months' tenure of office he introduced valuable reforms; extensive additions were made to the library; contacts were made with entomological specialists in many countries and large collections were distributed to them for identification. He designed an insectory and laboratories, which were erected, according to his plans after his departure. Dr. Imm's association with forest entomology at a time when a new Institute was projected and funds for expansion of entomological research had to be secured, was a most fortunate occurrence. In August, 1913, C.F.C. Beeson (IFS, 1911-42), an entomology student of H.M. Lefroy in London and K. Escherich in Tharandt, was transferred from the Punjab and took over the post of Forest Zoologist. In 1914 a new building of research institute was completed at Chandbagh.

In 1916 the Board of Forestry expressed its opinion that the Forest Department pay "Insufficient attention to the introduction of control measures for insect pests", and considered that "control measures to be of any use must be carried out continuously on a large scale and must be of a preventive rather than a remedial nature. As the necessary staff for such operations does not at present exist, progress must depend largely, if not entirely, on the appointment of provincial research officers working in connection with, and with assistance from, the Zoologist. The staff of Zoologist should be increased by the appointment of field assistants who would be employed in investigational work where prolonged local study is required and who would intimate and carry out control measures". Therefore, staff was increased and two assistants S.N. Chatterjee and B.M. Bhatia were added to the Zoologist's staff. The policy of extensive ecological surveys was accepted as essentially the function of the Forest Institute, and was incorporated in proposals for expansion of the Branch of Forest Zoology. The field of work in India and Burma was divided into four phytogeographical regions each of which was to be in charge of a Regional Zoologist (later renamed Divisional Forest Entomologist) with an assistant.

The erection of a new Forest Research Institute, and increase in the research staff in all branches was sanctioned in 1920. Beeson in 1920, arranged the recruitment of three officers to fill the new entomological posts. Dr. M. Cameron (Surgeon Commander, R. N., retired), J. C. M. Gardner (IFS., 1921, who studied entomology under Lefroy and Forestry at Cambridge), and D.J. Atkinson (IFS, 1921, then a probationer at Oxford) accepted the appointments. Cameron arrived in India in December, 1920 and his post was designated Systematic Entomologist, as it had been meanwhile decided to apply one of the posts of Divisional Forest Entomologist to Systematics until such time as the full regional strength could be recruited by training men in both entomology and forestry. A second assistant entomologist, S.K. Pillai who had been trained in Germany, was appointed in December, 1921. The two Divisional Forest Entomologists were posted to Bengal and Burma as Assistant Conservators of Forests in 1921. Ultimately, the Division was shifted to present FRI building in 1926.

Albeit enormous data generated on entomological aspects of timber research, yet the *ad hoc* Entomological Branch had to wait until 1927 when its future came under reconsideration and resolved. To meet its mandate the Forest Entomology Division researched on majority of insect species associated with different tree species which cause multiple injuries to the trees through extensive insect surveys, collection of insect species, laboratory rearing and identification. Subsequently, following its inclusion within the newly erected Indian Council of Forestry Research & Education, the objective of the entomology Branch also changed *vis-a-vis* forestry sector, emphasizing mainly on biodiversity conservation and ecological security through (i) Exploration, identification, and assessment of forest insect diversity, (ii) Conservation of rare and endangered insect species; identification and exploration of indicator species sensitive to the forest disturbances, (iii) Systematics of insects of forestry importance specially Coleoptera, Isoptera, Lepidoptera, Hymenoptera, etc., (iv) Strengthening, enrichment and digitization of National Forest Insect collection (NFIC), (v) Developing ecofriendly pest management system for important wood and non-wood forest insect species involving parasitoids, predators, microbials and plant based pesticides, (vi) Evaluation and identification of insect resistant clones of important tree species, (vii) Monitoring and control of insect defoliators and borer population of important tree species; exploration of insect pheromones and kairomones, (viii) Ecology of wood insect borers in forests, felling sites and timber yards for population monitoring and control, and (ix) Estimation of economic losses to the forest/plantation due to insect injuries/damages.

Dr. M. L. Roonwal (December, 1949 – July, 1956), ably assisted by his successors as Forest Entomologist, viz., Dr. R. N. Mathur (August, 1956 – October, 1961) and Dr. P. N. Chatterjee (November, 1961 – December, 1970), instituted many new research programmes. A major thrust area of research was

Isoptera (termites) on which many significant publications were brought about (Roonwal & Chhotani, 1989; Chhotani, 1997). Dr. Roonwal was also instrumental in bringing out the first complete catalogue of insects preserved in the insectary of the Forest Entomology Division, and continued to update it with his various dedicated colleagues (Roonwal et al., 1951a,b,c; 1953; 1963).

A truely colossal empire of entomological wisdom built up by the prolific Dr. Roonwal, the huge responsibility landed on the strong shoulders of the highly revered forest entomologist, Dr. P.K. Sen-Sarma (June, 1970 – May, 1973 and September, 1976 – August, 1982) who made great strides in termite research and produced slassical works on this important order of insects (Sen-Sarma, 1974). Following him many forest entomologists carried on his legacy of perseverance, focused attention and scientific documentation, such as Sh. Pratap Singh (June, 1973 – August, 1976 and October, 1983 – July, 1989), Sh. Avinash Chandra (September, 1982 – September, 1983), Dr. M.L. Thakur (August, 1989 – August, 1995), Dr. H. R. Khan (September, 1995 – October, 2002), Dr. M. Ahmad (November, 2002 – July, 2008), Shri R. S. Bhandari (August, 2008- June, 2009) and Dr. R. K. Thakur (July, 2009 – July, 2010). Dr. Neena Chauhan, a BM (NH) trained senior entomologist at the Forest Entomology Branch, healthily contributed to many insects' bioecology and described particularly the thrips of which she has described many new species such as that of *Pseudodendrothrips* (Vijay Veer & Chauhan Neena, 1992; Chauhan Neena & Vijay Veer, 2008). Dr. Sudhir Singh brought a new lease of life to the gradually stagnating entomology research at the FRI and not only elegantly narrated the history of forest entomology in India, particularly in the Dehra Dun Valley (Singh & Singh, 2012), but also helped restore to a great extent its past glory and the vintage collection of myriads of insect speciments in the insectary (Faisal et al., 2014).

Fig. 4. At the insectary of the FRI preserved collection of larvae of beetles and moths; termites are also stored in alcohol (*Source:* Singh & Singh, 2012).

For the nearly first five decades two institutions, viz., the erstwhile Forest Research Institute & Colleges, Dehradun and the Zoological Survey of India, headquartered in Kolkata with a regional centre in Dehradun, singularly played a major role in strengthening the datamining particularly on various different insect pest-host plant interrelationships through advancing knowledge on taxonomy, ecology, biology, dispersal and control. The Zoological Survey of India inventorized insect order-wise species occurring in India which helped the researchers all over India to concentrate on the dynamics of biological and dissemination processes of various insect species of economic importance (Table 1).

Of the 32 insect orders known globally, 31 are represented in India (Table 3). The insect order Mantophasmatodea, discovered recently in 2002 from the Namibian mountains in central-southern Africa, is yet to be reported from India. According to one estimate, 63,760 species of insects representing 658 families under 27 orders are reported from India. Eight orders, viz., the Lepidoptera, Coleoptera, Orthoptera, Diptera, Hemiptera, Odonata, Hymenoptera and Thysanoptera constitute the bulk (94%) of the insect fauna, while the remaining 19 orders are represented by small numbers (6%) of species. The order Coleoptera has the greatest diversity in terms of families (114) followed by Hemiptera (92 families), Diptera (87 families), Lepidoptera (84 families) and Hymenoptera (65 families) (*cf.* Kailash Chandra, 2011).

Table 3. Major orders of class Insecta in India

S.N.	Class/order	Common name	No. of species described (world) (Zhang, 2011)	Species in India (ZSI, 2012)	Genera	Families
	Class Insecta		10,29,741	63,760		658
1	Order Collembola	Springtails	8130	299	103	19
2	Order Protura	Proturans	804	20	10	3
3	Order Diplura	Diplurans	976	18	09	04
4	Order Zygentoma/ Thysanura	Jumping bristl-etails Silverfish	1074	38	21	5
5	Order Ephemeroptera	Mayflies	3240	124	46	12
6	Order Odonata	Dragonflies and damselflies	5899	463	139	19
7	Order Plecoptera	Stoneflies	3788	116	25	8
8	Order Orthoptera	Grasshoppers, crickets, katydids, etc.	24,276	1033	396	21
9	Order Phasmida	Walking-sticks	3029	144	41	8
10	Order Demaptera	Earwigs	1978	296	75	7
11	Order Embioptera	Web earwigs-spinners	464	31	5	2
12	Order Blattodea	Cockroaches	7314	186	58	12
13	Order Mantodea	Mantids	2400	174	72	11
14	Order Isoptera	Termites	2864	271	52	7
15	Order Psocoptera	Bark and book lice	5720	105	–	16
16	Order Phithiraptera	Lice	5102	400	–	8
17	Order Hemiptera	True bugs	1,03,590	6479	–	92
18	Order Thysanoptera	Thrips	6019	686	258	7

S.N.	Class/order	Common name	No. of species described (world) (Zhang, 2011)	Species in India (ZSI, 2012)	Genera	Families
19	Order Megaloptera	Alder flies, dobosonflies	354	25	7	1
20	Order Raphidiophtera	Snakeflies	254	5	2	1
21	Order Neuroptera	Lacewings, ant lions and others	5868	312	112	12
22	Order Coleoptera	Beeltes	3,87,100	17,455	–	114
23	Order Strepsiptera	Twisted-winged parasites	609	21	8	4
24	Order Mecoptera	Scorpion flies	757	23	2	2
25	Order Siphonaptera	Fleas	2075	46	24	8
26	Order Diptera	Flies, mosquitoes	1,59,294	6337	1180	87
27	Order Lepidoptera	Butterflies and moths	1,57,424	15,000	–	84
28	Order Trichoptera	Caddis flies	14,999	1046	94	27
29	Order Hymenoptera	Sawflies, ants, bees and wasps	1,16,861	12,605	65	57

Source: Adapted from Kailash Chandra (2011).

Uttrakhand within which the Doon Valley falls has an amalgamation of both apterygote (wingless) and pterygote (winged) insects. The Doon Valley provides a melange of varied ecosystems to its amazing insect fauna which is comparable in its richness to that of the Silent Valley in Western Ghat region in Kerala. India's insect fauna is distributed over a wide range of ecosystems, climatic regions and altitudes. Table 4 presents quantum diversity of insect fauna in the States and Union Territories of India (Kailash Chandra, 2011).

Many of the Indian species are endemic to a particular area or region. The diversity of insects is greater in the north-eastern states, the Western Ghats and the Andaman and Nicobar Islands, and these areas also have a high level of endemism. A high percentage of endemism is noted in the primitive insect groups, viz., Protura (85%), Diplura (66%) and Thysanura (60%), followed by Collembola (15 %). Among the exopterygotes, Thysanoptera has the highest percentage of endemism (75%), followed by Phasmida (68%), Ephemeroptera (58%), Plecoptera (57%), Orthoptera (54%), Embioptera (45%) and Isoptera (44%), and there is less than 40% endemism in the remaining orders. Among the endopterygotes, the endemism in species level is the highest in Mecoptera (86%), followed by Neuroptera (76%), Strepsiptera (71%), Hymenoptera (71%), Trichoptera (60%), Diptera (35%) and Coleoptera (17%), while the order Lepidoptera shows only 10% endemism since the moth fauna is widely distributed in the Indo-Pacific region (Table 5) (*cf.* Kailash Chandra, 2011).

Table 4. Insect fauna of states and union territories by group

Taxon	Andman and Nicobar islands	Andhra Pradesh	Arunachal Pradesh	Assam	Bihar	Chandigarh	Chhattisgarh	Delhi	Goa	Gujarat	Haryana	Himachal Pradesh	Jammu & Kashmir	Jharkhand	Karnataka	Kerala	Lakshadweep	Madhya Pradesh	Maharashtra	Manipur	Meghalaya	Mizoram	Nagaland	Odisha	Puducherry	Punjab	Rajasthan	Sikkim	Tamil Nadu	Tripura	Uttarakhand	Uttar Pradesh	West Bengal
Collembola	18	22	27	20								21	15	10		60		27	24	27	11	21	18	21			6	39	24	11	31	35	25
Protura																20																	1
Diplura													1			5			8										4				1
Archaeognatha	2	1										10	4			2													5		1	1	
Zygentoma	1	6						3				5	2						8	3							2	3	5	2	4	4	9
Ephemeroptera	1	1		4	3							25	8	1	5	5		12	19	6	5			5				2	24	5	1	3	17
Odonata	72	68	92	65	63	8	52	47	39	58	17	86	23	109	137	137		90	101	68	151	64	48	58		36	48	45	135	35	122	69	185
Dermaptera	14	3	80	18	5		6	4				30	14		32	19	3	16	13	28	49	15	3	14	4	3	3	50	75	14	36	12	79
Plecoptera				30								20	2		3	6		1	1	8	16					1		6	12			8	21
Embioptera	2	1		3							1				7	4		2			2							6	7		2		8
Orthoptera	85	84	149	221	123	36	81	94	37	22	30	166	77	14	156	129	10	121	148	120	161	94	46	146	19	62	87	123	341	108	144	100	278
Phasmida	4		4	21	3								1		3	6		1	1		11		8					16	20				28
Mantodea	7	13	20	29	14		3		9			20	4	10	23	14	1	31	56	10	25			27			9	17	35	2		55	35
Blattaria	12	14	20	5				9				1			2	2	4	10	12	8	26			12	4		4	17	47	10			23
Isoptera	40	12	44	26	9		6	14	5	60	8	12	16	18	47	56		35	17	22	31	15	11	31		9	61	16	46	27	47	28	55
Psocoptera				22									2			36			14	7	6			1						5			39
Phthiraptera	4	15		16	5		1	6		4	3	20	22		2	6		7	11	18	4		1	1		41	39	35	3		8	28	35
Thysanoptera	29	77	9	24	11	8		116	9	7	17	31	21		145	141		81	63	26	49	1	3	8		45	20	17	321	35	31	88	124
Coleoptera	580	137	73	810	156	17	250	102	4	126	280	205	99	52	144	302	9	355	420	325	911	50	45	196	21	296	239	994	750	283	1074	210	1570
Neuroptera	13	4	9	34	37		24		2			15	10		36	5		33	74	7	45	10		23	1	12	5	36	23	4		13	79
Mecoptera				3			1			1		2			1	3	1	1	2						1			4	7		1		3
Siphonaptera	2			3						1	2	5	8		9	4		2	7	2						2		6	11		1	8	13
Strepsiptera			2		1										2														1				14
Diptera	308	71	272	589	146	72	25	79	7	106		417	135		164	342	3	257	386	92	321	71	69	39		109	135	634	464	119	541	93	681
Lepidoptera	804	153	494	980	250	80	193	111	251	150	124	570	107	209	318	529	84	475	253	269	1775	145	493	141		340	216	2528	319	135	1523	190	1020
Trichoptera	20	4	160	129	36			1				8	22		53	40		24	43	131	168	16		7		24		179	65			154	79
Hymenoptera	176	201	117	227	180	3	20	248	32	62	61	319	83	43	669	997	3	77	339	52	754	7	7	199	31	102	116	619	612	61	181	806	430
Total	2439	1070	1592	3639	1077	243	692	986	400	609	557	2356	695	466	2084	3327	118	1793	2289	1522	5118	578	882	1106	96	1116	1049	5941	3609	1062	4160	1941	5818

(*Source:* Kailash Chandra, 2011)

Table 5. Numbers of endemic genera and species of insect in India (*Source*: Kailash Chandra, 2011)

	Class/order	No. of species known in India	Endemic genera	Endemic species	Percentage of species endemic to India
A	Class Collembola	299	22	45	15.05
B	Class Protura	20	4	17	85.00
C	Class Diplura	18	3	12	66.66
D	Class Insecta				
1	Order Thysanura	38	12	23	60.52
2	Order Ephemeroptera	124	–	72	58.06
3	Order Odonata	463	6	115	24.83
4	Order Plecoptera	116	–	66	56.89
5	Order Orthoptera	1033	77	563	54.50
6	Order Phasmida	144	–	99*	68.75
7	Order Dermaptera	298	3	117	39.26
8	Order Embioptera	31	–	14	45.16
9	Order Blattodea	186	14	60	32.25
10	Order Mantodea	174	24	77*	32.25
11	Order Isoptera	271	–	172*	44.25
12	Order Psocoptera	105	–	15	14.28
13	Order Phithiraptera	400	–	16	04.00
14	Order Hemiptera	6479	579	2421	37.36
15	Order Thysanoptera	686	92	520	75.80
16	Order Neuroptera	342	13	262	76.60
17	Order Coleoptera	17,455	923	3100	17.75
18	Order Strepsiptera	21	–	15	71.42
19	Order Mecoptera	23	–	20*	86.95
20	Order Siphonaptera	46	–	15	32.60
21	Order Diptera	6337	107	2183*	35.06
22	Order Lepidoptera	15,000	100	1500	10.00
23	Order Trichoptera	1046	5	650	62.14
24	Order Hymenoptera	12,605	516	9000	71.40
	Total	63,760		21,166	33.40

The Doon Valley, characteristic to its well preserved geography, is richly represented by several endemic species under different insect orders. For example, Odonata has *Burmagomphus siwalikensis* (Anisoptera) and *Agriocnemis corbeti* and *Calicnemia doonensis* (Zygoptera) as endemic taxa to the valley. Many species of thrips were endemically described from the Doon Valley (Kulshrestha &Vijay Veer, 1984, 1990). Likewise a large number of species, endemic to the Doon Valley, were described by Dr. Roonwal (*cf.* Roonwal & Sharma, 1988), Dr. Sen-Sarma (*cf.* Vijay Veer & Tyagi, 2008) and many others.

• • •

6

SELECTED MAJOR BIBLIOGRAPHIC REFERENCES USEFUL FOR ENTOMOLOGISTS OF THE DOON VALLEY

1. Gardner, J.C.M., 1947. On the larvae of the Noctuidae, Pt. III. Transactions of the Royal Entomological Society 98(4): 59-89.
2. Imms, A.D., 1925. A general textbook of entomology, including the anatomy, physiology, development and classification of insects. London: Methuen.
3. Beeson, C.F.C., 1941. The Ecology and Control of the Forest Insects of India and the Neighbouring Countries. Bishen Singh Mahendra Pal Singh, Dehra Dun, pp. viii + 1007, incl. 203 Figures.
4. Krishna, K. and Weesner, F. M., 1969. Biology of Termites, Vol. I. Academic Press, New York, pp. xiii + 598.
5. Krishna, K. and Weesner, F. M., 1970. Biology of Termites, Vol. II. Academic Press, New York, pp. xiv + 643.
6. Richards, O.W. & R.G. Davies, (Eds.), 1977. Imms' General Textbook of Entomology. Vol. 1: Structure, Physiology and Development. Springer Netherlands, pp. 418.
7. Richards, O.W. & R.G. Davies, (Eds.), 1997. Imms' General Textbook of Entomology. Vol. 2: Classification and Biology. Springer Netherlands, pp. 388.
8. Jha, L.K. and P.K. Sen-Sarma, 2008. Forest Entomology. APH Publishing Corporation, New Delhi, pp. 381.
9. Krishna, K., David, G.A., Krishna, V. and Engel, M.S., 2013. Treatise on the Isoptera of the World, Bulletin of the American Museum of Natural History, New York, vol. 377, pp. 2704.
10. Chhotani, O.B., 1997. Fauna of India-Isoptera (Termites), Vol. II. Zoological Survey of India Publication, Calcutta, pp. 800.
11. Roonwal, M.L. & Chhotani, O.B., 1989. Fauna of India-Isoptera (Termites). Vol. I. Zoological Survey of India Publication, Calcutta, pp. 672.
12. Sen-Sarma, P.K., 2000. Ecology and biogeography of the termites of India. In: Forest Entomology (Ed. M.L. Thakur). Sai Publication, Dehradun, pp. 422.
13. Kumar, A. & M. Prasad, 1981. Field ecology, zoogeography and taxonomy of the Odonata of Western Himalaya, India. Zool. Surv. India, Kolkata, pp. 118.
14. Tyagi, B.K. & Vijay Veer, 2011. Entomology: Ecology & Biodiversity. Scientific Publishers (India), Jodhpur, pp. 522.

15. Tyagi, B.K., Vijay Veer & Sri Prakash, 2008. Pests of Forests Importance and their Management. Scientific Publishers (India), Jodhpur, pp. 299.

16. Tyagi, B.K. 2007. Odonata: Biology of dragonflies. Scientific Publishers (India), Jodhpur, pp. 368.

Further non-scientific but interesting reading about the change in the environment of the Doon Valley:

1. Ganesh Saili, 2007. The Doon valley across the years. Rupa & Co., Dehra Dun, pp. 192.

2. Jaskiran Chopra, 2015. Memories of Another Day. Investcare Publications Pvt. Ltd., pp. 190.

• • •

7

INSTITUTIONS WITH FAMOUS RESEARCH HISTORY IN ENTOMOLOGY

"Learning is excellence of wealth that none destroy,
To man nought else affords reality of joy"
– Thirukkural

Even though quite some institutions have been known in past for carrying out remarkable research works in insects, yet three institutes, the erstwhile Forest Research Institute & Colleges, Zoological Survey of India and Dayanand Anglo Vedic College are of particular interest for not only history of entomological research but also the many milestones established by these institutes during past several decades.

Fig. 5. The world famous Forest Research Institute & Colleges (now known as Forest Research Institute, and is one of the premier institutions of the Indian Council of Forestry Research & Education, ICFRE).

1. FOREST RESEARCH INSTITUTE & COLLEGES (Fig. 5)

Set in the sylvan surroundings of Doon Valley, the Forest Research Institute is a proud testimony to the foresight and vision of foresters and administrators of long ago. Founded in 1878 as British Imperial Forest School, and later rechristened in 1906 as the Imperial Forest Research Institute under the British Imperial Forestry Service (IFS), the Forest Research Institute & Colleges (then popularly known as FRI & Colleges) is currently an institute of the Indian Council of Forestry Research and Education and is a premier institution in the field of forestry research in India and the world alike. It is located at Dehradun in Uttarakhand, and is one of the oldest institutions of its kind. In 1991, it was declared a deemed university by the University Grants Commission. The Forest Research Institute campus hosts the IGNFA (Indira Gandhi National Forest Academy), the staff college that trains officers selected for the Indian Forest Service (IFS). Insect ecology and biology as well as forestry related pest control are integral to the curriculum for the Indian Forest Service trainees at the Indian Forest College established in 1938; officers recruited to the Superior Forest Service by the states and provinces were trained there.

Fig. 6. Insect museum at thr Forest Entomology Division in the FRI, Dehradun (*Source:* Singh, 2008).

FRI Dehradun is one of the oldest institutions of its kind and acclaimed the world over. The Institute's history is virtually synonymous with the evolution and development of scientific forestry, not only in India, but over the entire sub-continent. Built over a lush green estate spread over 450 hectares, with the outer Himalaya forming its back drop, the institute's main building is an impressive edifice, marrying Greco-Roman and Colonial styles of architecture, with a plinth area of 2.5 hectares. The Institute has a developed infrastructure of all equipped laboratories, library, herbarium, arboreta, printing press and experimental field areas for conducting forestry research, quite in keeping with the best of its kind anywhere in the world (Fig. 6). It is 7 km west of the central city of Dehradun and has the biggest forest-based training institute in India. The F.R.I. with its majestic building houses one of the most authentically identified collection of insects of various orders, making the entomological reference collection and the zoological museum virtually a heaven for scholars and tourists alike. FRI and College Area campus is a census town, between Kashmir in the north and the Indian Military Academy to the south. The Tons River forms its Western Boundary.

Within its campus it hosts IGNFA (Indira Gandhi National Forest Academy), which is a separate organization of the Indian Ministry of Environment and Forests (MoEF). It also hosts the Central Academy for State Forest Services (CASFOS). The Wildlife Institute of India previously was on its campus but now has shifted to Chandrabani, and is now an independent organization under MoEF. IIFM is an independent autonomous organization of MoEF and has good liaisoning with FRI for forestry research and related activities. The deemed university of FRI runs four M.Sc. courses, *viz.,* Cellulose & Paper Technology, Environment Management, Forestry Management, Wood Science Technology and two P.G. Diploma courses in Natural Resource Management and Aroma Technology. It also enrolls large number of research scholars every year for Ph.D.

2. DAYANAND ANGLO VEDIC COLLEGE, DEHRADUN (Fig. 7)

Established in 1948, Dayanand Anglo Vedic College (*Website*: http://davpgcollege.com), or the DAV (PG) College, is the oldest academic institution in Dehradun. It is located in the heart of the city in its Karanpur area about 6 km from the railway Station and 30 km from the airport in Jolly Grant area. The institute has nearly all the academic courses for the UG and the PG. It allows doctoral research too in many of its departments, and the Department of Zoology, where MSc specialization in the disciplines of Ichthyology, Entomology and Helminthology is conducted, has always been in the forefront of high quality research.

The Department of Zoology has produced some outstanding academicians and scientists who had earned name and fame for their *alma mater*! Some of the most novel and innovative research investigations were carried out in insect orders of Odonata, Thysanoptera, Coleoptera, Hymenoptera, Homoptera, Heteroptera, Lepidoptera and Diptera.

Fig. 7. Dayanand Anglo Vedic College, Dehradun

3. ZOOLOGICAL SURVEY OF INDIA, NORTHERN REGIONAL CENTRE, DEHRA DUN

Founded in 1916, the Zoological Survey of India, a department under the Ministry of Environment and Forests, is a premier national institute exploring the faunal diversity of the country. It undertakes surveys in all the biogeographic zones to collect, identify, collate and inventorize faunal diversity of our country. The museum of this institute has a variety of Himalayan faunal diversity. One of its regional centres, the Northern Regional Centre, is situated in Dehra Dun, 5 kms away on Kaulagarh road.

The Northern Regional Centre is entrusted with exploration of faunal wealth in five North Indian states: Uttarakhand, Uttar Pradesh, Punjab, Haryana, Delhi and the union territory- Chandigarh, covering an area of c. 4,41,707 sq km. The region comprises ecologically diverse areas ranging from the western Himalayas in north with Shivaliks at the base, through Gangetic Plains with Terai in north and Bundelkhand in south, and the part of the Aravalis in southwest with equally diverse biodiversity in its one World Heritage Site (Nanda Devi), one Biosphere Reserve (Nanda Devi), two Tiger Reserves (Corbett and Dudhwa), nine National Parks, 46 Wildlife Sanctuaries, two Conservation Reserves and 48 Important Bird Areas. The Northern Regional Centre at Dehradun, Uttarakhand, was established in August 1960 under the second five year plan of the Government of India. It is one among the sixteen Regional Centres of the Zoological Survey of India (ZSI), the internationally acclaimed taxonomic organization of the country.

In as far as insects are concerned, ZSI-NRC has excelled in odonatological research particularly spearheaded by Dr. Arun Kumar, and ably supported by Dr. M. Prasad and Dr. Vinod Khanna.

• • •

8

INTERESTING SITES FOR ENTOMOLOGICAL EXPLORATIONS

Emperor Jahangir once said about Kashmir, "*Agar Firdaws ba roy-i zamin ast, hamin ast-u hamin ast-u hamin ast,*" meaning, "If there is Paradise on earth, it is this, it is this, it is this." The elegant Doon Valley is one of the most attractive ecosystems in the country, full of life with its highly rich fauna and flora, undulating landscapes, scores of water channels, fragrant forests *inter alia* dotted with villages that care the tourists most. The Doon Valley has a large number of sites which are abundantly rich with different kinds of specific insect orders, however, in the following pages, only a few of these habitats are described primarily with the viewpoint of encountering most of the insect fauna with likely endemic species.

1. SAHASTRADHARA (Fig. 8)

Sahastra means thousands and *Dhara* means streams, so Sahastradhara means thousands of streams. Around 14 km from Dehradun is the hot Sulphur water spring. The spring is believed to have exceptional medicinal value and the dripping caves and a bath in the Baldi River near the spring are said to rejuvenate the body and soul. Many people come to the spring in the hope of being cured of varied skin ailments.

Fig. 8. A stream flowing down from the mountain well (Source: Wikipedia, the free encyclopedia).

Set amidst the picturesque bounty of nature, Sahastradhara has become quite popular with tourists off late. The place makes an ideal picnic spot and is of immense attraction to visitors. The Baldi river and the caves provide a breathtaking view. The water here has a fall of about 9 m and leaves an incrustation of lime of all its touches.

Based on the banks of a tributary of the Song, the Kali Gad merges into the main Song river east of the Kalinga hill. After this confluence, the Song is a broad river valley, larger than its western counterpart - the Assan River. The river now flows through the valley in a south-eastern direction passing through the regions of Harrawala and Doiwala. It is a perennial river, however the water surges during the monsoon as can be seen from the bridge at Doiwala. The river proceeds southward till it merges with the Sooswa river, which in turn originates from the southern Shiwalik range separating the Doon valley from Saharanpur and Haridwar districts, collecting water from several small streams, including the once perennial Rispana. It flows westwards to Doiwala and can be seen through thick forest from the train running from Haridwar to Dehradun. The river then flows through the forest of Chilla before its confluence with the Ganga river at Raiwala.

The Sahastradhara or Sulpur Springs has been a natural haven for many a species of dragonflies, viz., *Neurobasis chinensis chinensis*, *Ceriagrion coromandelianum, Calicnemia doonensis, Rhodischnura nursei, Neurothemis fulvia,* many species of Orthetrum and Trithemis (particularly *T. festiva* and *T. aurora*).

2. MOTHRONWALA SWAMP (Fig. 9)

In Doon valley there are many patches of freshwater swamps, which are recognized as integral part of wetland ecosystems. The swampy zones are located in between the ridges and are composed of innumerable pools with characteristic bubbling and small intercommunicating streams. The northern portion however is drier than the southern, which is slushier and consists of loose soil. Kanjilal (1901) first emphasized on vegetation and botanical value of swamps. Ghildiyal & Srivastava (1989) studied the vegetation of other swamps of Doon valley that include Golatappar and Manu swamps.

Mothronwala, Dulhani-1 (new reserve) and Navada 10-14 (old reserve) of Lachiwala range are about 5 Km from main city of Dehradun, at an elevation of 600 m above sea level. It occupies an area of approximately 22 acres. The swamp lies at 300 15' 40" and 300 16' 45" N latitude and 780 1' and 780 2' 15" E longitude and lies to the South-East of Dehradun near the military township of Clement Town. On the East is the village of Mothronwala from which the swamp derives its name. On the north lies Banjarwala Tea Estate. On the West lies the Sushwa river. A stream coming out of the swampy zone drains into the river that ultimately discharges into the Ganga through Rispana River. On the South is the Clement Town water works.

Fig. 9. A marshy stream at Mothronwala (Source: B.K. Tyagi, 1978).

The swampy area of Mothronwala is humid and fairly green. The maximum rainfall ranges between 600 – 800 mm during the months of July-August and minimum is recorded during April - May. The maximum temperature reaches upto 400°C during the months of May and June whereas minimum of 2-30°C during December - January.

The ridge of Mothronwala swamp is about 10 - 11 m above the surrounding level. The slope along the ridge is approximately 200 - 300. The northern part of the ridge is drier than the southern area, which is slushy. Inside the swampy area, the sub-soil water level is quite high and remains so throughout the year. The slush in marshy place is knee deep. During rains the water infiltrates through the gravelly soil extending over a very large area of the terrain, oozes out here in a series of deep but narrow ravines giving rise to a number of streams which unite into a few main channels that pour into the Suswa River.

Mothronwala Swamp is very rich in varied invertebrate fauna including insects (Gupta et al., 2006).

3. DAAK-PATTHAR (Fig. 10)

45 km from Dehradun is located an excursion spot near Yamuna Hydel Scheme, Daakpatthar which has emerged as a lovely tourist spot in the Western Doon Valley. It encompasses lush green and blooming gardens. The vast stretch on the Yamuna river and the serrated Shivalik range add to the grandeur of the place. The place is connected by regular bus service from Parade Ground and Highway bus stand near Railway Station, Dehradun. A tourist Bungalow commands a beautiful view of the area with swimming pool facilities.

Fig. 10. A slow-running stream at Daak Patthar (Source: B.K. Tyagi, 1978).

4. ROBBER'S CAVE (Fig. 11)

An ideal place for picnic, Robber's Cave is just 8 Kms away from the City Bus Stand. Local bus services are available upto Anarwala Village, from where it is just a kilometer's trek to the spot. A magnificent spot that lies at the suburban areas of Dehradun, Robber's Cave is one of the best loved tourist locations. Popularly called as Guchhipani or Guchu Pani it is a perfect destination to spend the holiday in a pristine atmosphere.

Fig. 11. Robber's cave, well known for odonata such as *Burmagomphus siwalikensis* and several species of *Rhynocypha* (Source: Wikipedia, the free encyclopedia).

Beauty in the midst of adventure is the prime specialty of this scenic place. Bounded by majestic hills, this cave presents an amazing scenario of the Mother Nature. After the tedious trek of one kilometer, one would get the pleasant warmth of a refreshing stream. After a quite flow, this would vanish only to appear in a few kilometers away. Tourist-folks gather in large groups here to witness this strange and splendorous experience.

5. MALSI DEER PARK

The newly developed spot at the foot hills of Shivalik is situated about 10 kms from Dehradun, enroute to Mussoorie on Diversion road. A mini zoo with children's park and lovely natural surroundings make it ideal sightseeing-cum-picnic spot. Refreshment Tea and snacks are also available. It also is home to such fabulous creature like sambhar, deer, panther etc.

6. GULARGHATI (Fig. 12)

Tucked in the sprawling Doon Valley's cosy vale in its eastern slope, about 20 km away from the central town of Dehradun, Gularghati is a virtual paradise for abundance of a variety of insects. The vale is intersected by a perennial river, Song river, which supports breeding and aquatic development, apart from food, for a large number of insects. The Song river drains the central and eastern parts of the Doon Valley. A tributary of Sooswa river, which in turn is a tributary of the Ganges, it originates as spring-fed stream in the southern slopes of the Mussoorie ridge of the Himalayan range and runs from Dhanaulti towards Narendra Nagar. As is common in hilly areas, there are several streams running south from the mountains that merge into one river that emerges from the hills a little after Sahastradhara.

Fig. 12. A fast running stream at Raiwala (*Source:* B.K. Tyagi, 1978)

Song is one of the largest rivers that drain the Doon Valley, and its tributaries includes, Kali Gad, Sahastradhara, Assan River and Rispana River.

Many species of Odonata, Lepidoptera, Coleoptera, Hemiptera, Hymenoptera, Heteroptera, Diptera, Ephemeroptera etc. thrive in the Gularghati.

7. THANO-MALDEVTA-LACHHIWALA (Fig. 13)

The eastern part of the Dehradun valley is thickly forested with Sal-mixed forests of the Shivaliks that have multiple storeys, enhancing their appeal for birds and birdwatchers alike. These moist forests are interspersed with patches of scrub jungle, riverine forest, agricultural fields and orchards that further increase the diversity of insect species that may be seen aplenty. The area is easily accessible and a good day's insecting from Dehradun is possible. Thano is 24 Km from Dehradun.

Fig. 13. Song river at Lachhiwala (Source: B.K. Tyagi, 1978).

From the Lachhiwala entry gate, it is a good tarmac road down to the Song river, while passing through a variety of habitats, including Sal and riverine forest and scrub jungle. Total walking distance is about 2 km, one way. At Maldevta, follow the road past Kumalda village, across the iron bridge on the Bandal river and begin ascending into the hills. The road ascends along the river for some distance passing through scrubby hill forests with plentiful Lantana thickets, which attracts numerous insects. This road is worth exploring on foot for insects. The road that leads from Raiput to Thano, passes through protracted patches of Lantana scrub and follow agricultural fields, before moving into the reserved forest with rich stand of Sal.

8. RAIWALA (Fig. 14)

Raiwala is a village in Doiwala tehsil in Dehradun district of Uttarakhand State. It is located 46 km south of the city of Dehradun. Haripur Kalan (4 km) and Birpur Khurd (6 km) are the nearby villages to Raiwala which are surrounded by Haridwar Tehsil towards South, Yamkeshwar Tehsil towards East, Narendra Nagar Tehsil towards North, and Doiwala Tehsil towards West. Raiwala braces the borders of districts of both Dehradun and Tehri Garhwal. The main attraction is the Suswa river which engages in its flow myriads of insect species.

Fig. 14. Suswa river at Raiwala (Source: B.K. Tyagi, 1978).

While *Dysdercus singulatus* is very common, the vast stretches of the sylvan environs along the river is an abode for the uncommon gomphid dragonflies, *Onychogomphus saundersi duaricus* and *Burmagomphus pyramidalis*. The red Fire Ant, *Oecophylla smaragdina,* is abundantly present making nests in the trees.

• • •

9

INSECT ORDER-WISE TAXONOMY, BIOECOLOGY AND DISTRIBUTION, WITH MAJOR REFERENCES TO RESEARCH WORKS DONE IN THE DOON VALLEY

Insects are the most exuberant manifestations of earth's life. It defies imagination to understand how their simple unifying body plan has been modified to produce an enormous variety of species and how it has been able to exploit almost every conceivable type of environment from the equator to the arctic region and from sea level to the snowfields of the highest mountains, on land, in air and in water - almost everywhere! Estimates of the actual number of insects and related forms now living range from 3-30 millions; therefore, whatever the total, a great deal of scientific information on life comes essentially from the insects. The dominance of insects among the world's animal life is a fundamental scientific insight, yet one not widely appreciated. This dominance means that in numbers of species beyond our comprehension these animals permeate diverse and essential natural processes in Earth's terrestrial, aerial and freshwater ecosystems, contributing to the function of the natural world as a self-sustaining biological system.

No other class in the phylum Arthropoda, or for that matter, in the whole animal kingdom, has been taxonomically so complicated, unbridged, incomplete and subjected to repeated reviews and debates for a universally agreeable classification as that of the six-legged, the Insecta. The classification of Insecta which was considered fairly well defined in the early 20th century (Martinov, 1934; cf. Kiauta, 1974) is no more understood to be valid for many orders, notwithstanding appreciable amount of fossils available to support the classification. For good, the molecular tools and techniques have made many revelations of the insect's hitherto unknown characteristics and put many of them altogether at a different hierarchical level or sometimes completely detached from the rest of its old allies (e.g., Protura, Diplura and Collembola). Therefore, before we proceed with the individual insect orders we wish to emphasize to keep in mind a word of caution as to the classification of insects which will at times be discussed briefly as against the conventional system of classification!

The class Insecta or the higher taxa-subphylum Hexapoda (which additionally includes Collembola, Protura and Diplura) is the world's most species-rich group of organisms, with over 1 million described species, and possibly many millions awaiting to be described! They are most diverse in tropical forests, where the un-described fauna has been estimated to comprise 3-30 million species. Hexapods are a well-established monophyletic group, characterized by the presence of three major body divisions - head, thorax, and abdomen - and a single pair of locomotory appendages on each thoracic segment. Some primitive insects have retained appendages on the abdominal segments, but these are much smaller and less functional than those on the thorax. The more diverse groups of insects usually have a pair of wings on each of the mesothoracic and metathoracic segments, which have been lost or modified in some groups, especially the Diptera.

The fossil record of insects goes back to the earliest record of terrestrial life, with Collembola and lower insects recorded from the lower Devonian, almost 400 million years ago, and possibly, there are even earlier traces from the Silurian. Insects have been prominent members of the fossil record ever since, with most prominent major groups having been preserved from late Paleozoic or early Mesozoic formations (200-250 million years ago). There is ample proof that insects were systematically investigated many centuries before Aristotle. Hymns of Atharva Veda on the control of insects attacking crops reflect the variety of pests. Manudharma Sastra (1000 BC) identifies bees and biting insects such as mosquitoes and ants, while the treatise of Charaka (1200 BC) on bees and Shushmita's (100-200 AD) work on stings and his classification of ants (piplika) and flies (mahashikala) are of interest. Amarasimha coined the term shashpada for the Hexapoda nearly 1000 years ago (Ananthakrishnan, 2000).

The kingdom Animalia is represented by 15,52,319 species that have been described so far globally in 40 phyla in a new evolutionary classification. The phylum Arthropoda alone includes 12,42,040 species, constituting about 80% of the total number of species. The most successful group, Insecta, accounts for about 66% (10, 20,007 species in 39 orders) of all animals. The most successful insect order, Coleoptera, represents about 38% (3, 87,100 species) of the insect species of the world (Zhang, 2011). Compilations on the insect fauna of India have been produced from time to time. Maxwell and Howlett (1909) published the book 'Indian Insect Life', wherein 25,700 species of insect were reported from the Indian region, including adjacent countries. Beeson (1941) and Menon (1965) estimated the number of species from India to be 40,000 and 50,000, respectively. In recent times, Varshney (1997) reported 51,450 species under 589 families. Subsequently, Varshney (1998) reported the occurrence of 59,353 species of insect belonging to 619 families in the Indian subcontinent, particularly the hilly areas of the eastern and western Himalaya, the Western Ghats and popular places.

Out of 32 [the 32nd insect order of Mastophasmatoidea, discovered in 2002 by Klass (2002) and Adis (2002) from the Namibian desert in southern Africa is currently debatable as to its status] currently recognized orders of the class Insecta, a total of 31 have so far been reported from India, and about two dozen orders are represented in the Doon Valley. Every one of them is dealt with here under a separate caption. The main taxonomic and chorogeographic features that are important under a given Order are briefly discussed.

It goes without saying that the enormous amount of information available on various different aspects of taxonomy, geographic dispersal, biology, ecology, cytogenetics, cytotaxonomy, molecular biology, biotechnology, economics and control cannot be dealt with *in extenso* on the pages of this book. For each order, therefore, only general features are stated with mainstream reference(s), along with certain amount of data on Indian and, wherever possible, the Doon Valley fauna.

The classification of insects has passed through many changes and with the increased detailed knowledge an increasing number of orders has come to be recognized (e.g., Mastophasmatodea). The taxonomic status of some of these, however, is a matter of personal opinion of the respective research workers and is subject to continuous reorganizations and reclassifications. Where considered necessary, the general problems in this field are briefly stated at the respective places of treatment. Some five decades ago a group of scientists was of the opinion to reconsider the class Insecta subdivided into Apterygota and Pterygota and there were strong reasons in favour of the elevation of each of these to an independent class (*cf.* Kiauta, 1974). More recently, based on molecular phylogenetic studies, scientists have even ventured to exclude the orders Collembola, Protura and Diplura from the class Insecta, and independently upgraded to the class level (*cf.* Kailash Chandra, 2011).

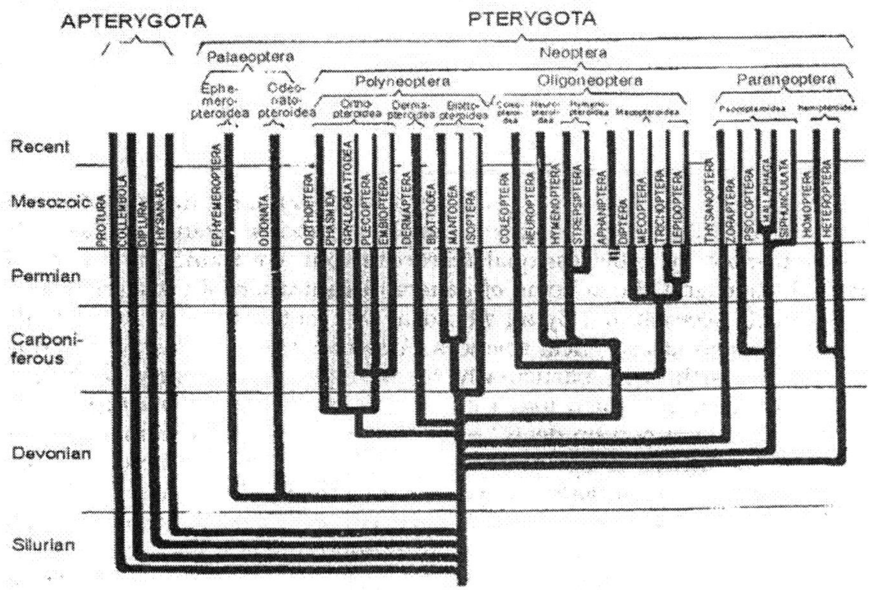

Fig. 15. Phylogenetic tree showing affinity amongst various orders of insects during the process of evolution (Source: Kiauta, 1974).

In India organizations like Zoological Survey of India, Forest Research Institute, Indian Council of Medical Research, Indian Agriculture Research Institute, National Centre for Diseases Control and their constituent instututes in various different regions of the country house remarkably rich depositories/ repositories of various forms of 'Types' of the species of insects.

Keeping in mind the unresolved complexities and intricacies of classification of various insect orders, it is considered indispersable to follow a most comprehensible and currently almost agreeable system. For the sake of convenience of readers, the Martinov's scheme of classification and phylogenetic affinities within the class Insecta, as adopted by Kiauta (1974), has been followed in this treatment (Fig. 15).

• • •

Phylum: Arthropoda

Class: Insecta

Subclass: Apterygota

1. Order Protura

The Protura, or proturans, and sometimes nicknamed coneheads, are very small, soil-dwelling animals, so inconspicuous they were not noticed until the 20th century (Fig. 16). Aparently they carry no known biomedical importance but are certainly a good player in developing a soil biomass especially in relation to food chain of certain macroinvertebrates.

Proturans, with entognathous, piercing mouthparts, are small, wingless, unpigmented hexapods measuring less than 2.5 mm in length. Antennae and eyes are absent. The anterior legs sensory, and all tarsi are one-segmented, with a single claw. The front legs serve as antennae.

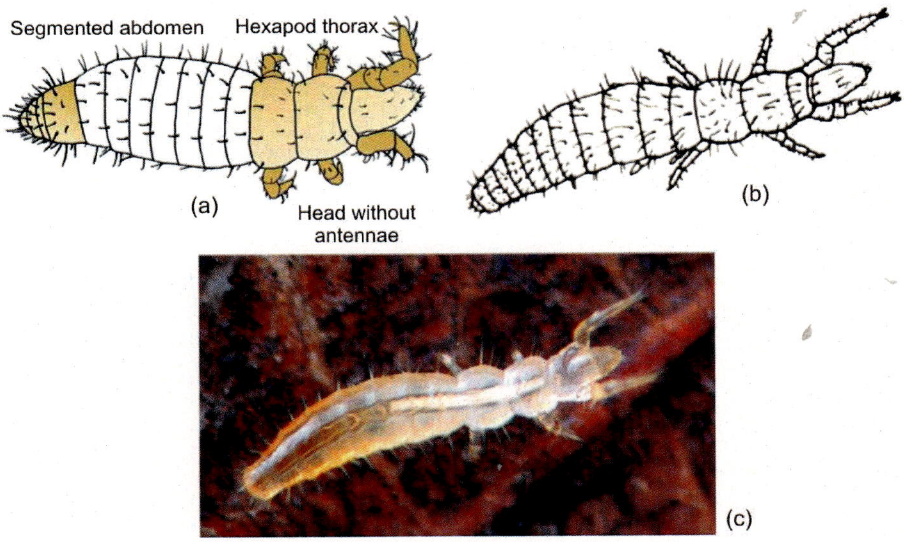

Fig. 16: (a-c). Protura (Source: Wikipedia, the free encyclopedia).

A unique feature is the telson tail, which is common in crustaceans but is absent in other insects. The telson tail is used for locomotion and for defense. There are 804 described species of Protura worldwide. From India, presently 20 species have been reported in 10 genera under 3 families (Prabhu, 1986), of which 17 species are endemic.

2. Order Collembola

Collembola or Springtails form the largest of the three lineages of modern hexapods that are no longer considered insects. They carry obviously no known biomedical importance but are of great significance in engendering soil biomass and food chain of macroinvertebrates.

Springtails are small, wingless, soft-bodied hexapods measuring usually between 0.2 and 5.0 mm (the largest species attaining a maximum length of 6 mm) in length (Fig. 17). The first abdominal segment bears a ventral pair of basally fused vesicles, the collophore; most species have a ventral bifurcate appendage, the tenaculum and furcular – the jumping organ - on the third and fourth abdominal segments. The mouthparts are entognathous (i.e., they are withdrawn into the head capsule). The antennae are usually four-segmented. Compound eyes may be present or absent.

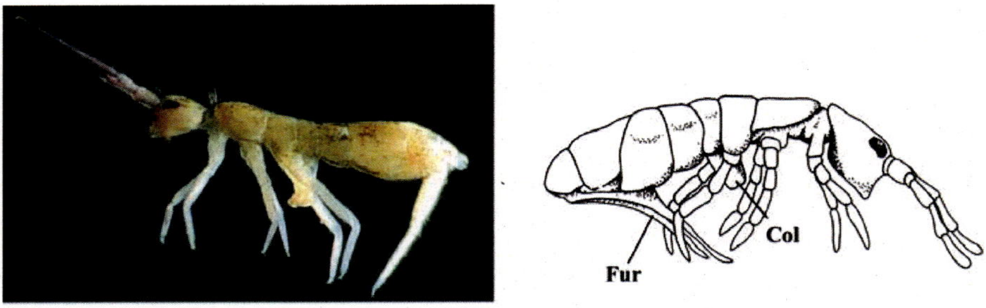

Fig. 17. Collembola (Source: Wikipedia, the free encyclopedia).

With very few exceptions (e.g., Tomocerus), juveniles closely resemble adults. Collembola are abundant and ubiquitous, probably exceeding many insect groups in number, but their small size and sensitivity to desiccation and light restrict most species to humid situations.

Currently there are 8130 known species of spring tails (Collembola) worldwide, classified into the suborders Arthropleona and Symphypleona, that contain 11 families (Janssens, 2012). These are cosmopolitan insects, and pre-eminently inhabitants of the highest alpine elevations. In the Himalaya they flourish still at elevations of 6300 m. The Indian Collembola fauna is represented by 299 species in 103 genera under 19 families (Mandal, 2009), of which 45 species are endemic. A first of its kind, the Zoological Survey of India made a detailed survey in Uttrakhand and reported 31 species under 16 genera of 3 families (ZSI, 2010).

Collembola differ from other apterygotes in more than one feature. The most striking character, however, is the presence of the antennal muscles running throughout from base to the top of the antenna. This very unique feature seems to be of great advantage to the 'Springtails' as it might quite well be functioning the role of the 'regenerator', re-forming the broken off parts of the antennae during the usual hopping manoevours. It appears that the antennal muscles help to regenerate the lost out parts of the antennae and make the latter workable again (Fig. 18; Tyagi & Tyagi, 1981).

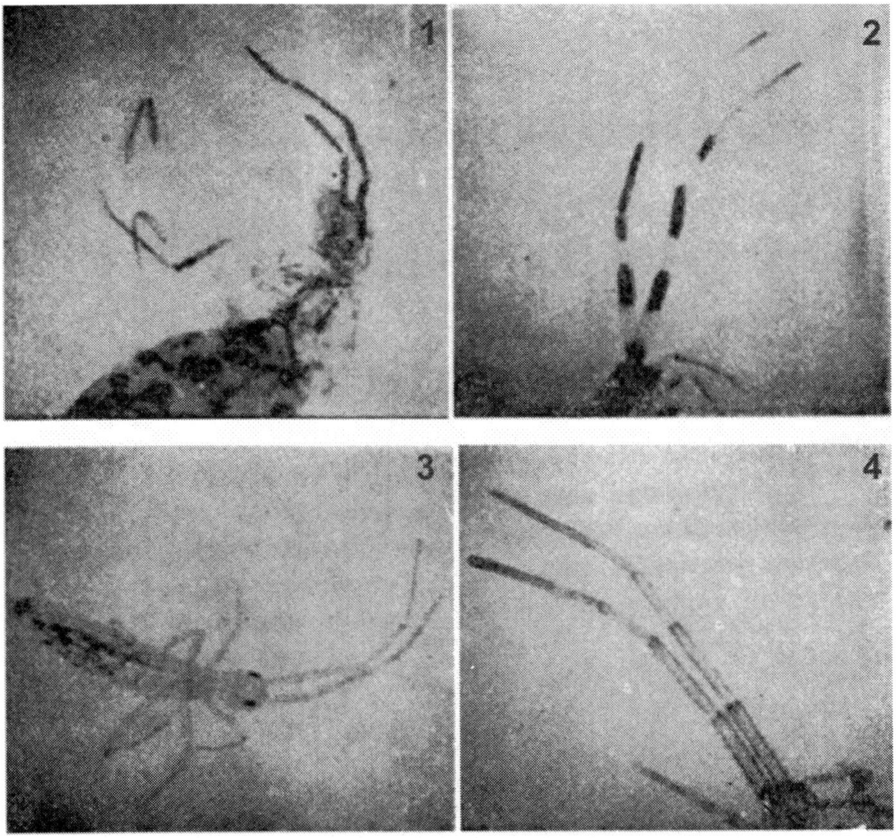

Figs. 18 (1-4). Assymmetrical antennae of some entomobryidae (Collembola) : 1 - 3, note the unequal and regenerated apical segment, while in 2, see also variation in the colour between the regenerated apical segment and its counterpart; 4, note the darkening and swelling of the broken antennal segment at the apex (Source: Tyagi & Tyagi, 1981).

Tyagi & Tyagi (1981) sampled four species of Collembola, e. g., *Salina indica* Imms, *Apysa indica* Handschin, *Entomobrya kali* Imms and *E. multifasciata* Tulberg (Entomobryidae, Arthropleona), from various places along the outer scarps of the Mussoorie Hills, Inner Siwaliks and the Dehra Dun Valley during January-March, 1979. The specimens collected following the technique prescribed by Oldroyd (1963) were authentically identified by Dr. H. N. Baijal of Agra College, Agra, an expert on Indian spring-tails. Dr. Baijal has described many endemic species from the Himalayan region particularly Uttarakhand, e.g., *Entomobrya kultinalensis* Baijal, 1958, *Entomobrya longisticta* Baijal, 1958, *Entomobrya nigrita* Baijal, 1958 and *Entomobrya rohtangensis* Baijal, 1958. (Baijal, 1958).

Although extensive bioecological, chorogeographical and taxonomical investigations have been made in other parts of the country (Pahari et al., 2014); Hazra, 2015), nevertheless the Doon Valley collembolan fauna is yet to receive proper attention as also there is no inventory available for the Doon Valley.

3. Order Diplura

Diplura still remains one of the least studied group of Indian fauna. These are soft bodied wingless insects. Mouthparts are retracted within the head capsule, eyes absent, body without scales (rarely with a few localised ones), cerei long and filiform or shorter and annular or foreep-like, and median tail appendages absent. These insects occur in large numbers in humus soil or in the forest floor throughout India. Knowledge on the Indian Diplura is available due to the contributions of Conde & Thomas (1957). Recently, Singh *et al.* (2008) have conducted some research on this group.

Diplurans are elongate, wingless hexapods usually found in moist soil, forest leaf litter and humus and measure 3-28 mm in length (Fig. 19). The japygids are easily identified by the pincers at the end of the abdomen (modified cerci, projecting backwards from the last of the 11 abdominal somites.), while in the campodeids the cerci are not pincer-like and many segments have long hair. These cerci give the group its name, from the Greek diplo ("two") and uros ("tail"). These cerci may be long and filamentous or short and pincer-like, leading to occasional confusion with earwigs. The mouthparts are entognathous and the antennae are many-segmented (10 or more bead-like segments projecting forward from the head). The flagellar segments are provided with muscles. They have no eyes, i.e., compound eyes and ocelli are absent. The tarsi have 1 or 2 segments. The abdomen is provided with appendages formed from lateral styli, and cerci are present.

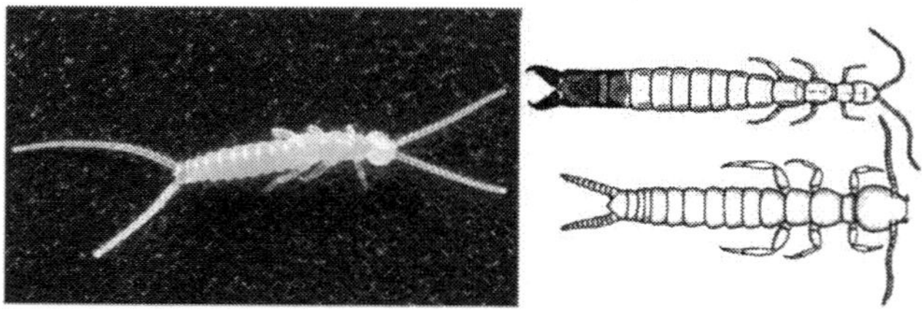

Fig. 19. Diplura (Source: Wikipedia, the free encyclopedia).

Some diplurans have the ability to shed their cerci if necessary (autotomy); of all terrestrial arthropods, only diplurans have been understood to possess the ability to regenerate these lost appendages over a series of moults (*cf.* Tyagi & Tyagi, 1981). Moulting occurs up to 30 times throughout the life of a dipluran, which is estimated to last up to one year.

The abdomens of diplurans bear eversible vesicles, which seem to absorb moisture from the environment and help with the animal's water balance.

Diplurans are common in moist soil, leaf litter or humus, but are rarely seen because of their size and subterranean lifestyles. They have biting mouthparts and feed on a variety of live prey and dead organic matter. Members of the family Japygidae are mainly predatory and use their pincer-like cerci to capture prey, including springtails, isopods, small myriapods, insect larvae, and even other diplurans, while members of the family Campodeidae feed on soil fungi, mites, springtails, and other small soil invertebrates, as well as detritus. Those species with long cerci are herbivorous.

Like other non-insect hexapods, diplurans have external fertilization. Males lay up to 200 spermatophores a week, which are held off the ground by a short stalk and probably only remain viable for about two days. The female collects the spermatophore with her genital opening, and later lays eggs in a cavity in the ground. The hatchlings do not undergo metamorphosis, but resemble the adults, apart from their smaller size, lesser number of setae and their lack of reproductive organs.

The relationships among the four groups of hexapods are not resolved, but most recent studies argue against a monophyletic Entognatha (Carapelli et al., 2006). The fossil record of the Diplura is sparse, but one apparent dipluran dates from the Carboniferous. This early dipluran, *Testa japyx*, had compound eyes, and mouthparts that more closely resembled those of true insects than those of modern diplurans do.

The world fauna of Diplura comprises 976 species, and the Indian fauna is represented by 18 species in 9 genera under 4 families (Mandal, 2010b), of which 12 species are endemic.

4. Order Thysanura

Thysanura is the now deprecated name of an order of the class insecta. The current name is Zygentoma and encompasses the silverfish or fishmoths, and firebrats. The superfamily Machiloidea (families Machilidae and Meinertellidae), the jumping bristletails, were included in the Thysanura until the late twentieth century, but then were allocated to their own order Archaeognatha. The remaining members of the Thysanura comprised the suborder Zygentoma; the suborder was raised to full order status, thereby rendering the term "Thysanura" redundant, leaving Zygentoma as the applicable name (Hoell et al., 1998). However, the momentum of nearly two centuries of nomenclature has preserved the name Thysanura in common use and in much published material. The name Thysanura is derived from the Greek θυσάνος, thysanos for fringe, tassel, bristle and οὐρά, oura for tail, a reference to the three fanned out caudal filaments. Members of the order are best known for their three long caudal filaments.

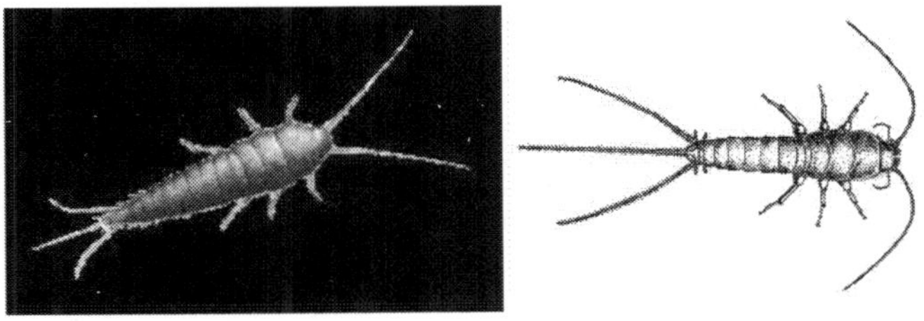

Fig. 20. Thysanura or silverfish (Source: Wikipedia, the free encyclopedia).

The order Thysanura (or Zygentoma) includes some of the most primitive, wingless insects. They are covered with silvery scales. The scales give rise to the common name "silverfish". The order is cosmopolitan. These insects are larger than the other members of the subclass Apterygota and are easily distinguishable from the other closely related groups by their long, many-segmented antennae, 2 anal cerci and single median telson at the terminal part of the abdomen. The mouthparts are ectognathous, adapted for biting (Fig. 20).

Silverfish may be found in moist, humid environments or dry conditions, both as free-living organisms or nest-associates. In domestic settings, they feed on cereals, paste, paper, starch in clothes, rayon fabrics and dried meats. In nature, they will feed on various dead plant matter and animals. Silverfish can sometimes be found in bathtubs or sinks at night, as they have difficulty moving on smooth surfaces and so become trapped. Wild species often are found in habitats such as caves, and some are commensals living in association with ant colonies, e.g., *Trichatelura manni*. There are no current species formally considered to be at conservation risk, though several are troglobites limited to one or a few caves or cave systems, and these species run an exceptionally high risk of extinction.

Lepismatidae is the largest family, widespread with more than 200 species, many living in human habitations. The Nicoletiidae are small and live in soil litter, humus and under stones. The Lepidotrichidae are represented by two species: *Lepidotrix pilifera* from Baltic amber and *Tricholepidion gertschi* from forests of northern California. Three species of Maindroniidae are found in the Middle East and in Chile. The Ateluridae (sometimes treated as a subfamily Atelurinae within the Nicoletiidae) live in nests of ants and termites and are small and blind.

Silverfish have an elaborate courtship ritual to ensure exchange of sperm. The male spins a silken thread between the substrate and a vertical object. He deposits a sperm packet (spermatophore) beneath this thread and then coaxes a female to walk under the thread. When her cerci contact the silk thread, she picks up the spermatophore with her genital opening. Sperms are released into her reproductive system, and then she ejects the empty spermatophore and eats it.

Silverfish continue to moult throughout their life, with several sexually mature instars, unlike the other pterygote insects. They are relatively slow growing, and lifespans of up to four years have been recorded.

Over 1200 species of Thysanura have been reported worldwide (Mendes, 1990). In India, the suborder Archaeognatha is represented by 10 species belonging to 6 genera and 2 families, and the suborder Zygentoma includes 28 species belonging to 15 genera and 3 families (Mandal, 2010c). Among the known taxa, 23 species are endemic to India.

• • •

Subclass: Pterygota

5. Order Ephemeroptera

Commonly known as mayflies, this ancient group of aquatic insects is believed to have evolved some 290 million years ago. Mayflies are classified into three superfamilies. Representatives of the order occur everywhere in the Temperate Zone and in the Alpine regions, flourishing almost up to the permanent snowline. In the Indian Himalaya mayflies are abundant in torrents up to elevations of 5000 m. For the high altitude fauna it is characteristic that it is composed of a few species, but these occur in immense numbers.

Aquatic insects were the theme for a series of six postage stamps issued by Poland on July 16, 1999. Immature and adult stages in the life cycle of a mayfly (*Ephemera vulgata*) are depicted on this stamp.

The immature stages of mayflies are aquatic. They generally live in unpolluted habitats with fresh, flowing water. Some species are active swimmers, others are flattened and cling to the underside of stones, a few are burrowers who dig U-shaped tunnels in the sand or mud. Most species are herbivorous. Their diet consists primarily of algae and other aquatic plant life scavenged from surrounding habitat. Some species mature quickly, in as little as four weeks, while others develop more slowly (one to four years per generation).

Once a mayfly completes development as a naiad, it leaves the aquatic environment, often rising to the water surface in a bubble of air. It quickly moults to a winged form (the subimago) and flies to a nearby leaf or stem. The subimago is a brief transitional stage that moults again into a sexually mature adult (imago). The imago usually has transparent wings and a smooth, shiny exoskeleton in contrast to the cloudy wings and dull, pubescent body of the subimago. Mayflies are the only living insects that moult again after they have wings.

Most adults are delicate insects with a very short lifespan. They do not feed (mouthparts are vestigial), and some species emerge, reproduce, and die in a single day. Males generally fly in swarms that undulate in the air 5-15 meters above the ground. Females fly into the swarm and are quickly grabbed by a male. Copulation takes place in flight, and the female usually lays her clutch of eggs within minutes or hours. Males die shortly after mating; females usually die soon after oviposition.

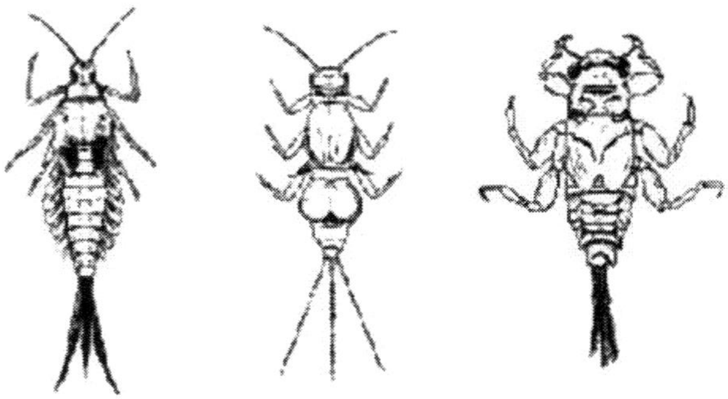

Fig. 21. Immature Ephemeroptera (Source: Wikipedia, the free encyclopedia).

Immatures have antennae short and bristle-like, four to nine pairs of leaf-like or fan-like gills along the sides of the abdomen, and three long filaments at rear of abdomen (Fig. 21).

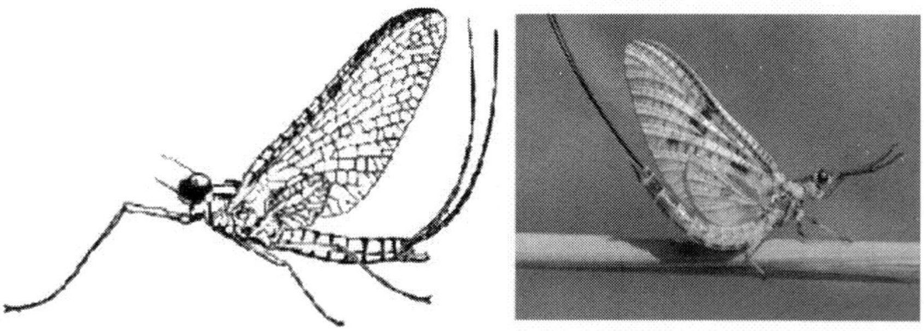

Fig. 22. Adult Ephemeroptera (Source: Wikipedia, the free encyclopedia).

Adults have antennae short and bristle-like; front legs long and often held out in front of body; Compound eyes large, usually covering most of the head; Wings: four membranous wings with many veins and crossveins; front wings large, triangular; hind wings smaller, fan-shaped; Abdomen slender, bearing two (or sometimes three) long terminal filaments (Fig. 22)

Many northern lakes and rivers (in both the United States and Europe) support unbelievably large populations of mayflies. The naiads are "ecological indicators" of good water quality and are an important source of food for fish and other aquatic wildlife. Anglers often use mayflies as bait, or tie "flies" that are made to resemble the imagos (spinners) or subimagos (dun). In some mayfly species, summer emergence of winged stages is a sudden and dramatic event that occurs almost simultaneously throughout the entire population. These mass emergences are often regarded as a major nuisance. The insects are attracted to city lights and blown inland by the wind. Their dead bodies pile up in drifts on porches and windowsills; they plaster car windshields and slicken highways. Europeans are often more sanguine than Americans about mass emergences - in some communities the dead insects are diligently gathered up, dried, and sold as bird food, fish bait, or fertilizer. Although mass emergences still occur, the populations are not as large as in the past. Urban development and water pollution in major lakes and streams has dramatically reduced mayfly populations over the past 50 years.

Mayflies have some atypical features of biology. The subimagos of mayflies are the only insects that moult when they have wings. Mayflies have paired genital openings. During copulation, the two penes of the male are inserted simultaneously into the two openings of the female. Sperm is transferred quickly (there is no spermatophore) and eggs are fertilized immediately. A few species of mayflies reproduce parthenogenetically -- no males have ever been found. Although most mayflies are herbivores, a few are predaceous, e.g., Siphloneuridae and Oligoneuriidae. Adult mayflies do not feed. Their digestive system is filled with air, making them light enough to float. In the Congo, there is one species of mayfly that excavates tunnels in fresh-water sponges. Some mayfly species require up to four years to complete development. In that time they may moult more than 20 times.

Recently mayflies were studied for their taxonomy and seasonal variation in the headwaters of Doon valley streams; Asan and Tawa, between October 2007 and September 2008. Total density of the macroinvertebrate fauna varied and differed significantly among the seasons in both the rivers. Insects were a major group followed by molluscs and annelids. The abundance of orders Ephemeroptera, Trichoptera and Diptera varied from summer to winter in both Asan and Tawa streams. In both the streams the abundance of Baetidae (Ephemeroptera) prevailed over Hydropsychidae (Trichoptera) during winter and summer in the Asan and during summer and monsoon in the Tawa. Gathering collector followed by scraper were the major functional feeding groups in both the streams of Doon region. The study indicated structural and functional similarity of the benthic macroinvertebrate fauna and follow 'Biome Dependence Hypothesis'. Appreciable work has been done on the benthic macroinvertebrate fauna in Garhwal region's rivers, such as Ganga (Joshi, et al., 2007; Nautiyal & Semwal, 2006; Nautiyal, 1996; Nautiyal, 2001, Nautiyal *et al.,* 2013).

More than 3000 species of mayfly belonging to 400 genera and 42 families are presently known globally. Of these, 124 species under 46 genera and 12 families have been reported from India, including 72 species endemic to India (Ramamoorthy *et al.,* 2009; Sivaramakrishnan *et al.,* 2009).

6. Order Odonata

Odonata (Dragonflies and Damselflies) are classified into three suborders (Zygoptera, Anisozygoptera and Anisoptera) and 28 families (Tillyard, 1917; Westfall & May, 1996; Needham et al., 2000). Although this classification was settled long ago for almost a century (Fraser, 1933, 1934, 1936; Fraser & Tillyard, 1957), yet the modern tools have added much to our revamped understanding about odonate classification and, though still debatable, offered a combined suborder Epiprocta (in which Anisoptera is an infraorder) when it was found that the "Anisozygoptera" was paraphyletic, i.e., composed of mostly extinct offshoots of dragonfly evolution. The four living species placed in that group are in the infraorder Epiophlebioptera, whereas the fossil taxa that were formerly there are now dispersed about the Odonatoptera (or Odonata *sensu lato*) (Lohmann, 1996; Bechly, 2002).

Odonata is an order of carnivorous insects, encompassing the dragonflies, a couple of relic dragonflies and the damselflies. The Odonata form a clade, which has existed since the Triassic. These are essentially and subtropical insects, though a considerable number is found in temperate zone and a few species penetrated far into the Subarctic. On higher alpine elevations they appear as occasional migrants. They occur in all faunal regions, and the Indian Subcontinent is famous for its rich and relatively well explored dragonfly fauna (*cf.* Kiauta, 1974). Although a large number of new species have been described in India following the Fraser's *magnum opuses* two new species, viz., *Agriocnemis corbeti* Kumar & Prasad, 1978 and *Calicnemia doonensis* Sangal & Tyagi, 1984 were described from the Doon Valley (Kumar & Prasad, 1978; Sangal & Tyagi, 1984; Hamalainen, 1989; Prasad and Varshney, 1995; Tyagi, 1997, 2007).

In Japan, this dragonfly (*Boninthemis insularis*) is known as "Shimaakane". The Japanese government officially designated this species a national landmark on March 12, 1969. This stamp (one of four in a set illustrating insects) was issued on September 14, 1977

These insects characteristically have large rounded heads covered mostly by well-developed, compound eyes, legs that facilitate catching prey (other insects) in flight, two pairs of long, transparent wings that move independently, and elongated abdomen. They have three ocelli and short antennae. The mouthparts are on the underside of the head and include simple chewing mandibles in the adult (Kumar & Khanna, 1983).

In most families there is a structure on the leading edge near the tip of the wing called the pterostigma. This is a thickened, hemolymph–filled and often colourful area bounded by veins. The functions of the pterostigma are not fully known, but it most probably has an aerodynamic effect and may also have a visual function. More mass at the end of the wing may also reduce the energy needed to move the wings up and down. The right combination of wing stiffness and wing mass could reduce the energy consumption of flying. A pterostigma is also found among other insects, such as bees.

The nymphs have stockier, shorter, bodies than the adults. In addition to lacking wings, their eyes are smaller, their antennae longer, and their heads are less mobile than in the adult. Their mouthparts are modified, with the labium being adapted into a unique prehensile organ for grasping prey. Damselfly nymphs breathe through external gills on the abdomen, while dragonfly nymphs respire through an organ in their rectum.

Fig. 23 (a-d). Damsel- and Dragonflies (a-c: Source; Wikipedia, the free encyclopedia; d: Source; B.K. Tyagi, *pers. comm.*).

Although generally fairly similar, dragonflies differ from damselflies in several, easily recognizable traits. Dragonflies are strong fliers with fairly robust bodies and at rest hold their wings either out to the side or out and downward (or even somewhat forward). Damselflies tend to be less robust, even rather weak appearing in flight, and when at rest most species hold their wings folded back over the abdomen (Fig. 23). Dragonfly eyes occupy much of the animal's head, touching (or nearly touching) each other across the face. In damselflies, there is typically a gap in between the eyes.

Odonates are aquatic or semi-aquatic as juveniles. Thus, adults are most often seen near bodies of water and are frequently described as aquatic insects. However, many species range far from water. They are carnivorous throughout their life, mostly feeding on smaller insects.

Male Odonata have complex genitalia, different from those found in other insects. These include grasping cerci for holding the female and a secondary set of copulatory organs on the abdomen in which the sperm are held after being produced by the primary genitals. To mate, the male grasps the female by the thorax or head and bends her abdomen so that her own genitalia can be grasped by the copulatory organs holding the sperm. Male odonates have a copulatory organ on the ventral side of abdominal segment 2 in which they store spermatozoa; they mate by holding the female's head (Anisoptera) or thorax (Zygoptera) with claspers located at the tip of the male abdomen; the female bends her abdomen forward to touch the male organ and receive sperm. This is called the "wheel" position.

Eggs are laid in water or on vegetation near water or wet places, and hatch to produce pronymphs which live off the nutrients that were in the egg. They then develop into instars with approximately 9–14 moults that are (in most species) voracious predators on other aquatic organisms, including small fishes. The nymphs grow and moult, usually in dusk or dawn, into the flying teneral immature adults, whose colour is not yet developed. These insects later transform into reproductive adults.

Odonates can act as bioindicators of water quality in rivers because they rely on good quality water for proper development in early life. Since their diet consists entirely of insects, odonate density is directly proportional to the population of prey, and their abundance indicates the abundance of prey in the examined ecosystem. They have also been exploited as biocontrol agents for vector mosquitoes (Venkatesh & Tyagi, 2013).

Odonata of the Doon Valley have been most extensively explored and described by a large number of workers prominent among which are Dr. Arun Kumar, Dr. M. Prasad, Dr. Vinod Khanna, and Dr. B.K. Tyagi (*cf.*, Kumar and Prasad, 1981; Tyagi, 2007).

Three independent catalogues of extant taxa of Odonata have so far been put forth after independence of India, with much change in its geographical limits than what was available in Fraser's classics published in the Fauna of British India (Prasad & Varshney, 1995; Tyagi, 1977, 1978a-c, 1986, 1997; Subramanian, 2014). Prasad and Varshney (1995) published a revised and updated check-list of 499 species and subspecies of Indian Odonata under 139 genera in 17 families, 32 subfamilies and 7 superfamilies. Tyagi (1997) has checklisted valid and extant taxa of Indian odonata, after consulting both Fraser (1933, 1934, 1936) and Prasad & Varshney (1995), and summed up the total of species and subspecies as 491. Singh & Prasad (1976) and Prasad & Singh (1976) collectively recorded 69 odonate species and subspecies from the DoonValley. Later, Tyagi et al. (1986) reported over 90 species, though they listed only 80 of these in their catalogue (due to a misprint only). Hamalainen (1989) listed 32 taxa, brought together during his short tour, and synonymized some of them. Kumar & Mitra (1998) recorded 42 species from the Sahastradhura region of Doon Valley, and added 3 new records. Recently, Mitra (1999) published 2 new records, which are new for the Western Himalaya as well.

7. Order Orthoptera

The name is derived from the Greek '*ortho*' meaning 'straight' and '*ptera*' meaning 'winged'. Grasshoppers are able to fold their wings, thereby placing them in the group Neoptera. The Order as currently understood to be divided into the suborders Ensifera and Caelifera. Locusts, crickets, and grasshoppers are essentially a thermophilous group, with its main distribution in the tropics, though members occur in all but the coolest zones. The highest elevation at which specifically identified Orthoptera seem to flourish and breed permanently is 4900 m on the Mt. Everest Massif, but larvae of an unidentified acridid were reported from an elevation of 5490 m in the Everest region, representing at present the highest known altitude record for the order.

The red locust (*Nomadacris septemfasciata*) is an important pest of field crops in central Africa. This stamp was issued by the government of Malawi on September 21, 1991.

Orthoptera is generally regarded as a dominant group in most terrestrial habitats. These insects feed on all types of plants and often cause serious economic damage. Swarms of grasshoppers (locusts) regularly appear in parts of Africa, Asia, and North America and destroy crops over wide land areas. Mole crickets are major pests in lawns and golf courses in the southern United States. Several species of field crickets are reared commercially as fish bait.

In many species of Orthoptera, the males use sound signals (chirping or whirring) in order to attract a mate. The sound is produced by stridulation -- rubbing the upper surface of one wing against the lower surface of another wing, or the inner surface of the hind leg against the outer surface of the front wing.

Each stridulating species produces a unique mating call. In fact, some species may be so similar to each other that they can only be distinguished by their mating calls. Many grasshoppers produce ultrasonic mating calls (above the range of human hearing). In some species, the sounds may be as high as 100 kHz (Human hearing extends to about 20 kHz). Species that produce sound also have auditory (tympanal) organs. In crickets and katydids, these "ears" are on the tibia of the front legs. In grasshoppers, they are on the sides of the first abdominal segment (Fig. 24).

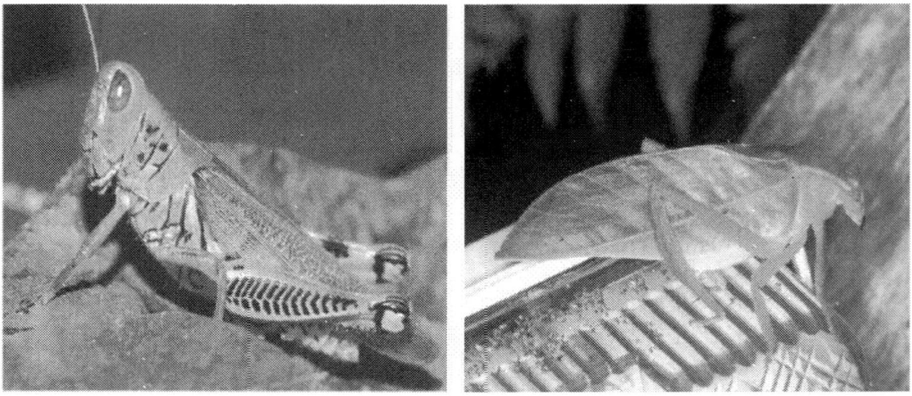

Fig. 24. (a) a cricket and (b) a Katitid (Source: Wikipedia, the free encyclopedia).

The snowy tree cricket, *Oecanthus fultoni* (family Gryllidae), is often called the temperature cricket. Adding 40 to the number of chirps it makes in 15 seconds will equal the ambient temperature in degrees Fahrenheit.

The redlegged grasshopper *Melanoplus femurrubrum* is not only a crop pest but also the intermediate host for a tapeworm *Choanotaenia infundibulum* that infests poultry.

Orthopterans have a generally cylindrical body, with hind legs elongated for jumping. They have mandibulate mouthparts for biting and chewing and large compound eyes, and may or may not have ocelli, depending on the species. The antennae have multiple joints and filiform type, and are of variable length (Hoell *et al.,* 1998). The first and third segments on the thorax are larger, while the second segment is much smaller. They have two pairs of wings, which are held overlapping the abdomen at rest. The forewings, or tegmina, are narrower than the hindwings and hardened at the base, while the hindwing is membranous, with straight veins and numerous cross-veins. At rest, the hindwings are held folded fan-like under the forewings. The final two to three segments of the abdomen are reduced, and have single-segmented cerci and their wing type is tegmina.

Orthopteroid species have a paurometabolous lifecycle or incomplete metamorphosis. The use of sound is generally crucial in courtship, and most species have distinct songs. Most grasshoppers lay their eggs in the ground or on vegetation. The eggs hatch and the young nymphs resemble adults, but lack wings and at this stage are often called 'hoppers'. They may often also have a radically different colouration from the adults. Through successive moults, the nymphs develop wings until their final moult into a mature adult with fully developed wings. The number of moults varies between species; growth is also very variable and may take a few weeks to some months depending on food availability and weather conditions.

Katydids, a kind of Tettigoniidae, offers an arcane type of prowess to hear sound even of ultralow frequency!

In many countries (Thailand, China etc.) many types of orthopterans are considered a delicacy being rich in proteins, mineral etc.

India has been a major research ground for various kinds of orthopteran particularly those belonging to Grasshoppers group as they are serious pests of agriculture. Classical work on Indian orthopterans exists by Roonwal (1936, 1037, 1938, 1940) and Pruthi (1954). Roonwal (1936) had proposed a new theory on migration of the locust, *Locusta migratoria migratoria.*

8. Order Phasmida

The Phasmatodea (also known as Phasmida or Phasmatoptera) are an order of insects, whose members are variously known as stick insects (in Europe and Australasia), stick-bugs or walking sticks (in the United States and Canada), phasmids, ghost insects and leaf insects (generally the family Phylliidae). Stick- and leaf insects are a prominently tropical group. Members of the order are found all over the world except for the Antarctic and Patagonia, though most species occur in Oriental region, being most diverse in Southeast Asia and South America, followed by Australia, Central America, and the southern United States. Over 300 species are known from the island of Borneo, making it the richest place in the world for Phasmatodea. They are divided into three families, of which a sizable number is found in India. The group's name is derived from the Ancient Greek φάσμα (phasma), meaning an apparition or phantom, referring to the resemblance of many species to sticks or leaves. Their natural camouflage makes them difficult for predators to detect, but many species have a secondary line of defence in the form of startle displays, spines or toxic secretions. The genus Phobaeticus includes the world's longest insects. Phasmids can be relatively large, ranging from 1.5 cm to over 30 cm in length. Females of the genus Phobaeticus are the world's longest insects, measuring up to 56.7 cm in total length in the case of *Phobaeticus chani*, including the outstretched legs. The heaviest species of phasmid is likely to be *Heteropteryx dilatata*, the females of which may weigh as much as 65 g (Fig. 25).

They are herbivorous with many species living unobtrusively in the tree canopy. They have a hemimetabolous life cycle with three stages: eggs, nymphs and adults. Many phasmids are parthenogenic, and do not require fertilized eggs for female offspring to be produced. In hotter climates, they may breed all year round; in more temperate regions, the females lay eggs in the autumn before dying, and the new generation hatches out in the spring. Some species have wings and can disperse by flying, while others are more restricted.

Fig. 25. Different phasmids (Source: Wikipedia, the free encyclopedia).

Phasmids have a range of shape and form, best to camouflage, deimatic behaviour and/or to attack. Phasmatodea species exhibit mechanisms for defense from predators that prevent an attack from happening in the first place (primary defense), and defenses that are deployed after an attack has been initiated (secondary defense). Some phasmids have cylindrical stick-like bodies, while others have flattened, leaflike shapes. Many species are wingless, or have reduced wings. The thorax is long in the winged species, since it houses the flight muscles, and is typically much shorter in the wingless forms. Where present, the first pair of wings is narrow and cornified (hardened), while the hind wings are broad, with straight veins along their length and multiple cross-veins. The body is often further modified to resemble vegetation, with ridges resembling leaf veins, bark-like tubercles, and other forms of camouflage. A few species, such as *Carausius morosus*, are even able to change their pigmentation to match their surroundings. The mouthparts project out from the head. Chewing mandibles are uniform across species. The legs are typically long and slender, and some species are capable of limb autotomy (shedding). Phasmids have long, slender antennae, as long as or longer than the rest of the body in some species (Hoell et al., 1998).

Phasmids have an impressive visual system that allows them to perceive significant detail even in dim conditions, which suits their typically nocturnal lifestyle. All phasmids possess compound eyes, but ocelli (light-sensitive organs) are only found in some winged males. They are born equipped with tiny compound eyes with a limited number of facets. As phasmids grow through successive moults, the number of facets in each eye is increased along with the number of photoreceptor cells. The sensitivity of the adult eye is at least tenfold that of the nymph in its first instar (developmental stage). As the eye grows more complex, the mechanisms to adapt to dark/light changes is also enhanced: eyes in dark conditions evidence less screening pigments, which would block light, than during the daytime, and changes in the width of the retinal layer to adapt to changes in available light are significantly more pronounced in adults. However, the larger size of the adult insects' eyes makes them more prone to radiation damage. This explains why fully grown individuals are mostly nocturnal. Lessened sensitivity to light in the newly emerged insects helps them to escape from the leaf litter wherein they are hatched and move upward into the more brightly illuminated foliage. Young stick insects are diurnal (daytime) feeders and move around freely, expanding their foraging range (Meyer-Rochow & Keskinen, 2003). Stick insects have two types of pad on their legs: sticky "toe pads" and non-sticky "heel pads" a little further up their legs. The heel pads are covered in microscopic hairs which create strong friction at low pressure, enabling them to grip without having to be peeled energetically from the surface at each step. The sticky toe pads are used to provide additional grip when climbing but are not used on a level surface.

A special mention is worth quoting here about the prowess of camouflaging, mostly in the form of mimicry, by the phasmids. Most phasmids are known for effectively replicating the forms of sticks and leaves, and the bodies of some species (such as *Pseudodiacantha macklotti* and *Bactrododema centaurum*) are covered in mossy or lichenous outgrowths that supplement their disguise. Remaining absolutely stationary enhances their disguise. Some species have the ability to change colour as their surroundings shift (*Bostra scabrinota, Timema californica*). In a further behavioural adaptation to supplement crypsis, a number of species perform a rocking motion where the body is swayed from side to side; this is thought to mimic the movement of leaves or twigs swaying in the breeze. Another method by which stick insects avoid predation and resemble twigs is by entering a cataleptic state, where the insect adopts a rigid, motionless posture that can be maintained for a long period. The nocturnal feeding habits of adults also help Phasmatodea to remain concealed from predators. In a seemingly opposite method of defense, many species of Phasmatodea seek to startle the encroaching predator by flashing bright colours that are normally hidden, and making a loud noise. When disturbed on a branch or foliage, some species, while dropping to the undergrowth to escape, will open their wings momentarily during free fall to display bright colours that disappear when the insect lands. Others will maintain their display for up to 20 minutes, hoping to frighten the predator and convey the appearance of a larger size. Some accompany the visual display with noise made by rubbing together parts of the wings or antennae.

Some species, such as the young nymphs of *Extatosoma tiaratum*, have been observed to curl the abdomen upwards over the body and head to resemble ants or scorpions in an act of mimicry, other defense mechanism by which the insects avoid becoming prey. The eggs of some species such as *Diapheromera femorata* have fleshy projections resembling elaiosomes (fleshy structures sometimes attached to seeds) that attract ants. When the egg has been carried to the colony, the ant feeds the elaiosome to a larva and the phasmid egg develops in the recesses of the nest in a protected environment. When threatened, some phasmids that are equipped with femoral spines on the metathoracic legs (*Oncotophasma martini, Eurycantha calcarata, Eurycantha horrida, Diapheromera veliei, Diapheromera covilleae*) respond by curling the abdomen upward and repeatedly swinging the legs together, grasping at the threat. If the menace is caught, the spines can draw blood and inflict considerable pain. Certain Phasmatodea, such as *Anisomorpha buprestoides*, sometimes form aggregations. These insects have been observed to congregate during the day in a concealed location, going their separate ways at nightfall to forage, and returning to their refuge before dawn. Such behaviour has been little studied, and how the insects find their way back is unknown.

Phasmids also use a kind of a chemical warfare as some species are equipped with a pair of glands at the anterior (front) edge of the prothorax that enables the insect to release defensive secretions, including chemical compounds of varying effect: some produce distinct odors, and others cause a stinging, burning sensation in the eyes and mouth of a predator. The spray often contains pungent-smelling volatile metabolites, previously thought to be concentrated in the insect from its plant food sources. Additionally, the chemistry of the defense spray from at least one species, *Anisomorpha buprestoides*, has been shown to vary based on the insect's life stage or the particular population it is part of. The spray from one species, *Megacrania nigrosulfurea*, is used as a treatment for skin infections by a tribe in Papua New Guinea because of its antibacterial constituents. Some species employ a shorter-range defensive secretion, where individuals bleed reflexively through the joints of their legs and the seams of the exoskeleton when bothered, allowing the blood (hemolymph), which contains distasteful additives, to discourage predators. Another ploy is to regurgitate their stomach contents when harassed, repelling potential predators.

Phasmids have a peculiar way to start life when the female deposits her eggs through one of these methods of oviposition: she will either flick her egg to the ground by a movement of the ovipositor or her entire abdomen, carefully place the eggs in the axils of the host plant, bury them in small pits in the soil, or stick the eggs to a substrate, usually a stem or leaf of the food plant. A single female lays from 100 to 1,200 eggs after mating, depending on the species. Parthenogenesis is common among many species of phasmids. *Acanthoxyla prasina* or the prickly stick insect, native to New Zealand, is believed to reproduce by parthenogenesis; no males have been found. One Australian species, the Lord Howe Island stick insect, is now listed as critically endangered. It was believed extinct until its rediscovery on the rock known as Ball's Pyramid. An effort is underway in Australia to rear this species in captivity. The best known of the stick insects is the Indian or laboratory stick insect (*Carausius morosus*). This insect grows to roughly 10 cm (4 in) and reproduces parthenogenically, and although males have been recorded, they are rare. Eggs from virgin mothers are entirely female and hatch into nymphs that are exact copies of their mothers. Stick insect species that are the product of hybridization are usually obligate parthenogens but non-hybrids are facultative parthenogens, meaning they retain the ability to mate and their sexual behavior depends on the presence and abundance of males. Phasmatodea eggs resemble seeds in shape and size, and have hard shells. They have a lid-like structure called an operculum at the anterior pole, from which the nymph emerges during hatching. The eggs vary in the length of time before they hatch which varies from 13 to more than 70 days, with the average around 20 to 30 days. Some species, particularly those from temperate regions, undergo diapause, where development is delayed during the winter months. Diapause is initiated by the effect of short day lengths on the egg-laying adults or can be genetically determined. Diapause is broken by exposure to the cold of winter, causing the eggs to hatch during the following spring. Among species of economic importance such as *Diapheromera femorata*, diapause results in the development of two-year cycles of outbreaks. Many species' eggs bear a fatty, knoblike capitulum that caps the operculum. This structure attracts ants because of its resemblance to the elaiosome of some plant seeds that are sought-after food sources for ant larvae, and usually contribute to ensuring seed dispersal by ants, a form of ant-plant mutualism called myrmecochory. The ants take the egg into their nest underground and can remove the capitulum to feed to their larvae without harming the phasmid embryo. There, the egg hatches and the young nymph, which initially resembles an ant (another instance of mimicry among Phasmatodea), eventually emerges from the nest and climbs the nearest tree to safety in the foliage.

The Phasmatodea life cycle is hemimetabolous, proceeding through a series of several nymphal instars. As is the case with hatching, if the nymph is caught in the encasing of a cast skin or egg capsule, it will likely die because it cannot free itself. Once emerged, the. nymphs will eat the cast skin. Adulthood is reached for most species after several months and many moults. The lifespan of Phasmatodea varies by species, but ranges from a few months to up to three years. Their role in the forest ecosystem is considered important by many scientists, who stress the significance of light gaps in maintaining succession and resilience in climax forests. The presence of phasmids lowers the net production of early successional plants by consuming them and then enriches the soil by defecation. This enables the late succession plants to become established and encourages the recycling of the tropical forest.

As phasmids are herbivorous and cause much damage to forests they are recognized as injurious to forest and shade trees by defoliation. The insects eat the entire leaf blade. In the event of heavy outbreaks, entire stands of trees can be completely ravaged. Continuous defoliation over several years often results in the death of the tree. Control efforts in the case of infestations have typically involved chemical pesticides. In New South Wales, research has been done into the feasibility of controlling stick insects using natural enemies such as parasitic wasps (Myrmecomimesis spp.) (Campbell, 1959).

Phasmids' mating behaviour is quite arcane. It is impressive because of the extraordinarily long duration of some pairings. A record among insects, the stick insect *Necroscia sparaxes*, found in India, is sometimes coupled for 79 days at a time. It is not uncommon for this species to assume the mating posture for days or weeks on end, and among some species (*Diapheromera veliei* and *D. covilleae*), pairing can last three to 136 hours in captivity. Overt displays of aggression between males over mates suggests that extended pairing may have evolved to guard females from sperm competition. It is known that tribes in Sarawak eat phasmids and their eggs.

9. Order Grylloblattodea

The name Grylloblattodea, members of which are commonly called Rock Crawlers and Icebugs, is derived from the Greek "*gryll*" meaning cricket and "*blatta*" meaning cockroach, refering to the blend of cricket-like and roach-like traits found in these insects. Grylloblattidae is the only family in this order with four genera and 25 species. They are hemimetabolous, i.e., have incomplete development (egg, nymph, adult), orthopteroid, i.e., closely related to Orthoptera and Dermaptera, and are in distribution being found in caves or near ice or snow at high elevations in mountains of Asia and North America. Rock crawlers are a small and obscure group of insects found only at high elevations in the mountains of China, Siberia, Japan, and western United States and Canada. No grylloblattids have ever been found in the Southern Hemisphere. Rock crawlers may require up to seven years to complete a single generation. Cave-dwelling species have been found in Korea and Japan. These omnivorous insects scavenge for food on the surface of snowfields, under rocks, or near melting ice. They are active only at cold temperatures and move downward toward permafrost during warm seasons. As their ordinal name implies, rock crawlers have a blend of physical characteristics from both crickets (gryllo-) and cockroaches (blatta-). The taxonomic status of this order is quite debatable since some taxonomists include these insects as a suborder or family within Orthoptera while others believe these insects are the only survivors of a primitive lineage that gave rise to other orthopteroid orders.

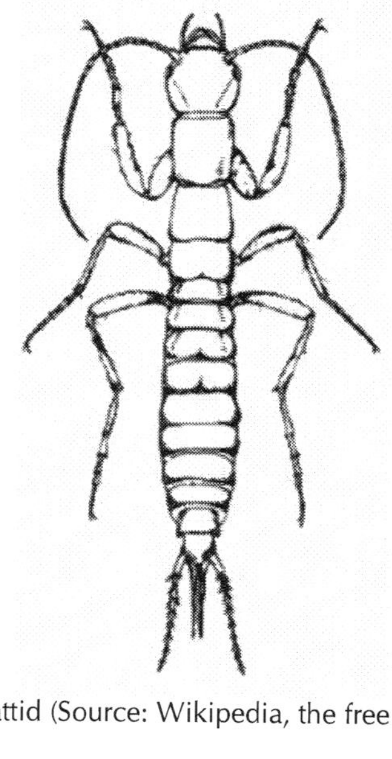

Fig. 26. A grylloblattid (Source: Wikipedia, the free encyclopedia).

These insects bear cylindrical body with slender and filiform antennae. They have mandibulate and hypognathous mouthparts. Tarsi on all legs are 5-segmented. They are secondarily wingless and their cerci are long and 8-segmented (Fig. 26).

10. Order Mantophasmatodea

For the first time in 90 years, a new order of insects has been discovered making the count to 32 insect oders. No new order of insects had been earlier identified since 1915!

Popularly known as gladiators, rock crawlers, heelwalkers, mantophasmids, and mantos, these insects live in the Brandberg Mountains of Namibia, on the west coast of Southern Africa. Entomologist Oliver Zompro of the Max Planck Institute of Limnology in Plön, Germany, who identified the creature as unique, claimed it to be uniquely placed between a stick insect and a grasshopper. The mantophasmatodan, however, differs from a stick insect because its first body segment is the largest. At the same hand, unlike a stick insect or mantid, it uses both its fore and mid legs to catch prey, and unlike a grasshopper, it can't jump. Growing up to 4 cm, "the gladiator" is carnivorous and nocturnal. It lives at the base of clumps of grass that grow in rock crevices. Their modern centre of endemism is western South Africa and Namibia (Brandberg Massif), although a relict population, Eocene fossils suggest a wider ancient distribution (Adis *et al.*, 2002) (Fig. 27).

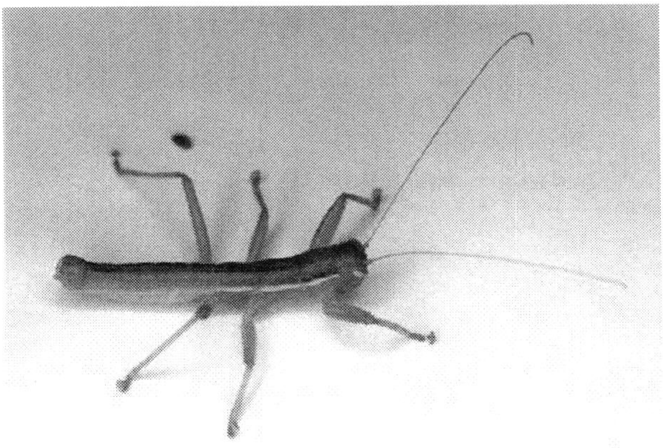

Fig. 27. *Mantophasma zephyrum* Zompro et al., 2002

Members of the order are wingless even as adults, making them relatively difficult to identify. They resemble a mix between praying mantids and phasmids, and molecular evidence indicates that they are most closely related to the equally enigmatic group, Grylloblattodea, with which they have now been debatably grouped together in the order Notoptera. The gladiators initially were described from old museum specimens that originally were found in Namibia

(*Mantophasma zephyrum*) and Tanzania (*M. subsolanum*), and from a 45-million-year-old specimen of Baltic amber (*Raptophasma kerneggeri*). Live specimens of *Mantophasma zephyrum* Zompro *et al.*, 2002 was found on the Erongoberg Massif in Namibia by an international expedition in early 2002, while another species of *Tyrannophasma* was found in the Brandberg Massif.

11. Order Plecoptera

Stoneflies or Plecoptera are classified by different workers into a different number of families. They occur particularly in cooler parts of the Temperate Zone of both the New and the Old world and are common in the alpine region of all mountain systems. There is a remarkable quantitative abundance of the individuals of relatively few species at high altitudes. Very little is known on the Indian fauna. Some 3,500 species are described worldwide, with new species still being discovered. Stoneflies are believed to be one of the most primitive groups of Neoptera, with close relatives identified from the Carboniferous and Lower Permian geological periods, while true stoneflies are known from fossils only a bit younger. The modern diversity, however, apparently is of Mesozoic origin. Their presence in a stream or still water is usually an indicator of good or excellent water quality since all species of Plecoptera are intolerant of water pollution (Fig. 28).

Fig. 28. (1) *Perlesta decipiens,* and (2) *Triznaka signata* (*Source*: C. Riley Nelson, 1996)

Stoneflies have a generalised anatomy, with few specialised features compared to other insects. They have simple mouthparts with chewing mandibles, long, multiple-segmented antennae, large compound eyes, and two or three ocelli. The legs are robust, with each ending in two claws. The abdomen is relatively soft, and may include remnants of the nymphal gills even in the adult. Both nymphs and adults have long, paired cerci projecting from the tip of their abdomens. The name "Plecoptera" literally means "braided-wings", from the Ancient Greek *plekein* (πλέκειν, "to braid") and *pteryx* (πτέρυξ, "wing"). This refers to the complex venation of their two pairs of wings, which are membranous and fold flat over their backs. Stoneflies are generally not strong fliers, and some species are entirely wingless. A few wingless species, such as the Lake Tahoe benthic stonefly (*Capnia lacustra*) or *Baikaloperla* sp. are the only known insects, perhaps with the exception of *Halobates* sp., that are exclusively aquatic from birth to death. Some true water bugs (Nepomorpha) may also be fully aquatic for their entire lives, but can leave the water to travel.

The females lay hundreds or even thousands of eggs in a ball which they initially carry about on their abdomens, and later deposit into the water. The eggs typically take two to three weeks to hatch, but some species undergo diapause, with the eggs remaining dormant throughout a dry season, and hatching only when conditions are suitable. Stoneflies usually live in areas with running water. The nymphs (technically, "naiads") are aquatic and live in the benthic zone of well-oxygenated lakes and streams. A few species found in New Zealand and nearby islands have terrestrial nymphs, but even these inhabit only very moist environments. The nymphs physically resemble wingless adults, but often have external gills, which may be present on almost any part of the body. Nymphs can acquire oxygen via diffusing through the exoskeleton, or through gills located on behind the head, on the thorax, or around the anus. Due to their nymph's requirement for well oxygenated water, the species is very sensitive to water pollution. This makes them important indicators for water quality. Most species are herbivorous as nymphs, feeding on submerged leaves and benthic algae, but many are hunters of other aquatic arthropods. The insects remain in the nymphal form for one to four years, depending on species, and undergo from 12 to 33 moults before emerging and becoming terrestrial as adults. The adults generally only survive for a few weeks, and emerge only during specific times of the year. Some do not feed at all, but those that do are herbivorous (Nelson, 1996; Nelson & Baumann, 1987).

12. Order Embioptera

Commonly known as webspinners, a few of them occur also in warmer parts of the Temperate Zone. Over 360 embiopteran species have been described, along with estimates of around 2000 species being in existence today. This is a small tropicosmopolitan order, organized into seven families. The name Embioptera ("lively wings") comes from Greek, εμβιος, *embios* meaning "lively" and πτερον, *pteron* meaning "wing", a name that has not been considered to be particularly descriptive for this group of fliers, perhaps instead referring to their remarkable speed of movement both forward and backward. The common name webspinner comes from the insects' unique ability to spin silk from structures on their front legs. They use the silk to make a web-like pouch or gallery in which they live.

All webspinners have a remarkably similar body form, although they do vary in colouration and size. The majority are brown or black in colour, ranging through to a pink or reddish shades in some species, and range in size from 15 to 20 mm. The body form of these insects is completely specialized for the silk tunnels and chambers in which they reside, being long, narrow and highly flexible. All the females and nymphs are wingless, whereas adult males can be either winged or wingless depending on species. The head has projecting mouthparts with chewing mandibles. The compound eyes are kidney-shaped, there are no ocelli, and the antennae are long, with up to 32 segments. The body is cylindrical in form, adapted for the tubular galleries within which the insects live. The first segment of the thorax is small and narrow, while the second and third are larger and broader, especially in the males, where they include the flight muscles. The wings, where present, occur as 2 pairs that are similar in size and shape: long and narrow, with relatively simple venation. These wings operate using basic hydraulics; pre-flight, chambers within the wings fill with hemolymph, making them rigid enough for use. On landing these chambers empty and wings become flexible, folding back against the body. Wings can also fold forwards over the body, and this, along with the flexibility, allows easy movement through the narrow silk galleries without any damage.

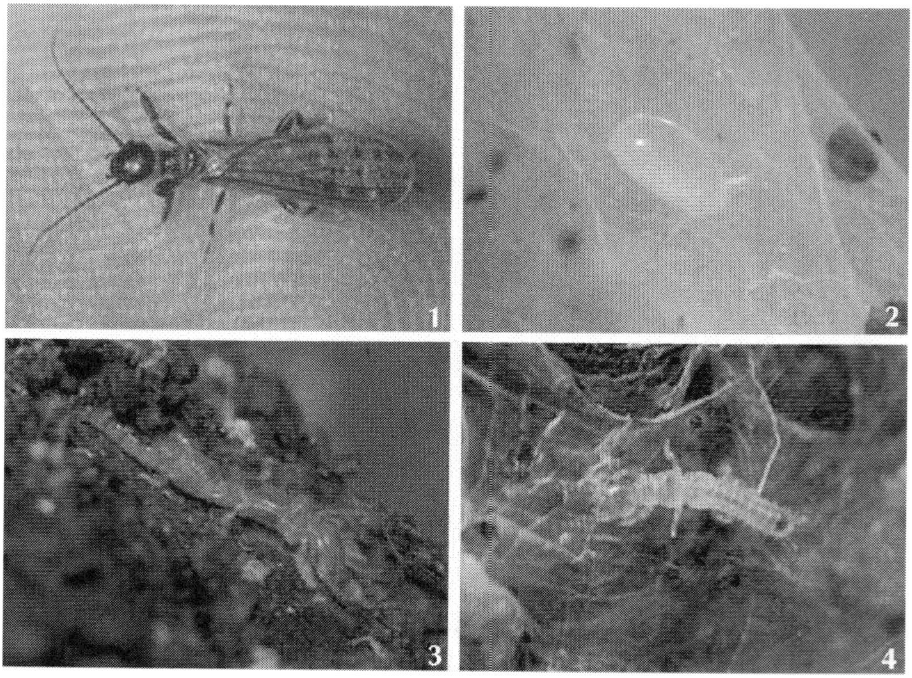

Fig 29. (1) Adult winged male *Oligotoma saundersii,* (2) Egg. (3) Nymph, and (4) 1st Larva (Source: Wikipedia, the free encyclopedia).

In both males and females the legs are short and sturdy, with an enlarged tarsomere on the first pair, containing the silk-producing glands. The abdomen has ten segments, with a pair of cerci on the final segment. These cerci are highly sensitive to touch, and allow the animal to navigate while moving backwards through the gallery tunnels, which are too narrow to allow the insect to turn round. Because morphology is so similar between species, it makes species identification extremely difficult. For this reason, the main form of taxonomic identification used in the past has been close observation of distinctive copulatory structures of males (although this method is now thought by some entomologists and taxonomists as giving insufficient classification detail). Although males never eat during their life span, they do have mouthparts similar to the females. These mouthparts are used to hold onto the female during copulation (Fig. 29).

After moulting, the female lays a single batch of eggs either within the existing gallery, or will find new territory to start a new colony. Here, the eggs hatch into nymphs that resemble small, wingless adults. After a short period of parental care, the nymphs undergo hemimetabolosis (moulting into several instars before emerging as a fully grown adult after the last moult), moulting a total of four times before reaching adult form. Adult males never eat, and leave the home colony almost immediately to find a female and mate. Those males that cannot fly will often mate with females in nearby colonies, meaning their chosen mates are often siblings or closely related. In some species, the female will eat the male after mating, but in any event, the male will not survive for long after mating. A few species are known to be parthenogenetic, meaning they are able to produce viable offspring without fertilization of eggs. This phenomenon occurs when a female is, for whatever reason, unable to find a male to mate with, thus giving her and her species reproductive security at all times. The nymphs and adult females are herbivorous, feeding on leaf litter, moss, bark and lichen. As stated above, adult males do not eat at all, meaning the majority die rapidly due to depleted energy resources (starvation) (Hoell, 1998).

13. Order Dermaptera

The scientific name for the order, *Dermaptera*, is Greek in origin, stemming from the words *derma*, meaning skin, and *pteron* (plural *ptera*), wing. Earwigs make up the insect order Dermaptera and are found throughout the Americas, Africa, Eurasia, Australia and New Zealand. With about 2,000 species in 12 families, they are one of the smaller insect orders. Earwigs have characteristic cerci, a pair of forceps-like pincers on their abdomen, and membranous wings folded underneath short forewings, hence the scientific order name, "skin wings." Some groups are tiny parasites on mammals and lack the typical pincers. Earwigs rarely use their flying ability. They are mostly nocturnal and often hide in small, moist crevices during the day, and are active at night, feeding on a wide variety of insects and plants. Damage to foliage, flowers, and various crops is commonly blamed on earwigs, especially the common earwig, *Forficula auricularia*.

Earwigs have five moults in the year before they become adults. Many earwig species display maternal care, which is uncommon among insects. Female earwigs may care for their eggs, and even after they have hatched as nymphs will continue to watch over offspring until their second moult. As the nymphs moult, sexual dimorphism such as differences in pincer shapes begins to show (Fig. 30).

Most earwigs are flattened (which allows them to fit inside tight crevices, such as under bark) with an elongated body generally 7-50 mm long. The largest certainly extant species is the Australian giant earwig (*Titanolabis colossea*) which is approximately 50 mm long while the possibly extinct Saint Helena earwig (*Labidura herculeana*) reached 78 mm. Earwigs are characterized by the cerci, or the pair of foreceps-like pincers on their abdomen; male earwigs generally have more curved pincers than females. These pincers are used to capture prey, defend themselves and fold their wings under the short tegmina. The antennae are thread-like with at least 10 segments or more. The forewings are short oblong leathery plates used to cover the hindwings like the elytra of a beetle, rather than to fly. Most species have short and leather-like forewings with very thin hindwings. The hindwing is a very thin membrane that expands like a fan, radiating from one point folded under the forewing. Even though most earwigs have wings and are capable of flight, they are rarely seen in flight. These wings are unique in venation and in the pattern of folding that requires the use of the cerci.

Fig. 30: (a) An earwig (*Forficula auricularia*), and (b) An earwig feeding on flower (Source: Wikipedia, the free encyclopedia).

Earwigs are hemimetabolous, meaning they undergo incomplete metamorphosis, developing through a series of 4 to 6 moults. The developmental stages between moults are called instrs. Earwigs live for about a year from hatching. They start mating in the autumn, and can be found together in the autumn and winter. The male and female will live in a chamber in debris, crevices, or soil 2.5 mm deep. After mating, the sperm may remain in the female for months before the eggs are fertilized. From midwinter to early spring, the male will leave, or be driven out by the female. Afterward the female will begin to lay 20 to 80 pearly white eggs in 2 days. Some earwigs, those parasitic in the suborders Arixeniina and Hemimerina, are viviparous (give birth to live young); they would be fed by a sort of placenta. When first laid, the eggs are white or cream-colored and oval-shaped, but right before hatching they become kidney-shaped and brown. Each egg is approximately 1 mm tall and 0.8 mm wide.

Earwigs are among the few non-social insect species that show maternal care. The mother will pay close attention to the needs of her eggs, such as warmth and protection, though studies have shown that the mother does not pay attention to the eggs as she collects them. The mother has been shown to pick up wax balls by accident, but they would eventually be rejected as they do not have the proper scent. The mother will also vigorously defend the eggs from predators, not eating unless an egg goes bad. Another distinct maternal care unique to earwigs is that the mother continuously cleans the eggs to protect them from fungi. Studies have found that the urge to clean the eggs persists for days after they are removed; when the eggs were replaced after hatching, the mother continued to clean them for up to 3 months (Fig. 31).

The eggs hatch in about 7 days. The mother may assist the nymphs in hatching. When the nymphs hatch, they eat the egg casing and continue to live with the mother. The nymphs look similar to their parents, only smaller, and will nest under their mother and she will continue to protect them until their second moult. The nymphs feed on food regurgitated by the other, and on their own moults. If the mother dies before the nymphs are ready to leave, the nymphs may eat her.-After five to six instars, the nymphs will moult into adults. The male's forceps will become curved, while the females' forceps remain straight. They will also develop their natural colour, which can be anything from a light brown (as in the Tawny earwig), or a dark black (as in the Ring-legged). In species of winged earwigs, the wings will start to develop at this time. The forewings of an earwig are sclerotized to serve as protection for the membranous hindwings.

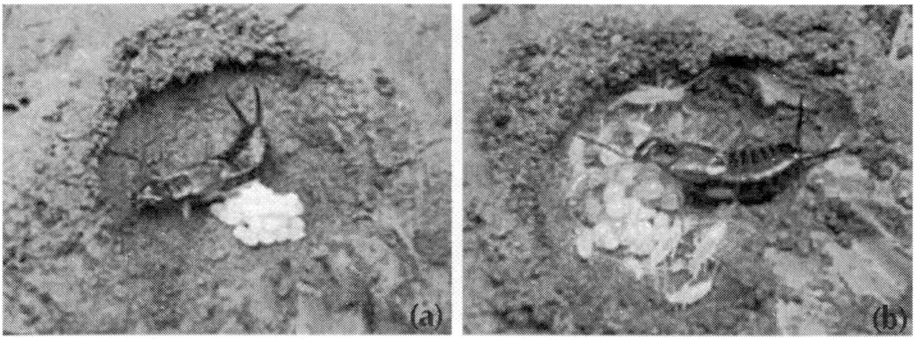

Fig. 31. Female earwig in her nest, with (a) eggs and (b) newly hatched youngones (Source: Wikipedia, the free encyclopedia).

Most earwigs are nocturnal and inhabit small crevices, living in small amounts of debris, in various forms such as bark and fallen logs. Species have been found to be blind and living in caves, or cavernicolous; reported to be found on the island of Hawaii and in South Africa. Food typically consists of a wide array of living and dead plant and animal matter. For protection from predators, the species *Doru taeniatum* of earwigs can squirt foul-smelling yellow liquid in the form of jets from scent glands on the dorsal side of the third and fourth abdominal segment. It aims the discharges by revolving the abdomen, a maneuver that enables it simultaneously to use its pincers in defense. Earwigs are mostly scavengers, but some are omnivorous or predatory. The abdomen of the earwig is flexible and muscular. It is capable of maneuvering as well as opening and closing the forceps. The forceps are used for a variety of purposes. In some species, the forceps have been observed in use for holding prey, and in copulation. The forceps tend to be more curved in males than in females.

14. Order Blattodea

There are more than 4000 species of cockroaches, occurring in all faunal regions and divided by different authors into different numbers of families. Blattodea is an order of insects that contains the cockroaches (Fig. 32). Ironically recent genetic and molecular studies put up enough evidence and suggest an intimate relationship of the termites (Isoptera) with the cockroaches; both cockroaches and termites having evolved from a common ancestor.

Cockroaches are a major representative of this group. India has many species of coackroaches, and so do have the Doon Valley. Some species of coackroaches, such as *Periplanata americana* and *Blatta orientalis* are strongly anthropophilic, and are serious nuisance and disease cerriers as well.

Fig. 32. *Periplanata americana* (Source: Wikipedia, the free encyclopedia).

Cockroaches are said to be the living fossil in a way that they have survived through ages without much changes in their structural characteristics. This also tends to mean that they have been obviously very successful in tiding over the calamitous vicissitudes of different geological periods, which saw even extinction of some of the most prominent animal groups of the Mesozoic era which followed much later to that when the insects had originated in the Paleozoic era some 400 million years ago. In course of time, however, cockroaches of certain species became close associates of humans, their pets and/or habitations. There, in the modern times, these cockroaches incidentally turned out to be the most common and obnoxious domestic pest, causing enormous material, psychological and health-related damages to the man. Cockroaches, besides being a potent source of allergy to several people, may also additionally carry pathogens responsible for diarrhea, dysentery, typhoid fever, food poisoning and cholera. Their filthy habits, grotesque appearance and obnoxious odour make them abhorred all over the world.

Cockroaches, or as popularly called "the red bird of kitchen" is dorsoventrally flattened, with the body clearly divided in three parts: head, thorax and abdomen. The head is bent downward and backward, with mouthparts equipped with a pair of very strong mandibles. The head has prominently a pair of long, cetaceous and segmented antennae, which help the insect to identify the nature of an object accessed, two large compound eyes and chewing mouthparts. The bent head is, in fact, tucked under the back plate shaped like a shield, called pronotum. This plate covers the bases of head and wings. Coackroaches have two pairs of wings; the forewings are coloured and leathery, while the hind wings are clear and membranous. Exceptionally, some coackroaches have either short wings or may be totally lacking wings. Their colour may be brown, tan or black and vary greatly in size; measuring from merely 2-3 mm to as gigantic as 80 mm.

Males can be easily identified from the female coackroaches by the presence of a pair of styli near cerci. The former also have a narrower and longer abdomen than compared to females who have a fatter and more rounded abdomen. Both males and females have a pair of large, segmented anal cerci.

Coackroaches are nocturnal by habit. In the day time they hide in a variety of sites imaginable, from cracks and crevices in walls, door hinges, bathroom sinks and drains points, cupboards, steam tunnels, animal houses, cellars, electric meters, radio, televisions, transistors and even refrigerators. They are gregarious in nature and are always sighted in groups. Some of the more important species are tabled below (Table 6).

Table 6. Some of the more important species of coacroaches.

Species	Common name	Distribution	Nature of public health hazard
Blatta orientalis	Oriental coackroach	Worldwide, but does not occur widely in the Orient (Eastern Asia)	Domestic pest
Blatella germanica	German coackroach	Ubiquitous	Domestic pest
Periplanata americana	American coackroach	Ubiquitous; most common cock-roach in India	Domestic pest
P. australasiae	Australian coackroach	Tropical and subtropical regions	Domestic pest
P. brunnea	Large brown coackroach	Warmer climates	Pest in foodstores, sewers etc.
P. fuliginosa	Smoky brown coackroach	Subtropical regions	Outdoor pest in garages, woodpiles
Supella longipalpa	Brown banded coackroach	Nearly worldwide	Domestic pest; electric appliances of all types
Leucophaea maderae	Madeira cockroach	Tropical	Pest in fruitstores, buildings with plantation

Cockroaches complete their life cycle in three stages: egg, nymph and adult. These are briefly described below.

(i) *The Egg:* Eggs are laid in groups and protectively covered in a hard case called ootheca. The egg case may be carried by the female internally or externally to some suitable place. These egg cases may tell quite much about the species which have laid these.

(ii) *The Nymph:* The tiny young cockroaches resemble adults. However, immediately after emergence from the ootheca, they are devoid of wings or the developed sexual organs. There may be 5-10 moultings in the development of nymphs to the adult stage. In general, males have fewer moults than the females.

(iii) *The Adult:* Males turn mature faster than females in view of less number of moultings. Although weak fliers initially, the adults, as are the young ones, are very quick in movement when disturbed. Both the nymphs and adults prawl for food in the night time, which may include a variety of material including raw and cooked food, small live and recently dead organisms. Females lay the eggs in a few days following the mating. In a single mating the male can pass sufficient sperm fluid to fertilize all the eggs of the female throughout her life.

Cockroaches carry tremendous public health significance. Not only they are potential contaminators of food items, they also emanate a highly detracting stench which can lead some people to psychologically depressive mood. But, perhaps the most noteworthy from medical point of view is the habit of cockroaches to vomit up or regurgitate some of their partly digested food and dropping faeces, often at the same place of feeding. They are a potential health risk to man as they carry disease pathogens from one filthy environment like human faeces to another like human food.

15. Oder Mantodea

The order of praying mantids (Mantodea), largely unsettled for its recognition in insect classification, is composed of some 2000 species under 430 genera in 15 families. The classification most commonly adopted is that proposed by Beier (1968). He divided the order into eight families. Klass (1997) studied the external male genitalia and postulated that the families Chaeteessidae and Metallytcidae diverged from the other families at an early date. However, the Mantidae and Thespidae are both polyphyletic, so the Mantodea will have to be revised. They are the largest family - the Mantidae. Mantises are distributed worldwide in temperate and tropical habitats. They have triangular heads with bulging eyes supported on flexible necks. Their elongated bodies may or may not have wings, but all Mantodea have fore legs that are greatly enlarged and adapted for catching and gripping prey; their upright posture, while remaining stationary with fore arms folded, has led to the common name praying mantis. They are mostly ambush predator but a few ground-dwelling species are found actively pursuing their prey. They normally live for about a year; in cooler climates, the adults lay eggs in autumn, and die. The eggs are protected by their hard capsule and hatch in the spring. Females sometimes practice sexual cannibalism, eating their mates after copulation. Occasionally, the females decapitate the males just before or during mating. The closest relatives of mantises are the termites and cockroaches, which are all within the superorder Dictyoptera. Mantises are sometimes confused with stick insects, other elongated insects such as grasshoppers, or other insects with raptorial fore legs such as mantisflies (Mantispidae) which however have their fore wings not leathery (Fig. 33).

Fig. 33: (a) Head of *Archimantis latistyla* showing the compound eyes and labrum, and (b) *Tenodera sinensis* feeding on a cricket (Source: Wikipedia, the free encyclopedia).

Mantises have large, triangular heads with a beak-like snout and mandibles. They have two bulbous compound eyes, three small simple eyes, and a pair of antennae. The articulation of the neck is also remarkably flexible; some species of mantis can rotate their heads nearly 180°. The mantis thorax consists of a prothorax, a mesothorax, and a metathorax. In all species apart from the genus *Mantoida*, the prothorax, which bears the head and fore legs, is much longer than the other two thoracic segments. The prothorax is also flexibly articulated, allowing for a wide range of movements of the head and fore limbs while the remainder of the body remains more or less immobile. The raptorial fore leg, showing the unusually long coxa, which, together with the trochanter, gives the impression of a femur. The femur itself is the proximal segment of the grasping part of the leg. Mantises have two spiked, grasping fore legs ("raptorial legs") in which prey items are caught and held securely. In most insect legs, including the posterior four legs of a mantis, the coxa and trochanter combine as an inconspicuous base of the leg; in the raptorial legs, however, the coxa and trochanter combine to form a segment about as long as the femur, which is a spiky part of the grasping apparatus. Located at the base of the femur is a set of discoidal spines, usually four in number, but ranging from none to as many as five depending on the species. These spines are preceded by a number of tooth-like tubercles, which, along with a similar series of tubercles along the tibia and the apical claw near its tip, give the fore leg of the mantis its grasp on its prey. The fore leg ends in a delicate tarsus used as a walking appendage, made of four or five segments and ending in a two-toed claw with no arolium. Mantises can be loosely categorized as being macropterous (long-winged), brachypterous (short-winged), micropterous (vestigial-winged), or apterous (wingless). If not wingless, a mantis has two sets of wings: the outer wings, or tegmina, are usually narrow and leathery. They function as camouflage and as a shield for the hind wings, which are clearer and more delicate. The abdomen of all mantises consist of 10 tergites, with a corresponding set of nine sternites visible in males and seven visible in females. The abdomen tends to be slimmer in males than females, but ends in a pair of cerci in both sexes.

Mantises have stereo vision. They locate their prey by sight; their compound eyes contain up to 10,000 ommatidia. A small area at the front called the fovea has greater visual acuity than the rest of the eye, and can produce the high resolution necessary to examine potential prey. The peripheral ommatidia are concerned with perceiving motion; when a moving object is noticed, the head is rapidly rotated to bring the object into the visual field of the fovea. Further motions of the prey are then tracked by movements of the mantis's head so as to keep the image centered on the fovea. The eyes are widely spaced and laterally situated, affording a wide binocular field and precise stereoscopic vision at close range. The dark spot on each eye that moves as it rotates its head is apseudopupil. This occurs because the ommatidia that are viewed "head-on" absorb the incident light, while those to the side reflect it. As their hunting relies heavily on vision, mantises are primarily diurnal. Many species, however, fly at night, and then may be attracted to artificial lights. Nocturnal flight is especially important to males in locating less-mobile females by detecting their pheromones. Flying at night exposes mantises to fewer bird predators than diurnal flight would. Many mantises also have an auditory thoracic organ that helps them avoid bats by detecting their echolocation calls and responding evasively. Mantises are generalist predators of arthropods (Fig. 34). The majority of mantises are ambush predators that only feed upon live prey within their reach. Either they camouflage themselves and remain stationary, waiting for prey to approach, or stalk their prey with slow, stealthy movements. Larger mantises sometimes eat smaller individuals of their own species, as well as small vertebrates such as lizards and frogs. The fore gut of some species extends the whole length of the insect and can be used to store prey for digestion later. This may be advantageous in an insect that feeds intermittently. Chinese mantises live longer, grow faster, and produce more young ones when they are able to eat pollen. Mantises are masters to disguise many a form as a anti-predation tactic (Beier, 1968; Klaus-Dieter, 1997).

Fig. 34: (a) Leaf mimicry: *Choeradodis* has leaf-like fore-wings and a widened green thorax, and (b) Adult female *Iris oratoria* performs a bluffing threat display (Source: Wikipedia, the free encyclopedia).

Mantids' mating season in temperate climates typically takes place in autumn, while in tropical areas, mating can occur at any time of the year. To mate following courtship, the male usually leaps onto the female's back, clasping her thorax and wing bases with his fore legs. He then arches his abdomen to deposit and store sperm in a special chamber near the tip of the female's abdomen. The female lays between 10 and 400 eggs, depending on the species.

Eggs are typically deposited in a frothy mass-produced by glands in the abdomen. This froth hardens, creating a protective capsule, which together with the egg mass is called an ootheca. Depending on the species, the ootheca can be attached to a flat surface, wrapped around a plant, or even deposited in the ground. Despite the versatility and durability of the eggs, they are often preyed on, especially by several species of parasitoid wasps. In a few species, mostly ground and bark mantises in the family Tarachodidae, the mother guards the eggs. The cryptic *Tarachodes maurus* positions herself on bark with her abdomen covering her egg capsule, ambushing passing prey and moving very little until the eggs hatch. A unique reproductive strategy is adopted by Brunner's stick mantis from the southern United States. No males have ever been found in this species, and the females breed parthenogenetically. In temperate climates, adults do not survive the winter and the eggs undergo a diapause, hatching in the spring (Fig. 35).

Fig. 35. (a) *Mantis religiosa* mating (brown male, green female), and (b) Mantis laying ootheca (Source: Wikipedia, the free encyclopedia).

Mantises go through three life stages: egg, nymph, and adult (mantises are among the hemimetabolous insects). For smaller species, the eggs may hatch in 3–4 weeks as opposed to 4–6 weeks for larger species. The nymphs may be coloured differently from the adult, and the early stages are often mimics of ants. A mantis nymph grows bigger as it moults its exoskeleton. Moulting can happen five to 10 times before the adult stage is reached, depending on the species. After the final moult, most species have wings, though some species remain wingless or brachypterous ("short-winged"), particularly in the female sex. The lifespan of a mantis depends on the species; smaller ones may live 4–8 weeks, while larger species may live 4–6 months.

16. Order Isoptera

Termites, most of which are Holarctic, are eusocial insects whose taxonomic status as a distinct order of insects is severely jeopardized after being reconsidered afresh by being classified at the taxonomic rank of infraorder Isoptera or as epifamily Termitoidae within the cockroach order Blattodea.

Termites were once classified in a separate order from cockroaches, but recent phylogenetic studies indicate that they evolved from close ancestors of cockroaches during the Jurassic or Triassic. It is possible, however, that the first termites emerged during the Permian or even the Carboniferous. Approximately 3,106 species are currently described, with a few hundred more left to be described. Although these insects are often called white ants, they are not ants. The infraorder name Isoptera is derived from the Greek words iso- (equal) and -ptera (winged), which refers to the nearly equal size of the fore-wings and hind-wings. Termite nests were commonly known as terminarium or termitaria (Fig. 36).

Like ants and some bees and wasps from the separate order Hymenoptera, termites divide labour among castes consisting of sterile male and female "workers" and "soldiers". All colonies have fertile males called "kings" and one or more fertile females called "queens". Termites mostly feed on dead plant material and cellulose, generally in the form of wood, leaf litter, soil, or animal dung. Termites are major detritivores, particularly in the subtropical and tropical regions. Termites are among the most successful groups of insects on Earth, colonising most landmasses except for Antarctica. Their colonies range in size from a couple of hundred individuals to enormous societies with several million individuals. Termite queens have the longest lifespan of any insect in the world, with some queens living up to 50 years. One can come across of those huge termite mounds, a wonderful engineering feat, across a forest on a morning walk (see below) which are aplenty in the Doon Valley inhabited generally by species like *Odontotermes redemanni.*

Fig. 36: (a) A termitarium or termite mound (Source: B.K. Tyag), (b) The external appearance of the giant northern termite *Mastotermes darwiniensis* is suggestive of the close relationship between termites and cockroaches (Source: Wikipedia, the free encyclopedia).

Unlike ants, which undergo a complete metamorphosis, each individual termite goes through an incomplete metamorphosis that proceeds through egg, nymph and adult stages. Colonies are described as superorganisms because the termites form part of a self-regulating entity: the colony itself. Their recycling of wood and plant matter is of considerable ecological importance. Termites are a delicacy in the diet of some human cultures and are used in many traditional medicines. Several hundred species are economically significant as pests that can cause serious damage to buildings, crops or plantation forests.

In their genetic structure termites essentially from that of social Hymenoptera, and sexualdimorphism is negligible. The main features can be summarized as follows: (1) the primary reproductive of termites are considered to constitute a single class, consisting of equal number of males and females, (2) termite colonies are gynandrarchtic, founded and headed by male and female, usually by a single pair, the male being fully functional in the social life, (3) the termite colony is completely bisexual, all immature stages and all castes consisting of approximately equal numbers of males and females, consequently, in termites sex determination has nothing to do with castes determination, (4) the males are diploid and almost certainly bisexually produced, parthenogenetically produced termites are exclusively females, and (5) there is nothing in the termites that corresponds to the haploid parthenogenesis of the male Hymenoptera.

Several aspects of biology and control of termites of the Doon Valley have been extensively studied by many distinguished scientists of the country (Roonwal & Chotani, 1989, Sen-Sarma, 1974, Tyagi, 1987; Tyagi et al., 1984).

17. Order Coleoptera

Coleoptera, of which beetles are main members, is among the most abundantly represented insect order. The beetles number 320,000 known species and are consequently the largest order of insects in the animal kingdom in general. They are the predominant insects of the present geological epoch, whose extraordinary adaptability has enabled them to invade land, air, and water to the same degree. The order includes some of the largest and some of the most minute of living insects and its members are found in all faunal regions and in most biotic environments. Taxonomically, though subject to debate, it is divided into three suborders, viz., Adephaga, Archostomata and Polyphaga. Beetles are the most important autochthonous insects in the high altitudes of all mountain systems; on the alpine-Himalayan system they amount to almost half of the total number of insects colonizing the biome above the timber line. The highest altitude record is at present 5600 m. The word "coleoptera" is from the Greek, *koleos*, meaning "sheath"; and *pteron*, meaning "wing", thus "sheathed wing", because most beetles have two pairs of wings, the front pair, the "elytra", being hardened and thickened into a shell-like protection for the rear pair and the beetle's abdomen. The order contains more taxa than any other order, constituting almost 25% of all known life-forms. The largest taxonomic family, the Curculionidae (the weevils or snout beetles), also belongs to this order. The controversy regarding evolution continues unabated and growing evidence suggest that possibly they possessed a polyphyletic evolution rather than conventionally thought monophyletic evolution. Though classification at the family level is a bit unstable, about 500 families and subfamilies are recognized.

Beetles exhibit an amazing range of diversity, being found in almost all types of habitats save for the sea or the polar regions. They interact with their ecosystems in several ways. They often feed on plants and fungi, break down animal and plant debris, and eat other invertebrates. Some species are prey of various animals including birds and mammals. Certain species are agricultural pests, such as the Colorado potato beetle, *Leptinotarsa decemlineata*, the boll weevil, *Anthonomus grandis*, the red floor beetle, *Tribolium castaneum*, and the mungbean or cowpea beetle *Callosobruchus maculatus*, while other species of beetles are important controls of agricultural pests. For example, beetles in the family Coccinellidae ("ladybirds" or "ladybugs") consume aphids, scale insects, thrips, and other plant-sucking insects that damage crops.

Species in the order are generally characterized by a particularly hard exoskeleton and hard forewings (elytra). These elytra distinguish beetles from most other insect species, except for a few species of Hemiptera. The beetle's exoskeleton is made up of numerous plates called sclerites, separated by thin sutures. This design creates the armored defenses of the beetle while maintaining flexibility. The general anatomy of a beetle is quite uniform, although specific organs and appendages may vary greatly in appearance and function between the many families in the order. Like all insects, beetles' bodies are divided into three sections: the head, the thorax, and the abdomen (Fig. 37). Coleopteran internal morphology is similar to other insects, although there are several examples of novelty. Such examples include species of water beetle which use air bubbles in order to dive under the water, and can remain submerged thanks to passive diffusion as oxygen moves from the water into the bubble.

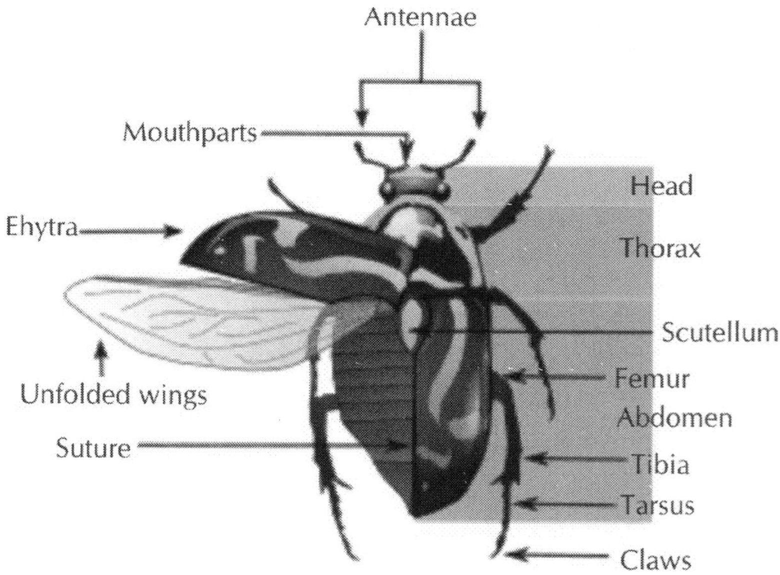

Fig. 37. The morphology of a beetle, with a Fiddler beetle as an example (Source: Wikipedia, the free encyclopedia).

Beetles are endopterygotes, which means that they undergo complete metamorphosis, a biological process by which an animal physically develops after birth or hatching, undergoing a series of conspicuous and relatively abrupt change in the its body structure. Coleopteran species have an extremely intricate behaviour when mating, using such methods as pheromones for communication to locate potential mates. Males may fight for females using very elongated mandibles, causing a strong divergence between males and females in sexual dimorphism.

Coleoptera are found in nearly all natural habitats, including freshwater and marine habitats, everywhere vegetative foliage is found, from trees and their bark to flowers, leaves, and underground near roots- even inside plants in galls, in every plant tissue, including dead or decaying ones.

The head, having mouthparts projecting forward or sometimes downturned, is usually heavily sclerotized and varies in size (Fig. 38). The eyes are compound and may display remarkable adaptability, as in the case of family Gyrinidae, where they are split to allow a view both above and below the waterline. Other species also have divided eyes – some longhorn beetles (family Cerambycidae) and weevils – while many have eyes that are notched to some degree. A few beetle genera also possess ocelli, which are small, simple eyes usually situated farther back on the head (on the vertex). Beetles' antennae are primarily organs of smell, but may also be used to feel a beetle's environment physically. They may also be used in some families during mating, or among a few beetle species for defence. Antennae vary greatly in form within the Coleoptera, but are often similar within any given family. Males and females sometimes have different antennal forms. Antennae may be clavate (flabellate and lamellate are subforms of clavate, or clubbed antennae), filiform, geniculate, moniliform, pectinate, or serrate. Beetles have mouthparts similar to those of grasshoppers. Of these parts, the most commonly known are probably the mandibles, which appear as large pincers on the front of some beetles. The mandibles are a pair of hard, often tooth-like structures that move horizontally to grasp, crush, or cut food or enemies. Two pairs of finger-like appendages, the maxillary and labial palpi, are found around the mouth in most beetles, serving to move food into the mouth. In many species, the mandibles are sexually dimorphic, with the males' enlarged enormously compared with those of females of the same species.

Fig. 38. Head of *Cephalota circumdata* showing the compound eyes (Source: https://en.wikipedia.org/wiki/Compound_eye and mouthparts).

The thorax is segmented into the two discernible parts, the pro- and pterathorax. The pterathorax is the fused meso- and metathorax which are commonly separated in other insect species, although flexibly articulate from the prothorax. When viewed from below, the thorax is that part from which all three pairs of legs and both pairs of wings arise. The abdomen is everything posterior to the thorax. When viewed from above, most beetles appear to have three clear sections, but this is deceptive: on the beetle's upper surface, the middle "section" is a hard plate called the pronotum, which is only the front part of the thorax; the back part of the thorax is concealed by the beetle's wings. This further segmentation is usually best seen on the abdomen.

Legs are multisegmented and have 2-5 tarsi generally bearing claws, usually one pair, on the end of the last tarsal segment of each leg. While most beetles use their legs for walking, legs may be variously modified and adapted for other uses. Among aquatic families, e.g., Dytiscidae, Haliplidae, many species of Hydrophilidae and others – the legs, most notably the last pair, are modified for swimming and often bear rows of long hairs to aid this purpose. Other beetles have fossorial legs that are widened and often spined for digging. Species with such adaptations are found among the scarabs, ground beetles, and clown beetles (family Histeridae). The hind legs of some beetles, such as flea beetles (within Chrysomelidae) and flea weevils (within Curculionidae), are enlarged and designed for jumping (Fig. 39).

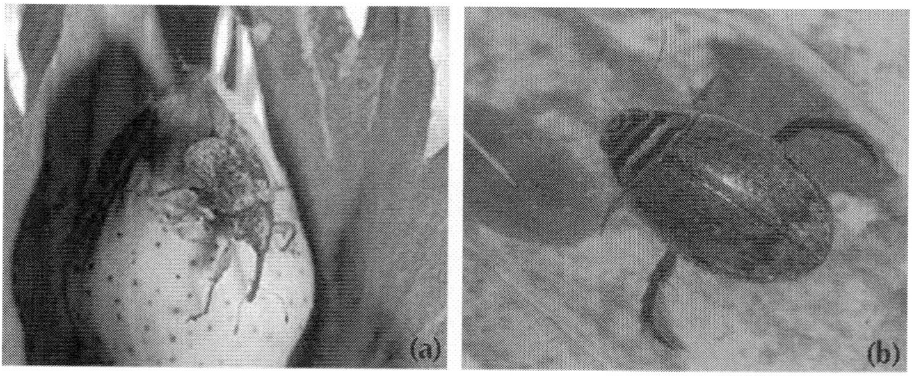

Fig. 39: (a) Cotton boll weevil, and (b) *Acilius sulcatus*, a diving beetle showing hind legs adapted for life in water (Source: Wikipedia, the free encyclopedia).

The elytra are connected to the pterathorax, so named because it is where the wings are connected. The elytra are not used for flight, but tend to cover the hind part of the body and protect the second pair of wings (*alae*). They must be raised to move the hind flight wings. A beetle's flight wings are crossed with veins and are folded after landing, often along these veins, and stored below the elytra. A fold (*jugum*) of the membrane at the base of each wing is a characteristic feature. In some beetles, the ability to fly has been lost. These include some ground beetles (family Carabidae) and some "true weevils" (family Curculionidae), but also desert- and cave-dwelling species of other families. Many have the two elytra fused together, forming a solid shield over the abdomen. In a few families, both the ability to fly and the elytra have been lost, with the best known example being the glow-worms of the family Phengodidae, in which the females are larviform throughout their lives.

The abdomen is made up of a series of rings, each with a spiracle, composing three different segmented sclerites: the tergum, the pleura, and the sternum. The tergum in almost all species is membranous, or usually soft and concealed by the wings and elytra when not in flight. The pleura are usually small or hidden in some species, with each pleuron having a single spiracle. The sternum is the most widely visible part of the abdomen, being a more or less sclerotized segment. The abdomen itself does not have any appendages, but some (for example, Mordellidae) have articulating sternal lobes.

A notable number of species have developed special glands to produce chemicals for deterring predators. The carabids' defensive glands, located at the posterior, produce a variety of hydrocarbons, aldehydes, phenols, quinones, esters, and acids released from an opening at the end of the abdomen. African carabid beetles employ the same chemicals as ants, i.e., formic acid. Bombardier beetles have well-developed, like other carabid beetles, pygidial glands that empty from the lateral edges of the intersegment membranes between the seventh and eighth abdominal segments. The gland is made of two containing chambers. The first holds hydroquinones and hydrogen peroxide, with the second holding just hydrogen peroxide plus catalases. These chemicals mix and result in an explosive ejection, forming temperatures of around 100 °C, with the breakdown of hydroquinone to $H_2 + O_2$ + quinone, with the O_2 propelling the excretion.

Most of beetles undergo complete metamorphosis, i.e., all stages of development are present - the egg, the larva, the pupa, and the imago or adult. The larvae are commonly called grubs and the pupa sometimes is called the chrysalis. In some species, the pupa may be enclosed in a cocoon constructed by the larva towards the end of its final instar. Going beyond "complete metamorphosis", however, some beetles, such as typical members of the families Meloidae and Rhipiphoridae, undergo hypermetamorphosis in which the first instar takes the form of a triungulin.

Fig. 40. Punctative flower chafers (*Neorrhina punctata*, Scarabaeidae) mating (Source: Wikipedia, the free encyclopedia).

Before mating, males and females may engage in various forms of behaviour. They may stridulate, or vibrate the objects they are on. In some species (for example, Meloidae), the male climbs onto the dorsum of the female and strokes his antennae on her head, palps, and antennae. In the genus *Eupompha* of said family, the male draws the antennae along his longitudinal vertex. They may not mate at all if they do not perform the precopulatory ritual (Figs. 40, 41).

A single female may lay from several dozen to several thousand eggs during her lifetime. Eggs are usually laid according to the substrate on which the larvae feed upon hatching. Among others, they can be laid loose in the substrate (for example, flour beetle), laid in clumps on leaves (for example, Colorado potato beetle), individually attached (for example, mungbean beetle and other seed borers), or buried in the medium (for example, carrot weevil). Parental care varies between species, ranging from the simple laying of eggs under a leaf to certain scarab beetles, which construct underground structures complete with a supply of dung to house and feed their young. Other beetles are leaf rollers, biting sections of leaves to cause them to curl inwards, then laying their eggs, thus protected, inside (Fig. 42).

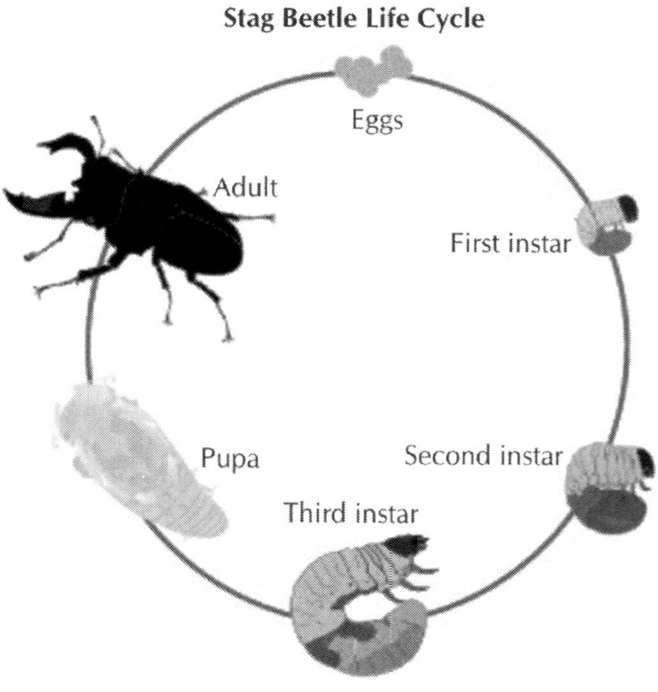

Fig. 41. The lifecycle of the stag beetle includes three instars (Source: Wikipedia, the free encyclopedia).

The larva is usually the principal feeding stage of the beetle lifecycle. Larvae tend to feed voraciously once they emerge from their eggs. Some feed externally on plants, such as those of certain leaf beetles, while others feed within their food sources. Examples of internal feeders are most Buprestidae and longhorn beetles. The larvae of many beetle families are predatory like the adults (ground beetles, ladybirds, rove beetles). The larval period varies between species, but can be as long as several years. The larvae are highly varied amongst species, with well-developed and sclerotized heads, and have distinguishable thoracic and abdominal segments (usually the tenth, though sometimes the eighth or ninth). Beetle larvae can be differentiated from other insect larvae by their hardened, often darkened heads, the presence of chewing mouthparts, and spiracles along the sides of their bodies. Like adult beetles, the larvae are varied in appearance, particularly between beetle families. Beetles with somewhat flattened, highly mobile larvae include the ground beetles, some rove beetles, and others; their larvae are described as campodeiform. Some beetle larvae resemble hardened worms with dark head capsules and minute legs. These are elateriform larvae, and are found in the family Elateridae, along with darkling beetles (Tenebrionidae). Some elateriform larvae of click beetles are known as wireworms. Beetles in the Scarabaeoidea have short, thick larvae described as scarabaeiform, more commonly known as grubs. All beetle larvae go through several instars, which are the developmental stages between each moult. In many species, the larvae simply increase in size with each successive instar as more food is consumed. In some cases, however, more dramatic changes occur. Among certain beetle families or genera, particularly those that exhibit parasitic lifestyles, the first instar (the planidium) is highly mobile to search out a host, while the following instars are more sedentary and remain on or within their host. This is known as hypermetamorphosis; examples include the blister beetles (family Meloidae) and some rove beetles, particularly those of the genus *Aleochara.*

Fig. 42. A scarabaeiform larva known as a curl grub (Source: Wikipedia, the free encyclopedia).

As with all endopterygotes, beetle larvae pupate, and from these pupae emerge fully formed, sexually mature adult beetles, or imagos. Adults have extremely variable lifespans, from weeks to years, depending on the species. In some species, the pupa may go through all four forms during its development, called hypermetamorphosis (for example, Meloidae). Pupae always have no mandibles (are adecticous). In most, the appendages are not attached to the pupae; ones that do have appendages are mostly obtect and the rest are exarate.

Beetles have a variety of ways to communicate, some of which include a sophisticated chemical language through the use of pheromones. From the host tree, mountain pine beetles have many forms of communication. They can emit both an aggregative pheromone and an anti-aggregative pheromone. The aggregative pheromone attracts other beetles to the tree, and the anti-aggregative pheromone neutralizes the aggregative pheromone. This helps to avoid the harmful effects of having too many beetles on one tree competing for resources. The mountain pine beetle can also stridulate to communicate, or rub body parts together to create sound, having a "scraper" on their abdomens that they rub against a grooved surface on the underside of their left wing cover to create a sound that is not audible to humans. Once the female beetles have arrived on a suitable pine tree host, they begin to stridulate and produce aggregative pheromones to attract other unmated males and females. New females arrive and do the same as they land and bore into the tree. As the males arrive, they enter the galleries that the females have tunneled, and begin to stridulate to let the females know they have arrived, and to also warn others that the female in that gallery is taken. At this point, the female stops producing aggregative pheromones and starts producing anti-aggregative pheromone to deter more beetles from coming. Since species of Coleoptera use environmental stimuli to communicate, they are affected by the climate. Microclimates, such as wind or temperature, can disturb the use of pheromones; wind would blow the pheromones while they travel through the air. Stridulating can be interrupted when the stimulus is vibrated by something else.

Some beetles display unique behavior of parental care, necessary for the survival of the larvae, protecting them from adverse environmental conditions and predators. One species, a rover beetle (*Bledius spectabilis*) displays both causes for parental care: physical and biotic environmental factors. Said species lives in salt marshes, so the eggs and/or larvae are endangered by the rising tide. The maternal beetle patrols the eggs and larvae and applies the appropriate burrowing behaviour to keep them from flooding and from asphyxiating. Another advantage is that the mother protects the eggs and larvae from the predatory carabid beetle *Dicheirotrichus gustavi* and from the parasitoid wasp *Barycnemis blediator*. Up to 15% of larvae are killed by this parasitoid wasp, being only protected by maternal beetles in their dens. Some species of dung beetle also display a form of parental care. Dung beetles collect animal feces, or "dung", from which their name is derived, and roll it into a ball, sometimes being up to 50 times their own weight; albeit sometimes it is also used to store food. Usually it is the male that rolls the ball, with the female hitch-hiking or simply following behind. In some cases the male and the female roll together. When a spot with soft soil is found, they stop and bury the dung ball. They then mate underground. After the mating, one or both of them prepares the brooding ball. When the ball is finished, the female lays eggs inside it, a form of mass provisioning. Some species do not leave after this stage, but remain to safeguard their offspring.

Besides being abundant and varied, beetles are able to exploit the wide diversity of food sources available in their many habitats. Some are omnivores, eating both plants and animals (Fig. 43). Other beetles are highly specialized in their diet. Many species of leaf beetles, longhorn beetles, and weevils are very host-specific, feeding on only a single species of plant. Ground beetles and rove beetles (family Staphylinidae), among others, are primarily carnivorous and catch and consume many other arthropods and small prey, such as earthworms and snails. While most predatory beetles are generalists, a few species have more specific prey requirements or preferences. Decaying organic matter is a primary diet for many species. This can range from dung, which is consumed by coprophagous species (such as certain scarab beetles of the family Scarabaeidae), to dead animals, which are eaten by necrophagous species (such as the carrion beetles of the family Silphidae). Some of the beetles found within dung and carrion are in fact predatory. These include the clown beetles, preying on the larvae of coprophagous and necrophagous insects.

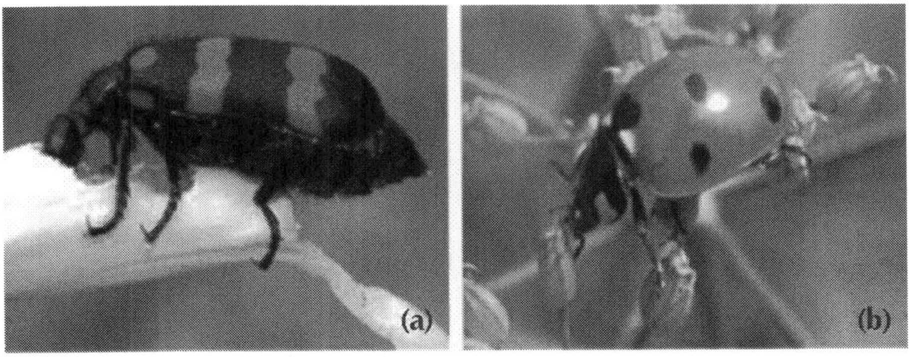

Fig. 43: (a) *Mylabris pustulata* (Meloidae) Feeding on the petals of Ipomoea carnea, (b) *Coccinella septempunctata*,. a beneficial beetle (Source: Wikipedia, the free encyclopedia).

More than 1000 species of beetles are known to be either parasitic, predatory, or commensals in the nests of ants. A few species of beetles are actually ectoparasitic on mammals. One such species, *Platypsyllus castoris*, parasitises beavers (*Castor* spp.). This beetle lives as a parasite both as a larva and as an adult, feeding on epidermal tissue and possibly on skin secretions and wound exudates. They are strikingly flattened dorsoventrally, no doubt as an adaptation for slipping between the beavers' hairs. They also are wingless and eyeless, as are many other ectoparasites. Other parasitic beetles include those that are kleptoparasites of other invertebrates, such as the small hive beetle (*Aethina tumida*) that infests honey bee hives. The larvae tunnel through comb towards stored honey or pollen, damaging or destroying cappings and comb in the process. Larvae defecate in honey and the honey becomes discolored from the feces, which causes fermentation and a frothiness in the honey; the honey develops a characteristic odour of decaying oranges. Damage and fermentation cause honey to run out of combs, destroying large amounts of it both in hives and sometimes also in honey extracting rooms. Heavy infestations cause bees to abscond; though the beetle is only a minor pest in Africa, beekeepers in other regions have reported the rapid collapse of even strong colonies.

Beetle-pollinated flowers are usually large, greenish or off-white in color, and heavily scented. Scents may be spicy, fruity, or similar to decaying organic material. Most beetle-pollinated flowers are flattened or dish-shaped, with pollen easily accessible, although they may include traps to keep the beetle longer. The plants' ovaries are usually well protected from the biting mouthparts of their pollinators. Beetles may be particularly important in some parts of the world such as semiarid areas of western India, southern Africa and southern California and the montane grasslands of Kwa Zulu- Nata in South Africa. *Austroplatypus incompertus* is eusocial, one of the few organisms outside Hymenoptera to do so, and the only species of Coleoptera.

Beetles are not only pests, but can also be beneficial, usually by controlling the populations of pests. One of the best, and widely known, examples are the ladybirds (family Coccinellidae). Both the larvae and adults are found feeding on aphid colonies. Other ladybugs feed on scale insects and mealybugs. If normal food sources are scarce, they may feed on small caterpillars, young plant bugs, or honeydue and nectar. Ground beetles (family Carabidae) are common predators of many different insects and other arthropods, including fly eggs, caterpillars, wireworms, and others. Dung beetles (Scarabidae) have been successfully used to reduce the populations of pestilent flies and parasitic worms that breed in cattle dung. The beetles make the dung unavailable to breeding pests by quickly rolling and burying it in the soil, with the added effect of improving soil fertility, tilth, and nutrient cycling. Dung beetles play a remarkable role in agriculture. By burying and consuming dung, they improve nutrient recycling and soil structure. They also protect livestock, such as cattle, by removing dung, which, if left, could provide habitat for pests such as flies.

Insects are used as human food in 80% of the world's nations. Beetles are the most widely eaten insects. About 344 species are known to be used as food, usually eaten in the larval stage. The mealworm is the most commonly eaten beetle species. The larvae of the darkling beetle and the rhinoceros beetle are also commonly eaten.

A huge mountain of information is amassed on the beetles of the Doon Valley and the readers are advised to consult ZSI and FRI publications on the Order.

18. Order Neuroptera

There is no agreement as to the classification of groups united under this classical name. While some authors, though admitting the heterogenous nature of the higher taxa, still regard the group as a single order, other workers have split it up by elevating the three suborders, Megaloptera (alder flies), Raphidio-idea (snake flies) Planipennia (lacewings and ant lions), to the rank of indepen-dent orders. From the cytological point of view the three taxa do not appear significantly distinct, and they have in common several characteristic and rather unusual cytological features. The three suborders include nearly 6000 known species, 90% of which only from Planipennia. Majority have a tropical or a subtropical distribution, but some penetrated deep into the Arctic and the Alpine zone.

Neuroptera, or net-winged insects, includes the lacewings, mantidflies, antlions, and their relatives. The order contains about 6,000 species.

They undergo complete metamorphosis and have chewing mouthparts. The adults of this order possess four membranous wings, with the forewings and hindwings about the same size, and with many veins. Neuropterans are soft-bodied insects with relatively few specialized features. They have large lateral compound eyes, and may or may not also have ocelli. Their mouthparts have strong mandibles suitable for chewing, and lack the various adaptations found in most other endopterygote insect groups. They have four wings, which are usually similar in size and shape, and a generalized pattern of veins. Some neuropterans have specialized sense organs in their wings, or have bristles or other structures to link their wings together during flight.

The larvae are specialized predators, with elongated mandibles adapted for piercing and sucking. The larval body form varies between different families, depending on the nature of their prey. In general, however, they have three pairs of thoracic legs, each ending in two claws. The abdomen often has adhesive discs on the last two segments (Fig. 44).

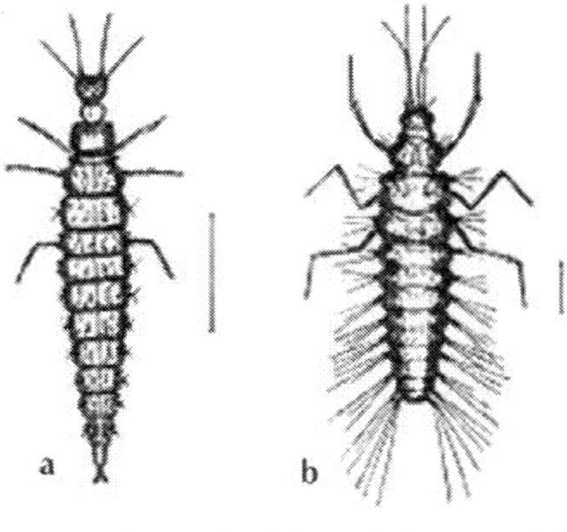

Fig. 44: (a) Larva of *Osmylus fulvicephalus*, Osmylidae, and (b) Larva of *Sycerus* sp., Sisyridae (Source: Wikipedia, the free encyclopedia).

The larvae of most families are predator. Many chrysopids eat aphids and other pest insects, and have been used for biological control (either from commercial distributors, but also abundant and widespread in nature). Larvae in various families cover themselves in debris (sometimes including dead prey insects) as camouflage, taken to an extreme in the ant lions, which bury themselves completely out of sight and ambush prey from "pits" in the soil. Larvae of some thonidae are root feeders, and larvae of Sysiridae are aquatic, and feed on freshwater sponges. A few mantispids are parasites of spider sacs. As in other holometabolic orders, the pupal stage generally is enclosed in some form of cocoon composed of silk and soil or other debris. The pupa eventually cuts its way out of the cocoon.

This is a very important order due to its high economic value in nature. For long it embodied dragonflies only to be found entirely different order to the dragonflies' Odonata which are hemimetabolous and paleoptrygotes.

Neuroptera research has a long tradition in the Doon Valley and a rich collection is housed in the FRI, Dehradun.

19. Order Hymenoptera

Of all the orders of insects Hymenoptera (Greek ὑμήν meaning *hymen*) is the third-largest, comprising the sawflies, wasps, and ants (Fig. 45). Over 200,000 species are recognized, with many more remaining to be described. Females typically have a special ovipositor for inserting eggs into hosts or otherwise inaccessible places. Hymenoptera originated in the Triassic, with the oldest fossils belonging to the family Xyelidae. Social hymenopterans appeared during the Cretaceous. The evolution of this group has been intensively studied. The ovipositor is often modified into a stinger. The young develop through holometabolism (complete metamorphosis), i.e., they have a worm-like larval stage and an inactive pupal stage before they mature. Hymenoptera continues to offer many an unresolved taxonomic query, and though it has been studied by examining the mitochondrial DNA, the ambiguities still exist (Gullan & Cranston, 2014; Grimaldi & Engel, 2005).

Fig. 45: (a) *Bombus muscorum* drinking nectar with its long proboscis, (b) Symphyta: the sawfly*Stethomostus fuliginosus* (Source: Wikipedia, the free encyclopedia).

Hymenoptera occur in all faunal regions, but the fauna is, of course, the richest in the tropics. Nevertheless, some groups have an essentially cool climate distribution, notably the Bombidae (Bumble bees). Hence, hymenopterans are but poorly represented above the timber line in the mountains, with the exception of ants (Formicidae) and bumble bees.

Hymenopterans range in size from very small to large insects, and usually have two pairs of wings. Their mouthparts are adapted for chewing, with well-developed mandibles (ectognathous mouthparts). Many species have further developed the mouthparts into a lengthy proboscis, with which they can drink liquids, such as nectar. They have large compound eyes, and typically three simple eyes (ocelli). The forward margin of the hind wing bears a number of hooked bristles, or "hamuli", which lock onto the fore wing, keeping them held together. The smaller species may have only two or three hamuli on each side, but the largest wasps may have a considerable number, keeping the wings gripped together especially tightly. Hymenopteran wings have relatively few veins compared with many other insects, especially in the smaller species. In the more ancestral hymenopterans, the ovipositor is blade-like, and has evolved for slicing plant tissues. In the majority, however, it is modified for piercing, and, in some cases, is several times the length of the body. In some species, the ovipositor has become modified as a stinger, and the eggs are laid from the base of the structure, rather than from the tip, which is used only to inject venom. The sting is typically used to immobilise prey, but in some wasps and bees may be used in defense.

The larvae of the more ancestral hymenopterans resemble caterpillars in appearance, and like them, typically feed on leaves. They have large chewing mandibles, three pairs of thoracic limbs, and, in most cases, a number of abdominal prolegs. Unlike caterpillars, however, the prolegs have no grasping spines, and the antennae are reduced to mere stubs. The larvae of other hymenopterans, however, more closely resemble maggots, and are adapted to life in a protected environment. This may be the body of a host organism, or a cell in a nest, where the adults will care for the larva. Such larvae have soft bodies with no limbs. They are also unable to defecate until they reach adulthood due to having an incomplete digestive tract, presumably to avoid contaminating their environment.

Among most or all hymenopterans, sex is determined by the number of chromosomes an individual possesses. Fertilized eggs get two sets of chromosomes (one from each parent's respective gametes), so develop into diploid females, while unfertilized eggs only contain one set (from the mother), so develop into haploid males; the act of fertilization is under the voluntary control of the egg-laying female. This phenomenon is called haplodiploidy. However, the actual genetic mechanisms of haplodiploid sex determination may be more complex than simple chromosome number. In many Hymenoptera, sex is actually determined by a single gene locus with many alleles. In these species, haploids are male and diploids heterozygous at the sex locus are female, but occasionally a diploid will be homozygous at the sex locus and develop as a male, instead. This is especially likely to occur in an individual whose parents were siblings or other close relatives. Diploid males are known to be produced by inbreeding in many ant, bee, and wasp species. Diploid biparental males are usually sterile but a few species that have fertile diploid males are known. One consequence of haplodiploidy is that females on average actually have more genes in common with their sisters than they do with their own daughters. Because of this, cooperation among kindred females may be unusually advantageous, and has been hypothesized to contribute to the multiple origins of eusociality within this order. In many colonies of bees, ants, and wasps, worker females will remove eggs laid by other workers due to increased relatedness to direct siblings, a phenomenon known as worker policing.

Parthenogenesis is another way the hymenopterans take advantage of. In this embryos are created without fertilization. Thelytoky is a particular form of parthenogenesis in which female embryos are created (without fertilization). The form of thelytoky in hymenopterans is a kind of automixis in which two haploid products (proto-eggs) from the same meiosis fuse to form a diploid zygote. This process tends to maintain heterozygosity in the passage of the genome from mother to daughter. It is found in several ant species including the desert ant *Cataglyphis cursor,* the clonal raider ant *cerapachys biroj,* the predaceous ant *Platythyrea punctata* and the electric ant (little fire ant) *Wasmannia auropunctata.* It also occurs in the Cape honey bee *Apis mellifera capensis.* Oocytes that undergo automixis with central fusion often have a reduced rate of crossover recombination. A low rate of recombination helps to maintain heterozygosity and avoid inbreeding depression. Species that display central fusion with reduced recombination include the ants *P. punctata* and *W. auropunctata* and the honey bee *A. m. capensis.* In *A. m. capensis*, the recombination rate during the meiosis is reduced by more than 10-fold. In *W. auropunctata* the reduction is 45-fold. Single queen colonies of the narrow headed ant *Formica exsecta* illustrate the possible deleterious effects of increased homozygosity. In this ant, colonies with more homozygous queens age more rapidly. The result is reduced colony survival.

Different species of Hymenoptera show a wide range of feeding habits. The most primitive forms are typically herbivorous, feeding on leaves or pine needles. Stinging wasps are predators, and will provision their larvae with immobilised prey, while bees feed on nectar and pollen. A number of species are parasitoid as larvae. The adults inject the eggs into a paralyzed host, which they begin to consume after hatching. Some species are even hyperparasitoid, with the host itself being another parasitoid insect. Habits intermediate between those of the herbivorous and parasitoid forms are shown in some hymenopterans, which inhabit the galls or nests of other insects, stealing their food, and eventually killing and eating the occupant (Hoell et al., 1998).

The Hymenoptera are divided into two groups, the Symphyta which have no waist, and the Apocrita which have a narrow waist (Aguiaret al., 2013).

The suborder Symphyta includes the sawflies, horntails and parasitic wood wasps. The group may be paraphyletic, as it has been suggested that the family Orussidae may be the group from which the Apocrita arose. They have an unconstricted junction between the thorax and abdomen. The larvae are herbivorous, free-living eruciforms, with three pairs of true legs, prolegs (on every segment, unlike Lepidoptera) and ocelli. The prolegs do not have crochet hooks at the ends unlike the larvae of the Lepidoptera.

The suborder Apocrita includes wasps, bees, and ants together, all of which are characterized by a constriction between the first and second abdominal segments called a wasp-waist (petiole), also involving the fusion of the first abdominal segment to the thorax. Also, the larvae of all Apocrita lack legs, prolegs, or ocelli. The hindgut of the larvae also remains closed during development, with feces being stored inside the body, with the exception of some bee larvae where the larval anus through developmental reversion has reappeared. In general, the anus only opens at the completion of larval growth.

20. Order Strepsiptera

The Strepsiptera, literally meaning "twisted wing", is a small order which comprise the insects commonly named, twisted-wing parasites. They are an endopterygote order of insects with nine extant families making up about 600 species. The early-stage larvae and the short-lived adult males are not sessile, but most of their lives are spent as endoparasites in other insects, such as bees, wasps, leafhoppers, silverfish, and cockroaches. They are extremely minute in size and are very anomalous in structure. The majority has a holarctic distribution, but a few species are met in other faunal regions as well.

The strepsipteran males have wings, legs, eyes and antennae, and superficially look like flies, though their mouthparts cannot be used for feeding. Many of their mouthparts are modified into sensory structures. Adult males are very short-lived, usually surviving less than five hours, and do not feed. Females, in all families except the Mengenillidae, are not known to leave their hosts and are neotenic in form, lacking wings, legs, and eyes. Virgin females release a pheromone which the males use to locate them. In the Stylopidia, the female's anterior region protrudes out of the host body and the male mates by rupturing the female's brood canal opening, which lies between the head and prothorax. Sperm passes through the opening in a process termed hypodermic insemination. The offspring consume their mother from the inside in a process known as haemocelous viviparity. Each female thus produces many thousands of triungulin larvae that emerge from the brood opening on the head, which protrudes outside the host body. These larvae have legs (which lack a trochanter, the leg segment that forms the articulation between the basal coxa and the femur), and actively search out new hosts. Their hosts include members of the orders Zygentoma, Orthoptera, Mantodea, Blatodea, Heteroptera, Hymenoptera and Diptera. In the strepsipteran family Myrmecolacidae, the males parasitize ants, while the females parasitize Orthoptera.

Strepsiptera eggs hatch inside the female, and the planidium larvae can move around freely within the female's haemocoal, which is unique to these animals. The female has a brood canal that communicates with the outside world, and the larvae escape through this. The larvae are very active, as they only have a limited amount of time to find a host before they exhaust their food reserves. These first-instar larvae have stemmata (simple, single-lens eyes), and once they latch onto a host, they enter it by secreting enzymes that soften the cuticle, usually in the abdominal region of the host. Some species have been reported to enter the eggs of hosts. Larvae of *Stichotrema dallatorreanum* Hofeneder from Palua New Guinea were found to enter their orthoptera host's tarsus (foot). Once inside the host, they undergo hypermetamorphosis and become a less mobile, legless larval form. They induce the host to produce a bag-like structure inside which they feed and grow. This structure, made from host tissue, protects them from the immune defences of the host. Larvae go through four more instars, and in each moult separation of the older cuticle occurs, but no discarding ("apolysis" without "ecdysis"), leading to multiple layers being formed around the larvae. Male larvae produce pupae after the last moult, but females directly become neotenous adults. The colour and shape of the host's abdomen may be changed and the host usually becomes sterile. The parasites then undergo pupation to become adults. Adult males emerge from the host bodies, while females stay inside. Females may occupy up to 90% of the abdominal volume of their hosts. Adult male Strepsiptera have eyes unlike those of any other insect, resembling the schizochroal eyes found in the trilobite group known as the Phacopida. Instead of a compound eye consisting of hundreds to thousands of ommatidia, that each produce a pixel of the entire image - the strepsipteran eyes consist of only a few dozen "eyelets" that each produce a complete image. These eyelets are separated by cuticle and/or setae, giving the cluster eye as a whole a blackberry-like appearance. Very rarely, multiple females may live within a single stylopized host; multiple males within a single host are somewhat more common. Adult males are rarely observed, however, although specimens may be lured using cages containing virgin females. Nocturnal specimens can also be collected at light traps. Strepsiptera of the Myrmecolacidae family can cause their ant hosts to linger on the tips of grass leaves, increasing the chance of being found by the parasite's males (in case of females) and putting them in a good position for male emergence (in case of males).

The Strepsiptera have two major groups: the Stylopidia and Mengenillidia. The Mengenillidia include three extinct families (Cretostylopidae, Protoxenidae, and Mengeidae) plus two extant families (Bahiaxenidae and Mengenillidae; the latter is not monophyletic, however). They are considered more primitive, and the known females (Mengenillidae only) are free-living, with rudimentary legs and antennae. The females have a single genital opening. The males have strong mandibles, a distinct labrum, and more than five antennal segments. The other group, the Stylopidia, includes seven families: the Corioxenidae, Halictophagidae, Callipharixenidae, Bohartillidae, Elenchidae, Myrmecolacidae, and Stylopidae. All Stylopidia have endoparasitic females having multiple genital openings. The Stylopidae have four-segmented tarsi and four- to six-segmented antennae, with the third segment having a lateral process. The family Stylopidae may be paraphyletic. The Elenchidae have two-segmented tarsi and four-segmented antennae, with the third segment having a lateral process. The Halictophagidae have three-segmented tarsi and seven-segmented antennae, with lateral processes from the third and fourth segments. The Stylopidae mostly parasitize wasps and bees, the Elenchidae are known to parasitize Fulgoroidea, while the Halictophagidae are found on leafhoppers, treehoppers, and mole cricket hosts.

Strepsipteran insects in the genus *Xenos* parasitize *Polistes carnifex*, a species of social wasps. These obligate parasites infect the developing wasp larvae in the nest and are present within the abdomens of female wasps when they hatch out. Here they remain until they thrust through the cuticle and pupate (males) or release infective first-instar larvae onto flowers (females). These larvae are transported back to their nests by foraging wasps (Hughes et al., 2003).

21. Order Aphaniptera

Order Aphaniptera (Siphonaptera) is represented by fleas. They have no wings and their mouthparts are adapted for piercing skin and sucking blood. Over 2,000 species have been described worldwide. Fleas are external parasites living by hematophagy off the blood of birds and mammals including human being and a cause of transmission of the deadly infection, plague. Flea systematics are not entirely fixed. While compared to many other insect groups, fleas have been studied and classified fairly thoroughly, yet the details still remain to be learned about the evolutionary relationships among the different flea lineages. Some important flea species include:

- Cat flea (*Ctenocephalides felis*)
- Dog flea (*Ctenocephalides canis*)
- Human flea (*Pulex irritans*)
- Moorhen flea (*Dasypsyllus gallinulae*)
- Northern rat flea (*Nosopsyllus fasciatus*)
- Oriental rat flea (*Xenopsylla cheopis*)

Fleas are wingless insects, 1.5 to 3.3 mm long, that are agile, usually dark coloured (for example, the reddish-brown of the cat flea), with tube-like mouth parts adapted to feeding on the blood of their hosts. They interestingly do not possess a mandible. Their legs are long, the hind pair well adapted for jumping; a flea can jump vertically up to 18 cm and horizontally up to 33 cm, making the flea one of the best jumpers of all known animals (relative to body size), second only to the froghopper. Their legs end in strong claws that are designed to grasp a host

Fleas take off from their tibiae and tarsi (the insect equivalent of feet) and not their trochanter, or knees. It has been known that fleas do not use direct muscle power, but instead use the muscle to store energy in a protein named resilin before releasing it rapidly.

Fleas' bodies are laterally compressed, permitting easy movement through the hairs or feathers on the host's body (or in the case of humans, under clothing). The flea body is covered with hard plates called sclerites. These sclerites are covered with many hairs and short spines directed backward, which also assist its movements on the host (Fig. 46). The tough body is able to withstand great pressure, likely an adaptation to survive attempts to eliminate them by mashing or scratching. Even hard squeezing between the fingers is normally insufficient to kill a flea.

Fleas lay tiny, white, oval-shaped eggs better viewed through a loupe or magnifying glass. The larva is small and pale, has bristles covering its worm-like body, lacks eyes, and has mouth parts adapted to chewing. The larvae feed on various organic matter, especially the feces of mature fleas. The adult flea's diet consists solely of fresh blood. In the pupal phase, the insect is enclosed in a silken, debris-covered cocoon. Fleas are holometabolous insects, going through the four lifecycle stages of egg, larva, pupa and imago (adult). Adult fleas must feed on blood before they can become capable of reproduction. Flea populations are distributed with about 50% eggs, 35% larvae, 10% pupae, and 5% adults. The flea life cycle begins when the female lays after feeding. Eggs are laid in batches of up to 20 or so, usually on the host itself, which means that the eggs can easily roll onto the ground. Because of this, areas where the host rests and sleeps become one of the primary habitats of eggs and developing fleas. The eggs take around two days to two weeks to hatch.

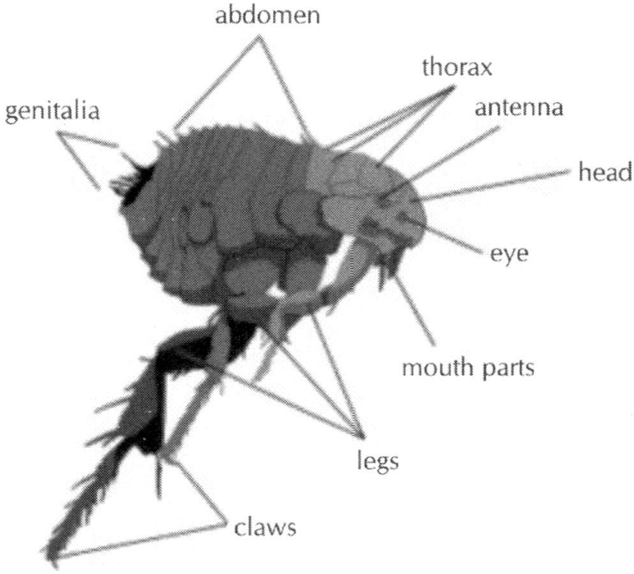

Fig. 46. Diagram of a Flea (Source: Wikipedia, the free encyclopedia).

Flea larvae emerge from the eggs to feed on any available organic material such as dead insects, feces, and vegetable matter. In laboratory studies, some dietary diversity seems necessary for proper larval development. Blood-only diets allow only 12% of larvae to mature, whereas blood and yeast or dog chow diets allow almost all larvae to mature. They are blind and avoid sunlight, keeping to dark places such as sand, cracks and crevices, and bedding. Given an adequate supply of food, larvae pupate and weave silken cocoons within 1–2 weeks after three larval stages. After another week or two, the adult fleas are fully developed and ready to emerge. They may remain resting during this period until they receive a signal that a host is near - vibrations (including sound), heat, and carbon dioxide are all stimuli indicating the probable presence of a host. Fleas are known to overwinter in the larval or pupal stages.

Once the flea reaches adulthood, its primary goal is to find blood and then to reproduce. Its total life span can be as long as one and one-half years in ideal conditions. Female fleas can lay 5000 or more eggs over their life, allowing for phenomenal growth rates averaging 30–90 days. These conducive conditions include the right temperature, food supply, and humidity. Generally speaking, an adult flea only lives for 2 or 3 months. Without a host for food a flea's life might be as short as a few days. With ample food supply, the adult flea will often live up to 100 days. Newly emerged adult fleas live only about one week if a blood meal is not obtained. However, completely developed adult fleas can live for several months without eating, so long as they do not emerge from their puparia. Optimum temperatures for the flea's life cycle are 21°C to 30°C, and optimum humidity is 70%.

Adult female rabbit fleas, *Spilopsyllus cuniculi*, can detect the changing levels of cortisol and corticosterone hormones in the rabbit's blood that indicate it is getting close to giving birth. This triggers sexual maturity in the fleas and they start producing eggs. As soon as the baby rabbits are born, the fleas make their way down to them and once on board they start feeding, mating, and laying eggs. After 12 days, the adult fleas make their way back to the mother. They complete this mini-migration every time she gives birth.

Fleas feed on a wide variety of warm-blooded vertebrates including dogs, cats, humans, chickens, rabbits, squirrels, rats, ferrets, and mice (Fig. 47). Fleas are a nuisance to their hosts, causing an itching sensation which in turn may result in the host attempting to remove the pest by biting, pecking, scratching, etc. in the vicinity of the parasite. Fleas are not simply a source of annoyance, however. Flea bites generally cause the formation of a slightly raised, swollen itching spot with a single puncture point at the center (similar to a mosquito bite). The bites often appear in clusters or lines of two bites, and can remain itchy and inflamed for up to several weeks afterwards. Fleas can also lead to hair loss as a result of frequent scratching and biting by the animal, and can cause anaemia in extreme cases.

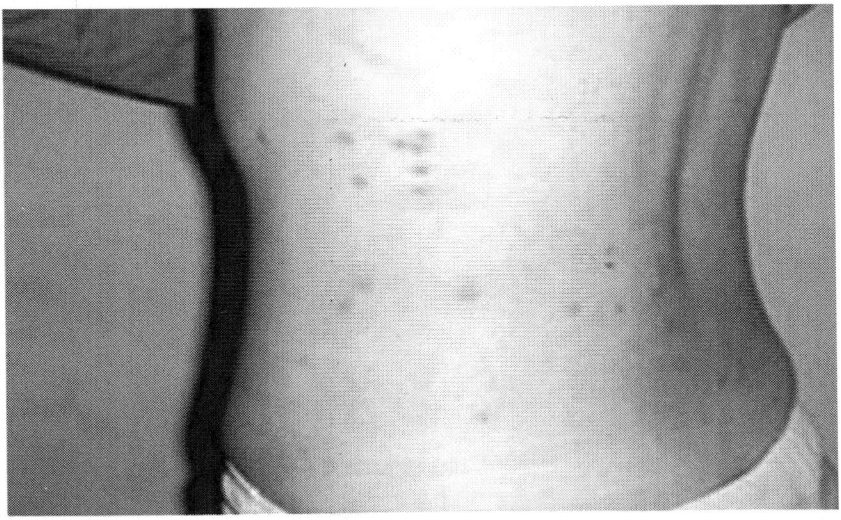

Fig. 47. Bite from a flea on the back of a man (Source: Wikipedia, the free encyclopedia).

Besides the problems posed by the creature itself, fleas can also act as a vector for disease. Fleas transmit not only a variety of viral, bacterial and ricketsial diseases to humans and other animals, but also protozoans and helminths.

- Bacteria – murine and endemic typhus. Fleas have helped cause epidemics by transmitting diseases such as the bubonic plague between rodents and humans by carrying *Yersenia pestis* bacteria. Fleas can also transmit *Ricketsial typhi, R. felis* and *Bartonella henselae.*

- Virus: myxamatosis.
- Helminth: infestation of *Hymenolepiasis* tapeworm.
- Protozoa: Trypanosome protozoans such as those of the subgenus, use a variety of flea species opportunistically as vectors.

Fleas that specialize as parasites on specific mammals may use other mammals as hosts; therefore, humans are susceptible to the predation of more than one species of flea. Fleas can settle in a person's hair in less than ten minutes, causing soreness and itching. The itching associated with flea bites can be treated with anti-itch cream, usually antihistamines or hydrocortisone. Calamine lotion has been shown to be effective for itching.

22. Order Diptera

The order Diptera (from the Greek *di* = two, and *ptera* = wings) comprise true flies. Diptera is a large order containing an estimated 1,000,000 species of mosquitoes, gnats, midges and others, although only 12% (about 120,000 species) have been described. It is one of the major insect orders, second only to Coleoptera, Hymenoptera and Lepidoptera, with ecological and human importance. Mosquitoes (Culicidae), are vectors of malaria, elephantiasis, dengue, chikungunya, Yellow Fever, West Nile Fever, Japanese encephalitis and other infectious diseases. The presence of a single pair of wings distinguishes most true flies from other insects with "fly" in their names. However, some true flies have become secondarily wingless. For example, many members of the superfamily Hippoboscoidea and some species that are inquilines live in social insect colonies. Species of Diptera occur in all faunal regions, in all climatic zones and in a great diversity of biotopes. They are among the dominant hypsobiont insects on nearly all high mountains, and are abundant even at the highest altitudes at which permanent insect life is still to be found. In the Himalaya the dipterans are common still at an elevation of 6300 m.

At least for three reasons the Diptera occupy a unique position in the field of evolutionary studies:

(1) The genus Drosophila, whose genetic system has been investigated more intensely than in any other organism (from the standpoint of formal genetics (1914-1930), and more recently, from phylogenetical and evolutionary viewpoints), belongs to this order.

(2) In Diptera the giant polytene salivary gland chromosomes are most beautifully developed and render possible a type of cytogenetic analysis that cannot be carried out in any other group of insects, though recently such studies were made out also on the polytene chromosomes of Collembola. The most favourable polytene chromosomes are found in the nematoceran genera Sciara and Chironomus, and in Drosophila.

(3) Diptera presents several entirely different types of genetic systems, whose relationships to one another present an evolutionary problem of considerable significance. Homopterans are the only other order in which a similar diversity of genetic systems is encountered. Dipterans are the only group in animal kingdom with very clear pairing of homologous elements in somatic metaphase. Various workers have claimed to have observed somatic pairing in other animal groups as well, but the phenomenon has never been conclusively demonstrated. The most primitive dipterans, viz., Tipulidae are characterized by the presence of chiasmata in both sexes and by the fact that the sex elements do not form a bivalent in the primary spermatocytes. From this group two main evolutionary lines arose. One led to Simuliidae, Culicidae and Chironomidae, in which the chiasmata were retained in the male, but the sex chromosomes disappeared as distinct elements.

Considerable revision in the taxonomy of the flies has taken place since the introduction of modern cladistic techniques, and much remains uncertain. The order Diptera is organized into two suborders, namely, Nematocera and Brachysera. The Nematocera are recognized by their elongated bodies and many-segmented, often feathery antennae as represented by mosquitoes and crane flies. The Brachycera have a more roundly proportioned body and much shorter antennae. Recently, in 1964, Nematocera was split into two suborders, the Archidiptera and the Eudiptera, by Boris Borisovitsch Rohdendorf:

1. Suborder Nematocera (77 families, 35 of them extinct) – long antennae, pronotum distinct from mesonotum. In Nematocera, larvae are either eucephalic or hemicephalic and often aquatic.
2. Suborder Brachysera (141 families, 8 of them extinct) – short antennae, the pupa is inside a puparium formed from the last larval skin, and they are generally robust flies with larvae having reduced mouthparts.
3. Infraorders Tabanomorpha and Asilomorpha – these comprise the majority of what were the Orthorrhapha under older classification schemes. The antennae are short, but differ in structure from those of the Muscomorpha.
4. Infraorder Muscomorpha – (largely the Cyclorrhapha of older schemes) Muscomorpha have three-segmented, aristate (with a bristle) antennae and larvae with three acephalic instars (maggots).

Most of the Muscomorpha are further subdivided into the subsections Acalyptratae and Calyptratae based on whether or not they have a calypter (a wing flap that extends over the halteres). The family Muscidae, that includes the house fly, is among the Calyptratae.

Diptera have one pair of fore wings on the mesothorax and a pair of halteres, or reduced hind wings, on the metathorax. Some species of flies are exceptional in that they are secondarily flightless. The only other order of insects bearing two true, functional wings in addition to any form of halteres are the Strepsiptera. In contrast to the flies, the Strepsiptera bear their halteres on the mesothorax and their flight wings on the metathorax.

Adult flies consume only liquids due to their sucking mouthparts, or haustellates. Female Tabanidae use knife-like mandibles and maxillae to make a cross-shaped incision in the hosts' skin and then lap up the blood. The gut includes large divrticulae, allowing the insect to store small quantities of liquid after a meal. Flies are adapted for aerial movement and typically have short and streamlined bodies. The first tagma of the fly, the head, consists of ocelli, antennae, compound eyes, and the mouthparts (the labrum, labium, mandible, and maxilla make up the mouthparts). The second tagma, the thorax, bears the wings and contains the flight muscles on the second segment, which is greatly enlarged; the first and third segments have been reduced to collar-like structures. The third segment of the thorax bears the halteres, which help to balance the insect during flight. A further adaptation for flight is the reduction in number of the neural ganglia, and concentration of nerve tissue in the thorax, a feature that is most extreme in the highly derived Muscomorpha infraorder.

Flies have a mobile head with large compound eyes on the sides of the head, and three small ocelli on the top. For visual course control, flies' optic flow field is analyzed by a set of motion-sensitive neurons. A subset of these neurons is thought to be involved in using the optic flow to estimate the parameters of self-motion, such as yaw, roll, and sideward translation. Other neurons are thought to be involved in analyzing the content of the visual scene itself, such as separating figures from the ground using motion parallax. The H1 Neuron is responsible for detecting horizontal motion across the entire visual field of the fly, allowing the fly to generate and guide stabilizing motor corrections midflight with respect to yaw. The antennae take a variety of forms, but are often short, which reduces drag while flying.

Fig. 48. Mating anthomyiid flies (Source: Wikipedia, the free encyclopedia).

Dipterans are holometabolous, i.e., they go through a complete metamorphosis (egg, larva, pupa, adult). In many flies, the larval stage is predominant. Most dipteran larvae have a sclerotized head capsule, or variably reduced ultimately to remnant mouth hooks and have no jointed, "true legs". Some dipteran larvae, such as species of Simulidae, Tabanidae, and Vermilionidae have prolegs adapted to hold onto a substrate in flowing water, host tissues, or prey. In some parasitic species, the larvae are not independent and therefore, are released when they mature and form pupae. Larvae have limited mobility.

Little demarcation is seen between thethorax and abdomen, though the demarcation may be visible in many Nematocera, such as mosquitoes; in the Brachycera, the head of the larva is not clearly distinguishable from the rest of the body, and few, if any, sclerites occur. Informally, such brachyceran larvae are called maggots, but the term is not technical and often applied indifferently to fly larvae or insect larvae in general. The eyes and antennae of brachyceran larvae are reduced or absent, and the abdomen also lacks appendages such as cerci. This lack of features is an adaptation to food such as carrion, decaying detritus, or host tissues surrounding endoparasites. Nematoceran larvae generally have visible eyes and antennae, though they are usually small and of limited function.

The pupae take various forms, and in some cases, develop inside a silk cocoon.

The adult stage is usually short, only to mate and lay eggs (Fig. 48). The genitalia of female flies are rotated to a varying degree from the position found in other insects. In some flies, this is a temporary rotation during mating, but in others, it is a permanent torsion of the organs that occurs during the pupal stage. This torsion may lead to the anus being below the genitals, or, in the case of 360° torsion, to the sperm duct being wrapped around the gut and the external organs being in their usual position. When flies mate, the male initially flies on top of the female, facing in the same direction, but then turns around to face in the opposite direction. This forces the male to lie on his back for his genitalia to remain engaged with those of the female, or the torsion of the male genitals allows the male to mate while remaining upright. This leads to flies having more reproduction abilities than most insects, and at a much quicker rate. Flies occur in large populations due to their ability to mate effectively and in a short period of time during the mating season.

The female lays her eggs as close to the food source as possible. Flies are oviviviparous, meaning eggs hatch immediately after being laid. Sometimes, the larvae hatch inside the mother. Development is rapid, allowing the larvae to consume as much food as possible in a short period of time before transforming into adults. Maggots found on corpses are useful to forensic scientists, specifically forensic entomologists. Maggot species can be identified by various means such as their anatomy and by matching their DNA. Maggots of various species of flies visit corpses and carcases at fairly well-defined times after the death of the victim, and so do their predators, such as beetles in the family Histeridae. Thus, the presence or absence of particular species of fly maggots and other scavenger insects provide evidence as a basis for estimation of the time since death, and sometimes other details such as the place of death.

Any description of the order Diptera withour mentioning about mosquitoes (Culicidae) is simply impossible (Tyagi, 2003, 2005; Pemola Devi & Jauhri, 2008). Many of these species are serious vectors and pests, and the Doon Valley has suffered by several ravages of malaria and dengue in past. It is important to understand these creatures in some more details on a wider horizon (Tyagi et al.,

2015). In addition to being severe pests to man in various different situations, mosquitoes are also very important vectors for the deadly diseases such as malaria, dengue, filariasis, Japanese encephalitis and a host of other infections, mostly viral. In fact, no other creature on earth has touched so directly and profoundly the lives of so many human beings as has been done by the mosquito. It has been said of the magnitude of ferocity of mosquitoes, merely by the fact that the diseases they transmitted had during oft-hailed epidemics in some distant past killed millions of people in the endemic countries, that they had felled great leaders, decimated armies, and decided the fates of nations! The recent epidemic of Zika Virus infection associated with microcephaly in the newborns in Pan America, Brazil in particular, and also other regions of the world including several countries in Europe and Asia, excluding so far India, has warranted an urgent need to develop effective control tools for the vectors, *Aedes aegypti* and *Ae. albopictus*, though a host of other various different mosquitoes have also been incriminated with Zika Virus.

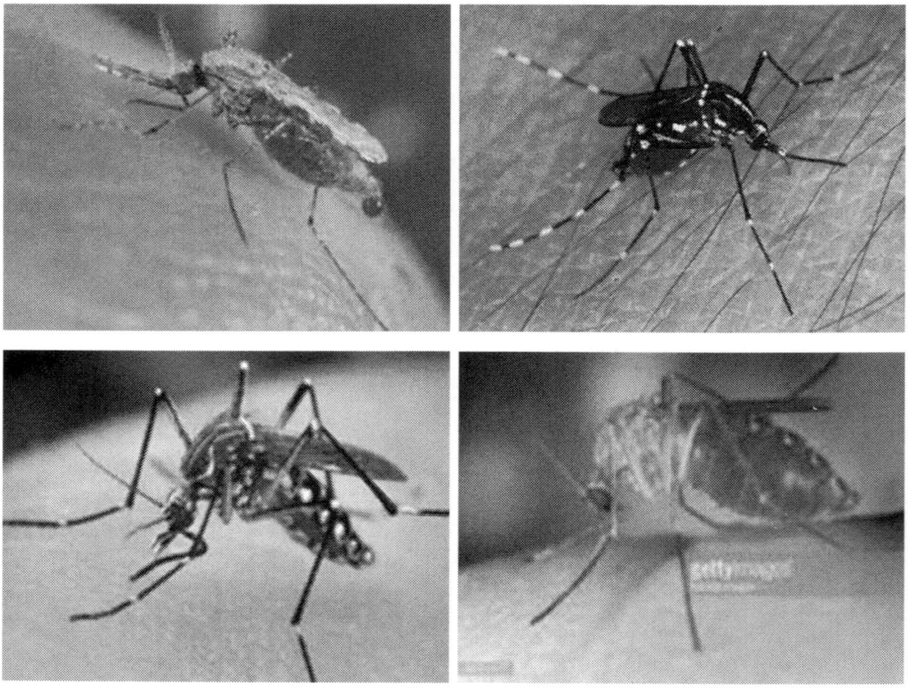

Fig. 49. Various vector mosquitoes: (a) *Anopheles stephensi*, (b) *Aedes albopictus*, (c) *Aedes aegypti*, and (d) *Culex quinquefasciatus* (Source; Wikipedia, the free encyclopedia).

Mosquitoes (Family Culicidae) are by far the most thoroughly researched insects. They probably have a much greater influence on human health and well-being throughout the world than any other arthropod, mainly because of their involvement in both transmitting a number of dreadful diseases (such as malaria, filariasis, dengue, Japanese encephalitis, Rift valley fever, Chikungunya, West Nile virus and O'Nyong-nyong) and creating nuisance of great public health importance. Consequently, among the many species of blood-sucking insects, mosquitoes belonging to genera Anopheles, Culex and Aedes are the most important arhropods from medical standpoint as they are responsible for nearly 10% of all the sickness of humans (Fig. 49). Although, there are more than 3500 species all over the world, with possibly the latest addition being *Anopheles pseudosundaicus* Tyagi et al., 2009, however, only fewer than 100 species under eight genera are vectors of diseases. These vectors are organized in two of the three subfamilies, Anophelinae and Culicinae, while the third subfamily Toxorhynchitinae has members with mouthparts suited to feed on plant sap only. Mosquitoes are cosmopolitan in distribution and are found in all climatic zones and zoogeographical regions.

To effect malaria, or for that matter any parasitic or viral infection, it is necessary that mosquito bites and blood feeds from humans. The whole blood feeding process by the vector mosquito is very intelligently executed exercise to ensure engorgement because the protein in blood is essential for the development of ova in females which only bloodfeed (Fig. 50). Male mosquito survive on plant sap etc.

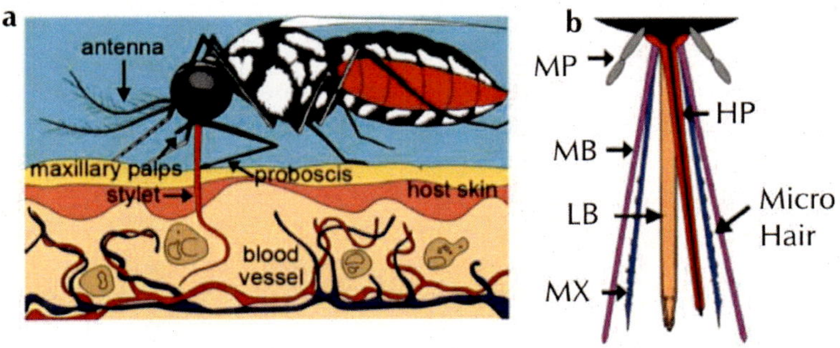

Fig. 50: (a) Mosquito's feeding process, and (b) mouth parts (Source: Jung et al., 2015. doi:10.1038/srep13444).

Mosquitoes have the capability to bite by day or by night, depending on species. They breed in all kinds of water fresh and salt, flowing or stagnant, foul or potable, water in tin cans, tree holes (phytotelmata) and salt marshes. Large coastal areas are occasionally uninhabitable because of salt marsh mosquitoes. Mosquitoes may render recreation centers unusable and make normal living in and around homes extremely unbearable and vexing. Ironically, real estate values may suffer where mosquitoes are present, and losses caused by reduced industrial efficiency can be considerable because of the mosquito menace.

As a principle, this is only the female of the vector mosquito which feeds on the blood of human or animal host, although females of some species are known to supplement their meals with plant sugars, too. Female mosquitoes may feed on the blood of a wide variety of mammals, birds, reptiles and amphibians. Mosquitoes are generally host-specific for their blood-feeding requirements but they may quickly diverse in case of any exigency.

Mosquitoes have their body clearly divided in three parts: head, thorax and abdomen. The head bears a pair of large compound eyes, a pair of antennae and mouthparts which are piercing and sucking type in females suited primarily for blood feeding, while in males they are developed to suck sap from the plants. Antennae too differ being shorter and sparsely haired in females while in males they are thickly haired and bushy. There is a pair of fully developed forewings, while the hind wings are reduced to rudimentary halteres or balancers. The males are equipped with external genitalia, called hypopygium. Following mating, males pass spermatozoa to the females where they are stored in spermatheca. Eggs are fertilized as they pass out of the ovary of the female. Females generally lay eggs in mildly moving or stagnant fresh (e.g., Anopheles) or polluted water bodies (e.g., Culex). Eggs are laid singly directly on water surface (Anopheles), in batches/groups cohesced in a boatraft-like shape (Culex, Culiseta, Mansonia and Uranotaenia) or along the moistened surfaces, like soil etc., close to water (Aedes). Eggs of the genus Toxorhynchites are also laid singly on the water surface. Some mosquitoes (e.g. Aedes, Toxorhynchites) are specialized to lay eggs in plant materials or tree-holes (phytotelmata). Some species may survive in the egg stage for 1-2 years before flooding occurs. Aedes, Orthopodomyia and Psorophora eggs are commonly laid in this manner. In general, development of mosquitoes is essentially the same.

There are four stages in the life cycle of mosquitoes; egg, larva, pupa and adult. The immature stages are aquatic in nature. The eggs are usually white when first deposited, becoming dark within an hour or two. Anopheline eggs are oval-shaped and float horizontally. They are usually pointed on one end and have a pair of lateral floats. The larvae of all mosquitoes get their food from the water where they live, but to respire they have to come to the surface. Larvae of the genus Mansonia are adapted to take oxygen underwater from the air-conducting parts of plants. The larval developmental period includes four instars which usually take 5-15 days to complete. At the end of the larval development pupa is formed by discarding the skin of the fourth instar larva. Only larvae feed while pupae do not feed. The larval body is characteristically divisible in three parts, but the pupa has only two parts; the fused cephalothorax and the curled and segmented abdomen. Pupae too occur in water and are actively motile. For respiring all pupae, except those of Mansonia and Coquilletidia, move to the surface where with the help of a pair of trumpets atop head region the oxygen is taken in. The pupal stage lasts for one day to several days. When the pupal growth is complete, the pupal skin or exuvium breaks, the adult works its way out, crawls on to the surface of the water and is soon ready to fly away. The adult mosquito, in most cases of species, is a small fragile insect with a slender abdomen, one pair of narrow and delicate wings, covered or thinly so with scales, and three pairs of long slender legs. As the case may be, species may greatly vary in their size varying from slightly over 4 mm to about 12 mm. The three body parts (i.e., head, thorax and abdomen) are clearly visible.

Most mosquitoes are nocturnal in their feeding behaviour (e.g., Anopheles), but some are diurnal or crepuscular (e.g., Aedes). When the female finds a suitable host, such as the man, it introduces the piercing mouthparts through skin deep into body and penetrates a capillary, to suck the blood directly from the capillary or from the pool of the blood oozed outside the punctured capillary. Females of certain mosquitoes are single-time feeder while many species tend to feed in shift or multiple occasions on a given night or day-time. The number of the egg formation and development in ovary of the female is directly dependent on the quantum and nature of supply of blood meal received by the female mosquito. Depending on a particular species, an engorged female may like to rest inside house or prefer outside environs. This characteristic of the vector species has been for long exploited for their control through indoor spraying of residual insecticides.

Mosquitoes are the sole vectors of human malaria, yellow fever and dengue, and they participate significantly in transmitting lymphatic filariasis and the encephalitides of various different kinds. Besides, their bites may itch for days, causing much loss of sleep and tranquility, and serious nervous irritation and restlessness. Their saliva contains proteins alien to the human body and repeated bites result in sensitivity to it. Mosquitoes often cause serious economic losses limiting outdoor movements.

Some of the important genera of mosquitoes are presented below:

(1) Anopheles mosquitoes

This is doubtlessly medically the most important of all the genera of mosquitoes. There are about 400 species worldwide, and in India a total of 57 species occur (Nagpal & Sharma, 1995; Tyagi, 2002). Species of this genus with varying numbers are found in all the zoogeographical regions of the world; the Oriental Region harbouring the maximum species.

Biology and Life Cycle of Anophelines

As stated above, there are four stages through which anopheline mosquitoes complete their life cycle (Fig. 51). Important taxonomic and biological features in each of these are described below:-

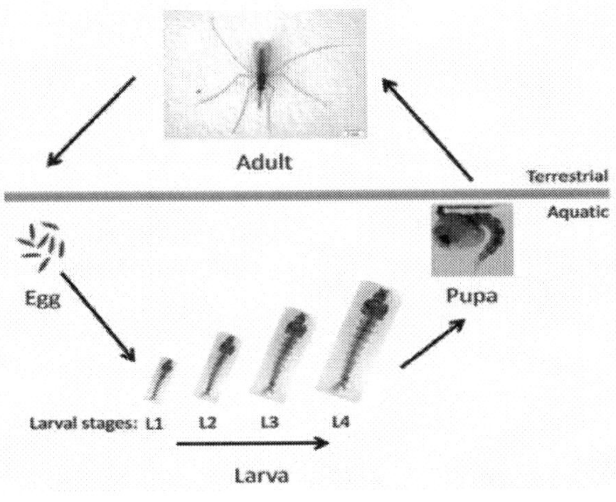

Fig. 51. Stages of the *Anopheles* mosquito life-cycle (Source: Wikipedia, the free encyclopedia).

(i) *The Egg:* Eggs of Anopheles mosquitoes are boat-shaped with an upper deck having a micropile at one end and a fringe all around, and usually with lateral floats with transverse striations called ridges. Eggs are laid singly. In optimal temperature it takes about 36-48 hours for eggs to hatch.

(ii) *The Larva:* This is very agile from the time it hatches out. There are four larval instars before it converts into pupal stage. It rests characteristically parallel to the surface of water. The head of a larva is capable to rotate at 180° to enable it feed more intensely and vigorously. Characteristic palmate hairs, present in many species, occur on both thorax and abdomen. Two (one on each side) reversible organs, called Nuttal's organs, occur on the first segment of the thorax. These are best seen in the live specimens as they are fleshy and protrusible. Larvae have a dorsal spiracular plate supporting two spiracles meant for bringing about respiration. The total duration taken by larvae under normal conditions of temperature ranges from 5-7 days.

(iii) *The Pupa*: This has a peculiar 'comma' form with highly active up and down vertical movements. The paddle is oval in outline. It may be either undivided or only slightly invaginated, with the tip having a pair of setae one of which is ventral and sub-apical. On the abdominal segments 3rd to 7th there are located spine-like seta No. 9 (Belkin's system) at the postero-lateral corners usually with branches on 8th abdominal segment. The trumpet, located on the top of the cephalothorax, often bears at least one cleft. Pupal development completes in about 48 hours.

(iv) *The Adult:* Soon after the adult emerges from pupae, it flies away to take shelter in some suitable place, where it rests for sometime before taking its first blood meal. Males generally emerge earlier than females, and have shorter lifespan than their counterparts.

Anopheline adults are identifiable from the rest of the mosquitoes due to their very characteristic colour patterns mostly on the palpi, wings and legs, beside other prominent traits as seen in their posture at rest etc. The palpi of both sexes are of similar length, almost equal to that of the proboscis. In males, however, the two distal segments in palpi are enlarged giving them a club-like appearance. The posterior edge of scutellum is only slightly curved and is continuous without any marked division into lobes. Hairs on these lobes are in continuous series.

The axes of head, thorax and abdomen are nearly in a straight line. Anopheline mosquitoes rest on an object in a characteristic posture. The angle of resting, varying from species to species, is close to 60°C with the tip of the abdomen directed away from the surface (e.g., *An. hyrcanus, An. barbirostris*). In a few species like *An. culicifacies, An. culiciformis, An. sintoni* etc. the body is almost parallel to the surface, giving adult a semblance of a Culex-like posture.

Of the three acini of the salivary glands of each side, the middle is sac-like instead of being tubular as is noticed in culicines.

Wings are spotted or unspotted; the former condition being very common, though. On the other hand, members of subgenus Anopheles, like *An. aitkenii,* have completely dark wings. Males have only one claw in the fore tarsi. Females have only one spermathecum.

Most vector species are endophilic and endophagic in behaviour (*An. culicifacies, An. subpictus, An stephensi*), although a few are sylvatic in nature, too (*An. dirus, An. pulcherrimus*). Although, there is no direct proof of hibernation exhibited by any Indian species, there are species in colder temperate regions elsewhere in the world which manifest overwintering. In India, *An. subpictus* is suspected to possess the potential to overwinter.

Some species exhibit "sibling species complexes", a phenomenon which carry tremendous medical significance in the sense that while some of them may be vectors, others are either less significantly important or are not able to transmit the pathogen at all. The most widely examined species-complexes belong to *An. gambiae* in Africa and *An. culicifacies* in India.

(2) Aedes mosquitoes

Several species of the genus Aedes are of tremendous medical and public health importance, for instance *Ae. aegypti*, *Ae. albopictus*, *Ae. sollicitans*, *Ae. nigramaculis*, *Ae. vexans* etc. In as far as India is concerned, the most important vector species are *Ae. aegypti* and *Ae. albopictus*, due to their involvement in transmission of dengue fever, chikungunya virus and Zika virus. *Aedes aegypti* is also the vector for the deadly Yellow Fever prevalent in Africa and South America.

Biology and life cycle of Aedes aegypti

Main biological and taxonomical features are described for the two important vector species (i.e., *Ae. aegypti* and *Ae. albopictus*). In general, the biology of *Ae. aegypti* follows the pattern of other culicid mosquitoes. The life cycle is completed through four distinct stages; egg, larva, pupa and adult. The larval development passes through four instars, each undergoing progressive increase in size till the pupal stage. Pupae and adults do not show any growth. Only female adults of *Ae. Aegypti* feed on blood, sugar and water for maturation of eggs, energy requirements and maintenance of the osmotic balance of the body.

(i) *The Egg:* Eggs are small, about 1 mm in length, torpedo shaped and deeply black in colour, although when freshly laid they are transluscent and pure white. The eggs are capable to remain viable for weeks or even months and can withstand prolonged desiccation. Eggs house fully developed larvae ready to hatch. The head of the larva occupies nearly one-third of the egg-volume, while the siphon and other terminal parts fill most of the remaining posterior pole. The egg has very characteristic polygonal markings, lined with a network of milky white membranous channels filled with air at the base of the exochorion. Beneath exochorion is a dense, tough and hard endochorion membrane which gives the egg its shape and rigidity.

(ii) *The Larva:* As in all other mosquitoes, there are four larval instars, each instar ending in a moult or ecdysis. The first instar larva hatches out from the egg by first widening the transverse crack at the dorsal side anteriorly. Finally the larva breaks the eggs by means of a thorne-shaped egg breaker. Initially the freshly hatched larva is unable to move but shortly the young larva wriggles its way to the water surface to make the contact with the free air. It takes about 2–3 days at 27-29°C for the larva to hatch out from the egg. However, the larva may take few minutes to several hours to emerge at the surface. The first instar larva is identifiable on the basis of the presence of the head of the egg breaker. The larva carries a characteristic siphon. The pectin spine arising on the soft membrane number 4-5. Siphon carries the trachea. The comb teeth, which are delicate, number only five (Table 7).

Table 7. Some identification characteristics of various larval instars as well as their time of development.

Larval instar	Transverse diameter of head (mm)	No. of pectin teeth	Period for completion of four instars (in hrs)
First	0.3	6	24
Second	0.45	7-10	24
Third	0.65	12-15	24
Fourth	0.95	12-20	96-240

The second and third instars can be identified by their size and general appearance. Little change occurs until fourth instar when the development of the future pupal structure begins. The compound eyes or the optic lobes begin to form in the late third instar. The imaginal buds at the bases of the pleural hairs become apparent in the second instar, and are quite conspicuous in the third, though their development is a feature of the fourth instar. The gonads are already visible in the 6th abdominal segment in the third instar, when also the buds of the developing gonoxites become apparent.

During the fourth instar period a condition is reached in which the larva is little more than the roughly sculptured body of the future adult disguised in the larval skin. Such a condition has been termed a prepupa. The duration of the larval stages is influenced by the water temperature. The most common period lies between 6 and 8 days.

(iii) *The Pupa:* The pupa of *Ae. aegypti* resembles most of other culicines. However, it differs from the anopeline pupa in having more rounded body and cylindrical breathing trumpets. The main characters used to identify the pupae of *Ae. aegypti* from other culicine pupae are: (a) form of the respiratory trumpets, (b) arrangement and characters of the hairs, particularly those in the abdominal region, and (c) characters of the tail fins or paddles. The pupal body consists of a large globular anterior region, the cephalothorax, and a narrower abdomen which is normally kept flexed under the former to propel the insect when swimming.

(iv) *The Adult:* It is a small dark structure easily identifiable by the lyre-shaped, silvery white lines on the thorax and the white bands on the tarsal segments. The adult is both vector as well as pest when present in large numbers, and therefore deserve special attention in medical entomology. This is thoroughly a domesticated mosquito and breeds almost exclusively in artificial containers in and around human habitations. The females lay their eggs singly on the water just at the margin, or on the sides of the container above the waterline. Given the opportunity, the adult prefer human blood to that of other animals. Adults readily enter houses to find suitable hosts. They bite mainly in the morning and in the evening. The adult quietly hails to settle for feeding, preferring to bite around the ankles, under shirt-sleeves, or in the back of the neck.

The adult of *Ae. aegypti* resembles markedly to that of *Ae. albopictus*, commonly called as the tiger mosquito, another mosquito of great medical significance, except that the lyre-shaped markings on the thorax are replaced by a stripe of silvery-white scales on the dorsum of thorax. It's a forest dwelling species and has adequately adapted itself to urban environments as well. A native of Asia, it has spread to most tropical parts of the globe. In Kerala State, southern India, it breeds abundantly in the latex collecting cups associated with rubber plantation.

(3) Culex mosquitoes

The genus Culex comprises several medically important species, such as *Cx. pipiens*, *Cx. quinquefasciatus, Cx. tarsalis, Cx. nigripalpus, Cx. tritaeniorhynchus, Cx. bitaeniorhynchus, Cx. pseudovishnui, Cx. restans and Cx. salinarius.* In Indian context, of greatest importance are *Cx. quinquefasciatus* and *Cx. tritaeniorhynchus* which, besides being serious pests, are in the first place responsible for transmitting bancroftian filariasis and Japanese encephalitis, respectively. These are brown coloured mosquitoes of medium size

Culex quinquefasciatus is fondly referred to as house mosquito due to its abundant occurrence in houses. It breeds in cesspools, storm-sewer catch basins, street gutters, open septic tanks and effluent from sewage disposal plants. Adults are active only at night. When disturbed, they can be easily spotted flying out of their hideouts in and around houses, outbuildings and other shelters. *Culex tritaeniorhynhus* breeds mainly in rice fields and it is both an exophilic and exophagic mosquito.

Culex pipiens and *Cx. quinquefasciatus* are found in most of the tropical countries. All culicines generally rest with their entire body more less close to the surface. The scales are narrow and not metallic coloured. There are no distinct patterns or markings on the body. The abdomen ends bluntly. Prespiracular and postspiracular areas are without bristles. Forelegs in males bear toothed claws. In females, the claws are there but simple. The tarsi in females have large pulvilli.

(4) Mansonia and Coquillettidia mosquitoes

Large sized, yellowish brown to dark brown mosquitoes with very painful bites. Mansonia occurs in all zoogeographical regions, while Coquillettidia occurs mainly in Afro-tropical region but also in Neartic, Neotropical, Oriental and Australasian regions.

Mansonia mosquitoes are characterized by possessing postspiracular setae, but no prespiracular setae are found. They are in the habit of biting and resting both inside and outside the houses. They breed in the heavily vegetated pondwater growing aquatic plants such as *Pistia* spp. etc. Larvae respire on air-conducting parts of the aquatic plants. Coquillettidia mosquitoes have neither pre- or postspiracular setae, but have an overall yellow colour. The adult body has a covering of broad scales, which may form fine speckled patterns of colour on the wings. The abdomen is blunt ended. The scales on the wings are very broad. The claws are simple and pulvilli are small or even absent. Some species like *Coquillettidia perturbans* breeds in marshes, ponds and lakes having a thick growth of aquatic vegetation. Larval development is generally slow, lasting several months. Just like that in Mansonia species, larvae in this species too breathe by penetrating air conducting plant tissues with their breathing tubes.

They readily enter the house and bite viciously. Some of them, such as *Ma. annulifera and Ma. uniformis*, are vectors of human lymphatic filariasis caused by *Brugia malayi.* In India, this type of lymphatic filariasis is restricted mainly in Kerala State.

(5) Haemagogus mosquitoes

These mosquitoes occur in Neotropical region only where they are of much importance in public health. These mosquitoes are identifiable due to certain specific characteristics. Scales on their wings are narrow, but they are fairly broad on the body. The anterior pronotal lobes are large and nearly meet at the top. The abdomen of these mosquitoes is blunt ended. Forelegs of the sexes possess toothed claws. There are no prespiracular or postspiracular areas. The palps of the males are shorter than the proboscis. These mosquitoes are not confirmed from India.

(6) Sabethes mosquitoes

Members of this genus occur only in Neotropical region. They are identifiable due to scutum and abdomen being covered with scales having a metallic colour. The scales of wings are narrow, but on the body they are fairly broad. The abdomen is blunt ended. The anterior pronotal lobes are very large. The mid or hindlegs have a conspicuous area of long scales in the shape of a paddle. Prespiracular bristles are present but there are no postspiracular bristles. There is a group bristles on the postnotum. The palps of males are shorter than the proboscis.

(7) Psorophora mosquitoes

This genus has several species which are of immense public health importance largely due to their nuisance value particularly when in large numbers. Many species like *P. ciliata* and *P. columbaie* are large sized with shaggy legs. The scales on body and wings are narrow and without metallic colour. Prespiracular and postspiracular bristles are present. The abdomen is pointed at the tip. Both sexes have forelegs with toothed claws. The male palps are longer than the proboscis. They are vicious biter and because of their gigantic appearance they often offer a frightening look. Some species do not lay eggs directly in water but on the ground that is subject to flooding from rainfall, overflow or irrigation. They breed in temporary pools, irrigation ditches and seepage pools, and larvae survive predating on other aquatic insects including mosquito larvae. Members of this genus occur only in the Neartic and Neotropical regions.

(8) Culiseta mosquitoes

Found in all zoogeographical regions in the world, they are identifiable due to presence of narrow scales on the body and wings which are non-metallic in colour; there are, however, of coloured patterns on the body. The abdomen is blunt ended. Prespiracular bristles alone are present. Forelegs of males have toothed claws; in female's forelegs they are simple. The stem vein (base of subcostal vein) of the wing usually has bristles on the ventral surface. Most species are painful biters.

In as far as Dehradun or the Doon Valley is concerned the mosquito research work carried out by Dr. R.K. Jauhari at his Department of Zoology of DAV (PG) College, Dehradun is most laudable (Pemola Devi & Jauhari, 2008). He has extensively investigated culicid fauna of Garhwal region, with emphasis on that of the Doon Valley (see Annex 3 for Dr. Jauhari's detailed mosquito-related research programmes and publications). Prof. Devinder Singh, Punjabi University, Patiala is an authority on Forensic Entomology including many dipteran species.

23. Order Mecoptera

Mecoptera (from the Greek: *meco-* = "long", *-ptera* = "wings") are an order of insects with about 550 species in nine families worldwide. Mecoptera are sometimes called scorpionflies after their largest family, Panorpidae, in which the males have enlarged genitals that look similar to the stinger of a scorpion. The Bittacidae, or hangingflies, are a prominent family of elongate insects known for their elaborate mating rituals, in which females choose mates based on the quality of gift prey offered by various males. They are overwhelmingly predators or consumers of dead organisms. The cosmopolitan scorpion flies are divided into two suborders, Protomecoptera and Eumecoptera.

Mecoptera are small to medium insects with slender, elongated bodies. They have relatively simple mouthparts, with long mandibles and fleshy palps, which resemble those of the more primitive true flies. Like many other insects, they possess compound eyes on the sides of their heads, and three ocelli on the top. Most Mecoptera feed on vegetation in moist environments; in hotter climates, they may therefore be active only for short periods of the year.

The wings are narrow in shape, with numerous cross-veins, and somewhat resemble those of primitive insects such as mayflies. A few genera, however, have reduced wings, or have lost them altogether. The abdomen is cylindrical, and typically curves upwards in the male, superficially resembling the tail of a scorpion, the tip containing an enlarged structure called the genital bulb (Fig. 52).

The female lays the eggs in close contact with moisture, and the eggs typically absorb water and increase in size after deposition. In species that live in hot conditions, the eggs may not hatch for several months, the larvae only emerging when the dry season has finished. More typically, however, they hatch after a relatively short period of time.

The larvae are usually quite catterpillar-like, with short, clawed, true legs, and a number of abdominal prolegs. They have sclerotised heads with compound eyes and mandibulate mouthparts. The tenth abdominal segment bears either a suction disc, or, less commonly, a pair of hooks. They generally eat vegetation or scavenge for dead insects, although some predatory larvae are known. The larva crawls into the soil or decaying wood to pupate, and does not spin a cocoon. The pupae are exarate, meaning the limbs are free of the body, and are able to move their mandibles, but are otherwise entirely nonmotile. In drier environments, they may spend several months in diapause, before emerging as adults once the conditions are more suitable. Scorpionflies in the *Bittacidae* provide a nuptial meal in the form of a captured insect prey. The size of the prey is correlated with the mating time allowed by the female.

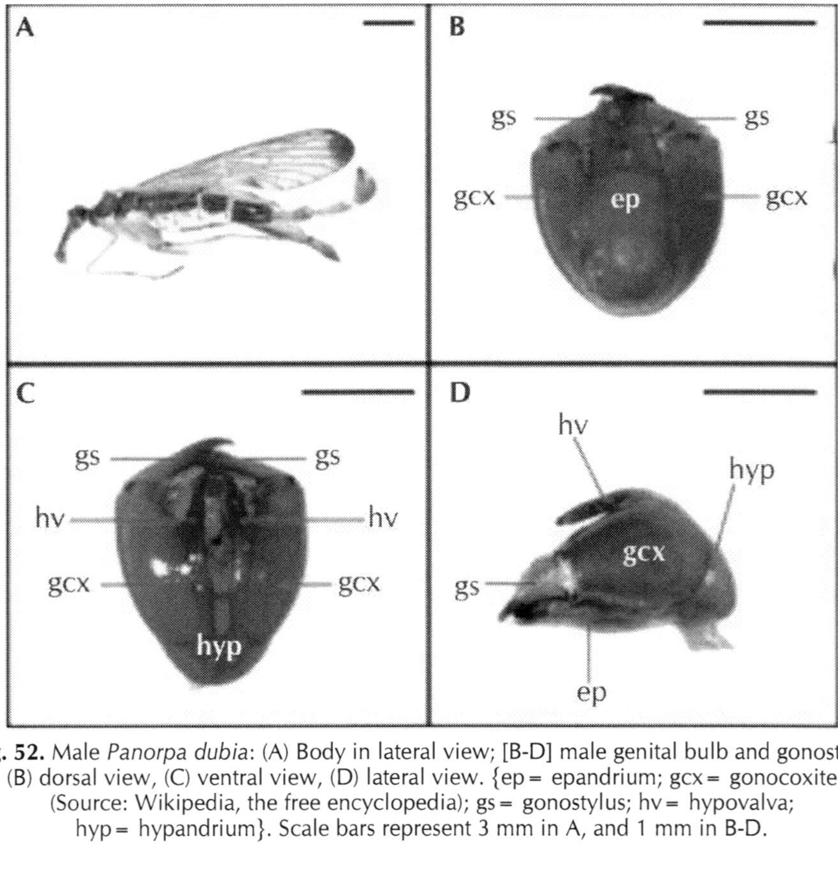

Fig. 52. Male *Panorpa dubia*: (A) Body in lateral view; [B-D] male genital bulb and gonostyli, (B) dorsal view, (C) ventral view, (D) lateral view. {ep = epandrium; gcx = gonocoxite (Source: Wikipedia, the free encyclopedia); gs = gonostylus; hv = hypovalva; hyp = hypandrium}. Scale bars represent 3 mm in A, and 1 mm in B-D.

24. Order Trichoptera

Trichoptera [from Greek: (*thrix*, "hair") + (*pteron*, "wings)] is an order of the caddisflies of which there are approximately 7,000 described species. Also called sedge-flies or rail-flies, they are small moth-like insects having two pairs of hairy membranous wings. They are closely related to Lepidoptera (moths and butterflies) which have scales on their wings, and the two orders together form the superorder Amphiesminoptera. Caddisflies have aquatic larvae and are found in a wide variety of habitats such as streams, rivers, lakes, ponds, spring seeps, and temporary waters (vermi- pools). The larvae of many species use silk to make protective cases of gravel, sand, twigs or other debris.

The order Tricoptera is subdivided into two suborders, Annulipalpia and Integripalpia, comprising together 37 families. They occur in all faunal regions, and the greatest bulk of the mountain autochthonous species is confined to the altitudes between 2500 and 4000 m. In the Himalaya they occur up to elevations of 5000 m, where the larvae develop in melt-water torrents and springs. In both suborders a few cases of facultative or obligatory parthenogenesis are known.

Although caddisflies may be found in waterbodies of varying qualities, species-rich caddisfly assemblages are generally thought to indicate clean water. Together with stoneflies and mayflies, caddisflies feature importantly in bioassessment surveys of streams and other water bodies. Caddisfly species can be found in all feeding guilds in stream habitats, with some species being predators, leaf shredders, algal grazers, and collectors of particles from the watercolumn and benthos. Most caddisfly larvae are underwater architects and use silk, excreted from salivary glands near their mouths, for building. Caddisflies can be divided loosely into three behavioural groups based on their use of silk: net-making caddisflies and case-making caddisflies, both of which may enlarge their structures throughout their larval lifespan; and free-living caddisflies, which only make such structures prior to pupation. Net-making Caddisflies usually live in running water, and their nets, often made amongst aquatic vegetation, serve both as a means to collect algae, detritus, and animal food and as retreats. Case-making caddisflies may build cases exclusively of silk, but more commonly the silk holds together substrate materials such as small fragments of rock, sand, small pieces of twig or aquatic plants. Caddisfly cases are open at both ends, the larva drawing oxygenated water through the posterior end, over their gills, and out of the wider anterior end. The anterior end is usually wider and it is to this end that they add material as they grow. Their abdomens are soft, but their tougher front ends project from their larval tubes, allowing them to walk while dragging their cases along with them. Caddisfly cases resemble bagworm cases, which are constructed by various terrestrial moth species. Free-living caddisflies do not build retreats or carry portable cases until they are ready to pupate, and their bodies tend to be tougher than the building.

Many species of caddisfly larvae enter a stage of inactivity called the pupa stage for weeks or months after they mature but prior to emergence. Their emergence is then triggered by cooling water temperatures in the fall, effectively synchronizing the adult activity to make mate-finding easier. In the

Northwestern US, caddisfly larvae within their gravel cases are called 'periwinkles. Caddisfly pupation occurs much like pupation of Lepidoptera. That is, caddisflies pupate in a cocoon spun from silk. Caddisflies that build the portable cases attach their case to some underwater object, seal the front and back apertures against predation though still allowing water flow, and pupate within it. Once fully developed, most pupal caddisflies cut through their cases with a special pair of mandibles, swim up to the water surface, cast off skin and the now-obsolete gills and mandibles, and emerge as fully formed adults. In a minority of species, the pupae swim to shore-either below the water or across the surface-and crawl out to emerge. Many of them are able to fly immediately after breaking from their pupal skin (Fig. 53).

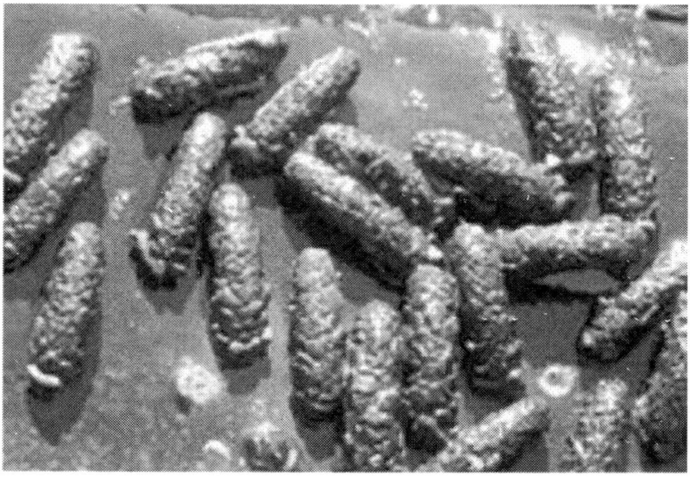

Fig. 53. Caddis pupae (Source: Wikipedia, the free encyclopedia)

The adult stage of caddisflies, in most cases, is very short-lived, usually only 1–2 weeks, but can sometimes last for 2 months. Most adults are non-feeding and are equipped mainly to mate. Once mated, the female caddisfly will often lay eggs (enclosed in a gelatinous mass) by attaching them above or below the water surface. Eggs hatch in as little as three weeks.

Caddisflies in most temperate areas complete their life cycles in a single year. The general temperate-zone lifecycle pattern is one of larval feeding and growth in autumn, winter, and spring, with adult emergence between late spring and early fall, although the adult activity of a few species peaks in the winter. Larvae are active in very cold water and can frequently be observed feeding under ice. In common with many aquatic insect species, many caddisfly adults emerge synchronously *en masse.* Such emergence patterns ensure that most caddisflies will encounter a member of the opposite sex in a timely fashion. Mass emergences of this nature are called 'hatches' by salmon and trout anglers, and salmonid fish species will frequently 'switch' to whatever species is emerging on a particular day. Anglers take advantage of this behaviour by matching their artificial flies to the appropriate fly.

25. Order Lepidoptera

The Lepidoptera (Ancient Greek '*lepis*' meaning scale and '*pteron*' meaning wing) is an order of insects that includes moths and butterflies (both called lepidopterans). About 180,000 species of Lepidoptera are described, in 126 families and 46 superfamilies 10% of the total described species of living organisms, making it the third largest order in insects. It is one of the most widespread and widely recognizable insect orders in the world, encompassing moths and the three superfamilies of butterflies, skipper butterflies, and moth butterflies. The modern Lepidoptera show many variations of the basic body structure that have evolved to gain advantages in lifestyle and distribution. Recent estimates suggest the order may have more species than earlier thought, and is among the four most speciose orders, along with the Hymenoptera, Diptera and Coleoptera.

Currently the order is classified into three suborders, viz., Zeugloptera, Monotrysia and Ditrysia. The more familiar divisions into Rhopalocera (butterflies) and Heterocera (moths), or into Microlepidoptera and Macrolepidoptera, are based on convenience rather than on structural criteria. The order has a worldwide distribution and the typical autochthonous high altitude species occur in the families Arctiidae, Geometridae, Lycaenidae, Noctuidae, Nymphalidae, Papilionidae, Parnassiidae, Pieridae and Satyridae. The highest records are from elevations of 6000 m in the Himalaya, but the majority of autochthonous alpine species breeds between 3000 and 4000 m, and a few are peculiar to elevations between 4000 and 5000 m.

Lepidopteran species are characterized by more than three derived features, some of the most apparent being the scales covering their bodies and wings, and a ptoboscis. The scales are modified, flattened "hairs", and give butterflies and moths their extraordinary variety of colors and patterns. Almost all species have some form of membranous wings, except for a few that have reduced wings or are wingless. Like most other insects, butterflies and moths are holometabolous, meaning they undergo complete metamorphosis (Fig. 57b). Mating and the laying of eggs are carried out by adults, normally near or on host plants for the larvae. The larvae are commonly called catterpillars, and are completely different from their adult moth or butterfly forms, having a cylindrical body with a well-developed head, mandible mouth parts, and from none to 11 (usually eight) pairs of prolegs (Fig. 54). As they grow, these larvae change in appearance, going through a series of stages called lnstars. Once fully matured, the larva develops into a pupa, referred to as a chrysalis in the case of butterflies and a cocoon in the case of moths. A few butterflies and many moth species spin a silk case or cocoon prior to pupating, while others do not, instead going underground.

Butterflies and moths play an important role in the natural ecosystem as pollinators and as food in the food chain; conversely, their larvae are considered very problematic to vegetation in agriculture, as their main source of food is often live plant matter. In many species, the female may produce from 200 to 600 eggs, while in others, the number may approach 30,000 eggs in one day. The caterpillars hatching from these eggs can cause damage to large quantities of crops. Many moth and butterfly species are of economic interest by virtue of their role as pollinators, the silk they produce, or as pest species.

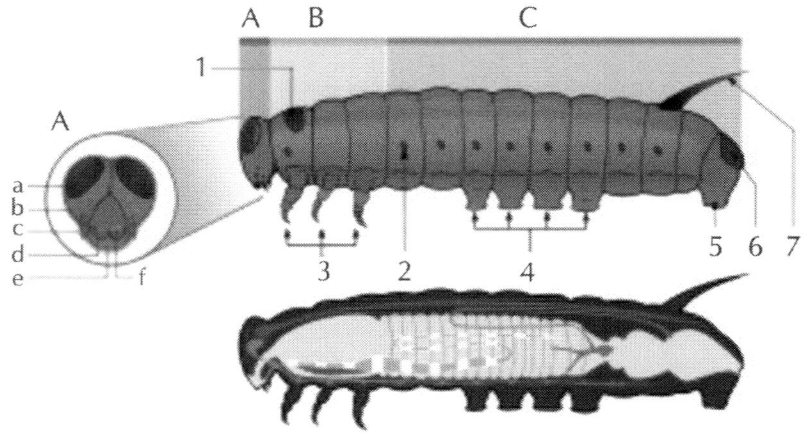

Fig. 54. Diagrammatic representation of a generalized body plan, including anatomy, of a lepidopteran (Source: Wikipedia, the free encyclopedia); A – head, B – thorax, C – abdomen, 1 – prothoracic shield, 2 – spiracle, 3 – true legs, 4 – midabdominal prolegs, 5 – anal proleg, 6 – anal plate, 7 – tentacle, a – frontal triangle, b – stemmata (ocelli), c – antenna, d – mandible, e – labrum.

The Lepidoptera are among the most successful groups of insects. They are found on all continents, except Antartica, and inhabit all terrestrial habitats ranging from desert to rainforest, from lowland grasslands to montane plateaus, but almost always associated with higher plants, especially angiosperms (flowering plants). Among the most northern dwelling species of butterflies and moths is the Arctic Apollo (*Parnassius arcticus*), which is found in the Arctic Circle in northeastern Yakutia, at an altitude of 1500 meters above sea level. In the Himalayas, various Apollo species such as *Parnassius epaphus* have been recorded to occur up to an altitude of 6,000 m above sea level.

There are over 174,250 lepidopteran species described, with butterflies and skippers estimated to comprise around 17,950, and moths making up the rest. The vast majority of Lepidoptera are to be found in the tropics, but substantial diversity exists on most continents.

Lepidoptera are morphologically distinguished from other orders principally by the presence of scales on the external parts of the body and appendages, especially the wings. Butterflies and moths vary in size from microlepidoptera only a few millimeters long, to conspicuous animals with a wingspan greater than 25 cm, such as the Monarch butterfly and Atlas moth (Gillot, 1995). Lepidopterans undergo a four-stage lifecycle: egg, larva or caterpillar, pupa or chrysalis, and imago (plural imagines), and show many variations of the basic body structure, which have evolved to gain advantages in lifestyle and distribution.

The Head houses many sensing organs and the mouth parts. Like the adult, the larva also has a toughened, or sclerotized head capsule. Here, two compound eyes, and *chaetosema*, raised spots or clusters of sensory bristles unique to

Lepidoptera, occur, though many taxa have lost one or both of these spots. The antennae have a wide variation in form among species and even between different sexes. The antennae of butterflies are usually filiform and shaped like clubs, those of the skippers are hooked, while those of moths have flagellar segments variously enlarged or branched. Some moths have enlarged antennae or ones that are tapered and hooked at the ends. The maxillary galeae are modified and form an elongated proboscis. The proboscis consists of one to five segments, usually kept coiled up under the head by small muscles when it is not being used to suck up nectar from flowers or other liquids. Some basal moths still have mandibles, or separate moving jaws, like their ancestors, and these form the family Micropterygidae.

The larvae, called catterpillars, have a toughened head capsule. Caterpillars lack the proboscis and have separate chewing mouthparts. These mouthparts, called mandibles are used to chew up the plant matter that the larvae eat. The lower jaw, or labium, is weak, but may carry a spinneret, an organ used to create silk. The head is made of large lateral lobes, each having an ellipse of up to six simple eyes.

The thorax is made of three fused segments, the prothorax, mesothorax and metathorax, each with a pair of legs. The first segment contains the first pair of legs. In some males of the butterfly family Nymphalidae, the fore legs are greatly reduced and are not used for walking or perching. The three pairs of legs are covered with scales. Lepidoptera also have olfactory organs on their feet, which aid the butterfly in "tasting" or "smelling" out its food. In the larval form there are 3 pairs of true legs, with up to 11 pairs of abdominal legs (usually eight) and hooklets, called apical crochets.

The two pairs of wings are found on the middle and third segments, or mesothorax and metathorax, respectively. In the more recent genera, the wings of the second segment are much more pronounced, although some more primitive forms have similarly sized wings of both segments. The wings are covered in scales arranged like shingles, which form an extraordinary variety of colours and patterns. The mesothorax is designed to have more powerful muscles to propel the moth or butterfly through the air, with the wing of this segment (fore wing) having a stronger vein structure. The largest superfamily, the Noctuidae, has their wings modified to act as tympanal or hearing organs.

The caterpillar has an elongated, soft body that may have hair-like or other projections, three pairs of true legs, with none to 11 pairs of abdominal legs (usually eight) and hooklets, called apical crochets. The thorax usually has a pair of legs on each segment. The thorax is also lined with many spiracles on both the mesothorax and metathorax, except for a few aquatic species, which instead have a form of gills (Fig. 55).

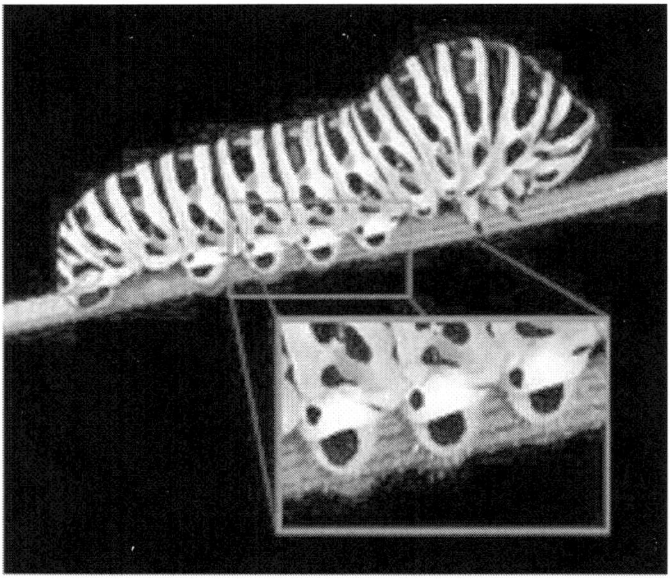

Fig. 55. Caterpillar prolegs on *Papilio machaon* (Source: Wikipedia, the free encyclopedia).

Fig. 56. Electron microscopy images of scales; (a) A patch of wing (×50), (b) Scales close up (×200) (c)A single scale (×1000), and (d) Microstructure of a scale (×5000) (Source: Wikipedia, the free encyclopedia).

The abdomen, which is less sclerotized than the thorax, consists of 10 segments with membranes in between, allowing for articulated movement. The sternum, on the first segment, is small in some families and is completely absent in others. The last two or three segments form the external parts of the species' sex organs. The genetilia of Lepidoptera are highly varied and are often the only means of differentiating between species. Male genitals include a valva, which is usually large, as it is used to grasp the female during mating. Female genitalia include three distinct sections. The females of basal moths have only one sex organ, which is used for copulation and as an ovipositor, or egg-laying organ. About 98% of moth species have a separate organ for mating, and an external duct that carries the sperm from the male.

The abdomen of the caterpillar has four pairs of prolegs, normally located on the third to sixth segments of the abdomen, and a separate pair of prolegs by the anus, which have a pair of tiny hooks called crotchets. These aid in gripping and walking, especially in species that lack many prolegs (e.g. larvae of Geometridae). In some basal moths, these prolegs may be on every segment of the body, while prolegs may be lost completely in other groups, which are more adapted to boring and living in sand (e.g., Prodoxidae and Nepticulidae, respectively).

Scales on wings form the colour and pattern on wings (Fig. 56). The wings, head, and parts of the thorax and abdomen of Lepidoptera are covered with minute scales, a feature from which the order derives its name. Most scales are lamellar, or blade-like and attached with a pedicel, while other forms may be hair-like or specialized as secondary sexual characteristics. The lumen or surface of the lamella has a complex structure. It gives colour either by coloured pigments it contains, or through structural coloration with mechanisms that include photonic crystals and diffraction gratings. Scales function in insulation, thermoregulation, producing pheromones (in males only), and aiding gliding flight, but the most important is the large diversity of vivid or indistinct patterns they provide, which help the organism protect itself by camouflage or mimicry, and which act as signals to other animals including rivals and potential mates.

In the reproductive system of butterflies and moths, the male genitalia are complex and unclear (Fig. 57). In females the three types of genitalia are based on the relating taxa: 'monotrysian', 'exoporian', and 'ditrysian'. In the monotrysian type is an opening on the fused segments of the sterna 9 and 10, which act as insemination and oviposition. In the exoporian type (in Hepialoidea) and Mnesarchaeoidea) are two separate places for insemination and oviposition, both occurring on the same sterna as the monotrysian type, i.e., 9 and 10. The ditrysian groups have an internal duct that carries sperm, with separate openings for copulation and egg-laying. In most species, the genitalia are flanked by two soft lobes, although they may be specialized and sclerotized in some species for ovipositing in area such as crevices and inside plant tissue. Hormones and the glands that produce them run the development of butterflies and moths as they go through their lifecycles, called the endocrine system. The first insect hormone prothoracicotropic hormone (PTTH) operates the species lifecycle and diapause. This hormone is produced by corpora allata and corpora cardiaca, where it is also stored. Some glands are specialized to perform certain task such as producing silk or producing saliva in the palpi. While the corpora cardiaca produce PTTH, the corpora allata also produces juvenile hormones, and the prothorocic glands produce moulting hormones.

(a) (b)

Fig. 57: (a) Mating pair of *Laothoe populi* (popular hawk-moth) showing two different colour variants, (b) The four stages of the lifecycle of ananise swallowtail (Source: Wikipedia, the free encyclopedia).

In the digestive system, the anterior region of the foregut has been modified to form a pharyngeal sucking pump as they need it for the food they eat, which are for the most part liquids. An oesophagus follows and leads to the posterior of the pharynx and in some species forms a form of crop. The midgut is short and straight, with the hindgut being longer and coiled. Ancestors of lepidopteran species, stemming from Hymenoptera, had midgut ceca, although this is lost in current butterflies and moths. Instead, all the digestive enzymes, other than initial digestion, are immobilized at the surface of the midgut cells. In larvae, long-necked and stalked goblet cells are found in the anterior and posterior midgut regions, respectively. In insects, the goblet cells excrete positive potassium ions, which are absorbed from leaves ingested by the larvae. Most butterflies and moths display the usual digestive cycle, but species with different diets require adaptations to meet these new demands (Fig. 58).

In Lepidoptera, polymorphism can be seen not only between individuals in a population, but also between the sexes as sexual dimorphism, between geographically separated populations in geographical polymorphism, and between generations flying at different seasons of the year (seasonal polymorphism or polyphenism). In some species, the polymorphism is limited to one sex, typically the female. This often includes the phenomenon of mimicry when mimetic morphs fly alongside nonmimetic morphs in a population of a particular species. Polymorphism occurs both at specific level with heritable variation in the overall morphological design of individuals, as well as in certain specific morphological or physiological traits within a species. Geographical isolation causes a divergence of a species into different morphs. A good example is the Indian white admiral *Limenitis procris*, which has five forms, each geographically separated from the other by large mountain ranges. An even more dramatic showcase of geographical polymorphism is the Apollo (*Parnassius apollo*). Because the Apollos live in small local populations, thus having no contact with each other, coupled with their strong stenotopic nature and weak migration ability, interbreeding between populations of one species practically does not occur; by this, they form over 600 different morphs, with the size of spots on the wings of which varies greatly.

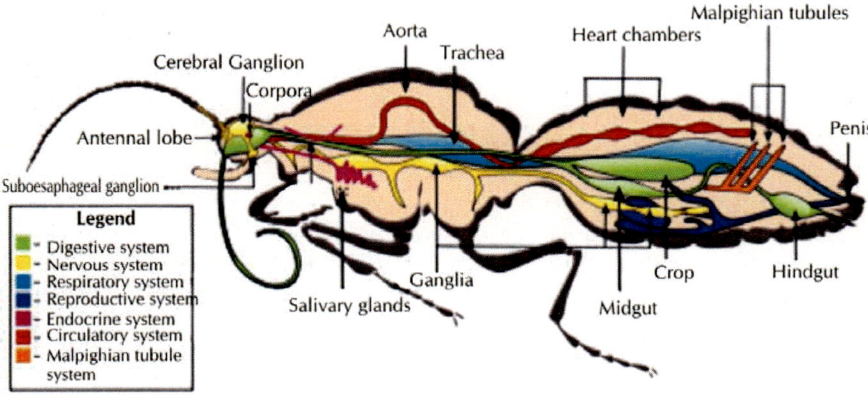

Fig. 58. Internal morphology of adult male in the family Nymphalidae, showing most of the major organ systems, with characteristic reduced forelegs of that family: The corpora include the corpus allatum and the corpus cardiaca (Source: Wikipedia, the free encyclopedia).

Males usually start eclosion (emergence) earlier than females and peak in numbers before females. Both of the sexes are sexually mature by the time of eclosion. Butterflies and moths normally do not associate with each other, except for migrating species, staying relatively asocial. Mating begins with an adult (female or male) attracting a mate, normally using visual stimuli, especially in diurnal species like most butterflies. However, the females of most nocturnal species, including almost all moth species, use pheromones to attract males, sometimes from long distances. Some species engage in a form of acoustic courtship, or attract mates using sound or vibration such as the polka-dot wasp moth, *Syntomeida epilais*.

Adaptations include undergoing one seasonal generation, two or even more, called voltinism (univoltism, bivoltism, and multivism, respectively). Most lepidopterans in temperate climates are univoltine, while in tropical climates most have two seasonal broods. Some others may take advantage of any opportunity they can get, and mate continuously throughout the year. These seasonal adaptations are controlled by hormones, and these delays in reproduction are called diapause. Many lepidopteran species, after mating and laying their eggs, die shortly afterwards, having only lived for a few days after eclosion. Others may still be active for several weeks and then overwinter and become sexually active again when the weather becomes more favourable, or diapause. The sperm of the male that mated most recently with the female is most likely to have fertilized the eggs, but the sperm from a prior mating may still prevail.

Lepidoptera usually reproduce sexually and are oviparous (egg-laying), though some species exhibit live birth in a process called ovoviviparity. A variety of differences in egg-laying and the number of eggs laid occur. Some species simply drop their eggs in flight (these species normally have polyphagous larvae, meaning they eat a variety of plants, e.g., hepialids and some nymphalids) while most lay their eggs near or on the host plant on which the larvae feed (Fig. 59). The number of eggs laid may vary from only a few to several thousand. The females of both butterflies and moths select the host plant instinctively, and primarily, by chemical cues. The eggs are derived from materials ingested as a larvae and in some species, from the spermataphores received from males during mating. An egg can only be 1/1000th the mass of the female, yet she may lay up to her own mass in eggs. Females lay smaller eggs as they age. Larger Females lay larger eggs. The egg is covered by a hard-ridged protective outer layer of shell, called the chorion. It is lined with a thin coating of wax, which prevents the egg from drying out. Each egg contains a number of micropyles, or tiny funnel-shaped openings at one end, the purpose of which is to allow sperm to enter and fertilize the egg. Butterfly and moth eggs vary greatly in size between species, but they are all either spherical or ovate. The egg stage lasts a few weeks in most butterflies, but eggs laid prior to winter, especially in temperate regions, go through diapause, and hatching may be delayed until spring. Other butterflies may lay their eggs in the spring and have them hatch in the summer. These butterflies are usually temperate species (e.g. *Nymphalis antiopa*).

Fig. 59. Larval form typically lives and feeds on plants (Source: Wikipedia, the free encyclopedia).

The larvae or caterpillars are the first stage in the lifecycle after hatching. Caterpillars, are "characteristic polypod larvae with cylindrical bodies, short thoracic legs, and abdominal prolegs (pseudopods)". They have a toughened head capsule with an adfrontal suture formed by medial fusion of the sclerites, mandibles for chewing, and a soft tubular, segmented body, that may have hair-like or other projections, three pairs of true legs, and additional prolegs (up to five pairs). The body consists of thirteen segments, of which three are thoracic and ten are abdominal. Most larvae are herbivores, but a few are carnivores (some eat ants or other caterpillars) and detritivores. Different herbivorous species have adapted to feed on every part of the plant and are normally considered pests to their host plants; some species have been found to lay their eggs on the fruit and other species lay their eggs on clothing or fur (e.g., *Tineola bisselliella*, the common clothes moth). Some species are carnivorous and others are even parasitic. Some lycaenid species such as *Maculinea rebeli* are social parasites of *Myrmica* ants nests. A species of Geometritidae from Hawaii has carnivorous larvae that catch and eat flies. Some pyralid caterpillars are aquatic.

The larvae develop rapidly with several generations in a year; however, some species may take up to 3 years to develop, and exceptional examples like *Gynaephora groenlandica* take as long as seven years. The larval stage is where the feeding and growing stages occur, and the larvae periodically undergo hormone-induced ecdysis, developing further with each instar, until they undergo the final larval-pupal moult. The lepidopteran pupa, known as a chrysalis in the case of butterflies, has functional mandibles with appendages fused or glued to the body in most species, while the pupal mandibles are not functional in others.

The larvae of both butterflies and moths exhibit mimicry to deter potential predators. Some caterpillars have the ability to inflate parts of their heads to appear snake-like. Many have false eye-spots to enhance this effect. Some caterpillars have special structures called osmeteria (family Palilionidae), which are exposed to produce smelly chemicals used in defense. Host plants often have toxic substances in them and caterpillars are able to sequester these substances and retain them into the adult stage. This helps make them unpalatable to birds and other predators. Such unpalatability is advertised using bright red, orange, black, or white warning colours. The toxic chemicals in plants are often evolved specifically to prevent them from being eaten by insects. Insects, in turn, develop counter measures or make use of these toxins for their own survival. This "arms race" has led to the coevolution of insects and their host plants.

No form of wing is externally visible on the larva, but when larvae are dissected, developing wings can be seen as disks, which can be found on the second and third thoracic segments, in place of the spiracles that are apparent on abdominal segments. Wing disks develop in association with a trachea that runs along the base of the wing, and are surrounded by a thin peripodial membrane, which is linked to the outer epidermis of the larva by a tiny duct. Wing disks are very small until the last larval instar, when they increase dramatically in size, are invaded by branching tracheae from the wing base that precede the formation of the wing veins, and begin to develop patterns associated with several landmarks of the wing. Near pupation, the wings are forced outside the epidermis under pressure from the haemolymph, and although they are initially quite flexible and fragile, by the time the pupa breaks free of the larval cuticle, they have adhered tightly to the outer cuticle of the pupa (in obtect pupae). Within hours, the wings form a cuticle so hard and well-joined to the body that pupae can be picked up and handled without damage to the wings.

After about five to seven instars, or moults, certain hormones, like PTTH, stimulate the production of ecdysis, which initiates insect moulting. Then, the larva puparium, a sclerotized or hardened cuticle of the last larval instar, develops into the pupa. Depending on the species, the pupa may be covered in silk and attached to many different types of debris, or may not be covered at all. The pupa stays attached to the leaf by silk spun by the caterpillar before it spins the silk for the full pupa. Features of the imago are externally recognizable in the pupa. All the appendages on the adult head and thorax are found cased inside the cuticle (antennae, mouthparts etc.), with the wings wrapped around, adjacent to the antennae. While encased, some of the lower segments are not fused, and are able to move using small muscles found in between the membrane. Moving may help the pupa, for example, escape the sun, which would otherwise kill it. The pupa of the Mexican Jumping Bean moth (*Cydia deshaisiana*) does this. The larvae cut a trapdoor in the bean (species of *Sebastiania*) and use the bean as a shelter. With a sudden rise in temperature, the pupa inside twitches and jerks, pulling on the threads inside. Wiggling may also help to deter parasitoid wasps from laying eggs on the pupa. Other species of moths are able to make clicks to deter predators.

The length of time before the pupa ecloses (emerges) varies greatly. The monarch butterfly may stay in its chrysalis for two weeks, while other species may need to stay for more than 10 months in diapause. The adult emerges from the pupa either by using abdominal hooks or from projections located on the head. The mandibles found in the most primitive moth families are used to escape from their cocoon (e.g., Micropterigoidea).

Most lepidopteran species do not live long after eclosion, only needing a few days to find a mate and then lay their eggs. Others may remain active for a longer period (from one to several weeks), or go through diapause and overwintering as monarch butterflies do, or waiting out environmental stress. Some adult species of Microlepidoptera go through a stage where no reproductive-related activity occurs, lasting through summer and winter, followed by mating and oviposition in the early spring. While most butterflies and moths are terrestrial, many species of Pyralidae are truly aquatic with all stages except the adult occurring in water. Many species from other families such as Arctiidae, Nepticulidae, Cosmopterygidae, Oletheutidae, Noctuidae, Cossidae and Sphingidae are aquatic or semiaquatic (James et al., 1992).

Flight is an important aspect of the lives of butterflies and moths, and is used for evading predators, searching for food, and finding mates in a timely manner, as lepidopteran species do not live long after eclosion. It is the main form of locomotion in most species. In Lepidoptera, the fore wings and hind wings are mechanically coupled and flap in synchrony. Flight is anteromotoric, or being driven primarily by action of the fore wings. Although lepidopteran species reportedly can still fly when their hind wings are cut off, it reduces their linear flight and turning capabilities. Lepidopteran species have to be warm, about 25 to 26°C, to fly. They depend on their body temperature being sufficiently high and since they cannot regulate it themselves, this is dependent on their environment. Butterflies living in cooler climates may use their wings to warm their bodies. They will bask in the sun, spreading out their wings so that they get maximum exposure to the sunlight. In hotter climates butterflies can easily overheat, so they are usually active only during the cooler parts of the day, early morning, late afternoon or early evening. During the heat of the day, they rest in the shade. Some larger thick-bodied moths (e.g. Sphingidae) can generate their own heat to a limited degree by vibrating their wings. The heat generated by the flight muscles warms the thorax while the temperature of the abdomen is unimportant for flight. To avoid overheating, some moths rely on hairy scales, internal air sacs, and other structures to separate the thorax and abdomen and keep the abdomen cooler. Some species of butterflies can reach fast speeds, such as the southern dart, which can go as fast as 48.4 km/h. Sphingids are some of the fastest flying insects, some are capable of flying at over 50 km/h, having a wingspan of 35–150 mm. In some species, sometimes a gliding component to their flight exists. Flight occurs either as hovering, or as forward or backward motion. In butterfly and moth species, such as hawk hovering is important as they need to maintain a certain stability over flowers when feeding on the nectar.

Many lepidopterans take to migration. Lepidopteran migration is typically seasonal, as the insects moving to escape dry seasons or other disadvantageous conditions. Most lepidopterans that migrate are butterflies, and the distance travelled varies. Some butterflies that migrate include the mourning cloak, painted lady, American lady, red admiral and the common buckeye. The most well-known migrations are those of the eastern population of the monarch butterfly from Mexico to northern United States and southern Canada, a distance of about 4,000–4,800 km. Other well-known migratory species include the painted lady and several of the danaine butterflies. Spectacular and large-scale migrations associated with the monsoons are seen in peninsular India (Williams, 1927). Migrations have been studied in more recent times using wing tags and stable hydrogen isotopes. Moths also undertake migrations, an example being the Uraniids, *Urania fulgens*, undergoes population explosions and massive migrations that may be not surpassed by any other insect in the Neotropics. In Costa Rica and Panama, the first population movements may begin in July and early August and depending on the year, may be very massive, continuing unabated for as long as five months.

Some larvae of both moths and butterflies have a form of hair that has been known to be a cause of human health problems. Caterpillar hairs sometimes have toxins in them and species from approximately 12 families of moths or butterflies worldwide can inflict serious human injuries (Urticarial dermatitis and atopic asthma to osteochondritis, consumption coagulopathy, renal failure, and intracereberal haemorrhage). Skin rashes are the most common, but there have been fatalities. *Lonomia* is a frequent cause of economization in humans in Brazil, with 354 cases reported between 1989 and 2005. Lethality ranging up to 20% with death caused most often by intracranial hemorrhage. These hairs have also been known to cause kerato-conjunctivitis. The sharp barbs on the end of caterpillar hairs can get lodged in soft tissues and mucus membranes such as the eyes. Once they enter such tissues, they can be difficult to extract, often exacerbating the problem as they migrate across the membrane. This becomes a particular problem in an indoor setting. The hairs easily enter buildings through ventilation systems and accumulate in indoor environments because of their small size, which makes it difficult for them to be vented out. This accumulation increases the risk of human contact in indoor environments.

Lepidoptera, as an order, is thoroughly investigated in the Doon Valley since the time of Dr. Stebbing. All Forest Entomologists investigated lepidopera, and many scientists from outside valley got interested in the fauna of the Doon Valley, e.g., Dr. H.S. Rose and Dr. Jagbir Singh from Punjab, both highly distinguished entomologists, with the latter particularly on Lepideoptera. Both FRI and ZSI have marvelous collections of catterpillars and dried and preserved specimens of adults from whole of Garhwal region, with adequate documentation.

26. Order Thysanoptera

Order Thysanoptera [Greek *thysanos* ("fringe") + *pteron* ("wing")], represented by minute, slender insects with fringed wings, called thrips, is an important group from agriculture and horticulture points-of-view. Other common names for thrips include thunderflies, thunderbugs, storm flies, thunderblights, storm bugs, corn flies and corn lice. There are currently over 6000 species of thrips recognized, grouped into 777 extant genera and sixty fossil genera. Essentially thermophilous, they are present in all faunal regions. Several species are adapted to life in high alpine elevations.

Thrips species feed on a large variety of plants and animals by puncturing them and sucking up the contents (Ananthakrishnan, 1993). A large number of thrips species are considered pests, because they feed on plants with commercial value. Some species of thrips feed on other insects or mites and are considered beneficial, while some feed on fungal spores or pollen. Approximately 6,000 species have been described. Thrips are generally tiny (1 mm long or less) and are not good flyers, although they can be carried long distances by the wind. In the right conditions, like indoor grow rooms or greenhouses, many species can exponentially increase in population size and form large swarms because of a lack of natural predators, making them an irritation to humans.

Thrips are small hemimetabolic insects with a distinctive cigar-shaped bauplan. They are elongated with transversely constricted bodies. They range in size from 0.5 to 14 mm in length for the larger predatory thrips, but most thrips are about 1 mm in length. Flight-capable thrips have two similar, strap-like pairs of wings with a ciliated fringe (from which the order derives its name). Their legs usually end in two tarsal segments with a bladder-like structure known as an "arolium" at the pretarsus. This structure can be everted by means of hemolymph pressure, enabling the insect to walk on vertical surfaces (Fig. 60).

Fig. 60: (a) *Franklinothrips uttarakhandiensis*, (b) *Kulshresthothrips moundi*, (c) *Streothrips anshumani* (*Courtesy:* Dr. Vijay Veer)

Thrips have asymmetrical mouthparts that are also unique to the group. Unlike the Hemiptera, the right mandible of thrips is reduced and vestigial – and in some species completely absent. The left mandible is larger, and forms a narrow stylet used to pierce the cell wall of tissues. Some species may then inject digestive enzymes as the maxillary stylets and hypopharynx are inserted into the opening to drain cellular fluids. This process leaves a distinctive silvery or bronze scarring on the surfaces of the stems or leaves where the thrips feed.

Thysanoptera is divided into two suborders: Terebrantia, and Tubulifera; these can be distinguished by morphological, behavioural, and developmental characteristics. Members of Tubulifera can be identified by their characteristic tube-shaped apical abdominal segment, egg-laying atop the surface of leaves, and three "pupal" stages. Females of the eight families of the Terebrantia all possess the eponymous saw-like ovipositor on the anteapical abdominal segment, lay eggs singly within plant tissue, and have two "pupal" stages.

Even through globally the order Thysanoptera is largely recognized to be organized into two suborders: Terebrantia and Tubulifera, yet Dr. Bhatti provides an alternative classification based on morphological characters that would split the group into two orders and forty different families (Bhatti, 1994, 1998, 2006).

Thrips are believed to have descended from a mycetophilic (fungus-feeding) ancestor during the Mesozoic and many groups still feed upon and inadvertently redistribute fungal spores, but most research has focused on those species feeding on or in association with economically significant crops. Some thrips are predatory, but the majority are phytophagous insects feeding on pollen and the chloroplasts harvested from the outer layer of plant epidermal and mesophyll cells. These species are minute organisms that prefer to feed within the tightly packed apical buds of new growth. Feeding usually occurs along the main vein or ribs of leaves and petals. Flower-feeding thrips may be responsible for pollination while feeding, but their most obvious contribution to their ecosystem remains the damage they can cause during feeding. This impact may fall across a broad selection of prey items, as there is considerable breadth in host affinity across the order, and even within a species, varying degrees of fidelity to a described host remain. Family Thripidae is particularly notorious for members with broad host ranges, and the majority of pest thrips come from this family.

While poorly documented, chemical communication is believed to be important to the group. Anal secretions are produced in the hindgut, and released along the posterior setae as predator deterrents. Many thrips form galls on plants when feeding or laying their eggs. Some of the gall-forming Phlaeothripidae, such as genera *Kladothrips* and *Oncothrips* form eusocial groups similar to ant colonies, with reproductive queens and nonreproductive soldier castes.

The rate at which thrips move through their developmental cycles is highly dependent upon environmental conditions, including the temperature and nutrient quality of their food sources. Thrips begin their lives as eggs. These are extremely small (about 0.2 mm long) and kidney-shaped. Hatching may take from as little as a day to several weeks. The females of the suborder Terebrantia are equipped with an ovipositor, which they use to cut slits in plant tissue and then insert their eggs, one per slit. Females of the suborder Tubulifera lack an ovipositor and lay their eggs singly or in small groups on the outside surfaces of plants. Thrips then pass through two wingless instars of nymph.

As hemimetabolous insects, the Thysanoptera do not actually undergo complete metamorphosis, but pass through a similar stage in which they do not feed and are mostly immobile. Both suborders of thrips will first enter a short prepupal stage lasting a day at most, during which they will seek out dark crevices on plants, hiding in the tightly packed flower buds or bark – or drop off of the plant entirely, burrowing into leaf litter or loose soil. Some thrips will then construct a pupal cell or cocoon. In Terebrantian thrips, a single pupal instar follows, whereas in the Tubulifera, two pupal stages will follow. During these stages, wing-buds and reproductive structures will grow and mature into their adult forms.

All described genera of thrips are haplodiploid organisms capable of parthenogenesis, with some favouring arrhenotoky and others displaying thelytoky. In some cases the sex-determining bacterial endosymbiont *Wolbachia* plays a role in defining the reproductive mode for some populations of thrips.

Several normally bisexual species have become established in the United States with only members of a single sex present.

When mating occurs, it may last from minutes to hours. Most female thrips have a preoviposition period lasting from a day to a week during which their eggs mature, and before which they cannot mate.

Many thrips are pests of commercial crops due to the damage caused by feeding on developing flowers or vegetables, causing discoloration, deformities, and reduced marketability of the crop. Thrips may also serve as vectors for plant diseases, such as Tospoviruses. Over 20 plant-infecting viruses are known to be transmitted by thrips. These enveloped viruses are considered among some of the most damaging of emerging plant pathogens around the world. Virus members include the tomato spotted wilt virus and the impatiens necrotic spot viruses. The western flower thrips, *Franliniella occidentalis*, now has a worldwide distribution and is considered the primary vector of plant diseases caused by Tospoviruses.

Flower-feeding thrips, like *Thrips flavus*, are routinely attracted to bright floral colours (including white, blue, and especially yellow), and will land and attempt to feed. It is not uncommon for some species (e.g., *Frankliniella tritici* and *Limothrips cerealium*) to "bite" humans under such circumstances. Although no species feed on blood and no known animal disease is transmitted by thrips, some skin irritation has been described. Although many thrips are considered to be pests, they have been found to aid in pollination of plants as well. It has been found that a species of thrips *Thrips setipennis* is the sole pollinator of *Wilkiea huegeliana*, which is a small, uni-sexual annually flowering tree or shrub that is found in the rainforests of eastern Australia. In Australia, *T. setipennis* serves as an obligate pollinator for other rainforest plant species, including *Rapanea howittiana* and *R. variabilis*. Due to their small sizes and high rates of reproduction, thrips are difficult to control using classical biological control. All predators must be small and slender enough to penetrate the crevices where thrips hide while feeding, and then prey extensively on eggs and larvae. Only two families of parasitoid hymenoptera are known to parasitize eggs and larvae, the Eulophidae and the Trichogrammatidae. Other biocontrol agents of adults and larvae include aphid wasps, anthocorid bugs of genus *Orius*, and phytoseiid mites. For this reason, many growers are occasionally forced to make limited use of pesticides to control thrips populations in the field and in greenhouses. Another effective strategy for pest thrips are biological insecticides, including *Beauveria bassiana* or *Verticillium lecanii*. These demonstrate a clear effect on eggs, larvae and adults of thrips. Insecticidal soap spray is effective against thrips. It is commercially available or can be made of certain types of household soap. Scientists in Japan report that significant reductions in larva and adult melon thrips occur when plants are illuminated with red light.

On national level thrips were extensively investigated into by Prof. T.N. Ananthakrishnan and Prof. J.S. Bhatti, and in the Doon Valley the classical research work conducted for many years by Dr. Vijay Veer, supported ably by Dr. Neena Chauhan and Dr. A.K. Srivastava, is much appreciated and lauded internationally. Besides the international treatments available by Dr. L.A. Mound and Dr. Trevor Lewis, the Indian masterpieces by Ananthakrishnan (1979),

Ananthakrishnan & Sen (1980) and Bhatti (2006) are essential for reading by the budding thysanopterists in India. However, a large quantum of information mined by Vijay Veer & Neena Chauhan (1990) and Neena Chauhan & Vijay Veer (1992) is inevitably indispensable for all researchers engaged in thrips research in the Garhwal region. For more details see Annex 1(2).

27. Order Zoraptera

The insect order Zoraptera contains a single family, the Zorotypidae, which in turn contains one extant genus with 39 species, *Zorotypus* as well as 9 extinct species. Living species are found worldwide, mainly in tropical and subtropical regions around the world. Four species occur north of the Tropic of Cancer, two in the USA and two in Tibet.

The phylogenetic relationship of the order remains controversial and elusive. At present the best supported position based on morphological traits recognizes the Zoraptera as polyneopterous insects related to the webspinners of the Order Embioptera. However, molecular analysis of 18s ribosomal DNA supports a close relationship with the Superorder Dictyoptera.

The members of this order are small insects, 3 mm or less in length, that resemble termites in appearance and in their gregarious behaviour. They are short and swollen in appearance. They belong to the hemimetabolous insects. They possess mandibulated biting mouthparts, short cerci (usually 1 segment only), and short antennae with 9 segments. The abdomen is segmented in 11 sections. The maxillary palps have five segments, labial palps three, in both the most distal segment is enlarged. Immature nymphs resemble small adults (Fig. 61).

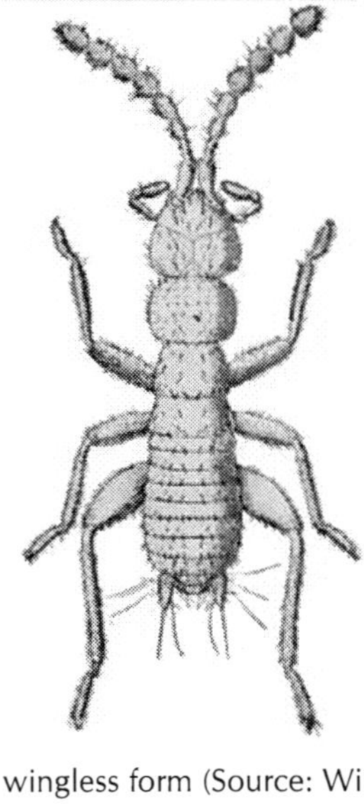

Fig. 61. *Zorotypus guineensis*, wingless form (Source: Wikipedia, the free encyclopedia).

Each species shows polymorphism. Most individuals are the apterous form or "morph", with no wings, no eyes, and no or little pigmentation. A few females and even fewer males are in the alate form with relatively large membranous wings that can be shed at a basal fracture line. Alates also have compound eyes and ocelli, and more pigmentation. This polymorphism can be observed already as two forms of nymphs. Wingspan can be up to 7 mm, and the wings can be shed spontaneously. When observed, wings have simple venation. Under good conditions the blind and wingless form predominates, but if their surroundings become too tough, they produce offspring which develop into winged adults with eyes. The wings are paddle shaped, and have reduced venation.

They live in small colonies beneath rotting wood, eating fungal spores and detritus but also can hunt smaller arthropods like mites and collembola. *Zorotypus gurneyi* lives in colonies consisting of up to several hundred of individuals. Most commonly the colonies have a size of around 30 individuals, of which about 30% are nymphs, the remainder adults. Zoraptera spend most of their time grooming one another. The grooming process is thought to be a way of removing fungal pathogens. When two colonies of *Z. hubbardi* are brought together experimentally, there is no difference in behaviour towards members of the own or new colony. Therefore, colonies in the wild might merge easily. Winged forms are rare. The males in such average colonies establish a linear dominance hierarchy in which age or duration of colony membership is the prime factor determining dominance. Males appearing later in colonies are at the bottom of the hierarchical ladder, regardless of their body size. By continually attacking other males, the dominant male monopolizes a harem of females. The members of this harem stay clumped together. There is a high correlation between rank and reproductive success of the males. *Z. barberi* lack such a dominance structure but display complex courtship behavior including nuptial feeding. The males possess a cephalic gland that opens in the middle of their head. During courtship they secrete a fluid from this gland and offer it to the female. Acceptance of this droplet by the female acts as behavioural releaser and immediately leads to copulation. In *Z. impolitus*, copulation does not occur, but fertilization is accomplished instead by transfer of a spermatophore from the male to the female. This 0.1 mm spermatophore contains a single giant sperm cell, which unravels to about the same length as the female herself, 3 mm. It is thought that this large sperm cell prevents fertilization by other males, by physically blocking the female's genital tract.

Zoraptera are thought to provide some important supporting service in ecosystems. By consuming detritus, such as dead arthropods, they assist in decomposition and nutrient cycling.

28. Order Psocoptera

Psocoptera (Greek word, *psokos* meaning gnawed or rubbed and *ptera* meaning wings) are an order of insects that are commonly known as booklice, barklice or barkflies. They first appeared in the Permian period, 295-248 million years ago.

They are often regarded as the most primitive of the hemipteroids. The book lice have mainly the tropical distribution and are organized into three suborders, viz., Troglomorpha, Troctomorpha and Psocomorpha. More recently morphological and molecular evidences have shown that the parasitic lice (Phthiraptera) evolved from within the psocopteran suborder Troctomorpha. In modern systematics, Psocoptera and Phthiraptera are therefore, treated together in the order Psocodea. There are more than 5,500 species in 41 families in three suborders. Many of these species have only been described in recent years.

The species known as booklice received their common name because they are commonly found amongst old books — they feed upon the paste used in binding. The barklice are found harmlessly on trees, feeding on algae and lichen. No member of this order is currently considered endangered; in fact, in 2007, *Atlantopsocus adustus*, a species native to Madeira and the Canary Islands, was found to have colonized the mild Cornish coast of southwest England.

Psocids are small, scavenging insects with a relatively generalized body plan. They feed primarily on fungi, algae, lichen, and organic detritus. They have chewing mandibles, and the central lobe of the maxila is modified into a slender rod. This rod is used to brace the insect while it scrapes up detritus with its mandibles. They also have a swollen forehead, large compound eyes, and three ocelli. Their bodies are soft with a segmented abdomen. Some species can spin silk from glands in their mouth. They may festoon large sections of trunk and branches in dense swathes of silk.

Some psocids have forewings that are up to 1.5 times as long as the hindwings, and all four wings have a relatively simple venation pattern, with few cross-veins. The wings, if present, are held tent-like over the body. The legs are slender and adapted for walking, rather than gripping, as in the true lice Abdomen has nine segments, and no cerci.

There is often considerable variation in the appearance of individuals within the same species. Many have no wings or ocelli, and may have a different shape to the thorax. Other, more subtle, variations are also known, such as changes to the development of the setae. The significance of such changes is uncertain, but their function appears to be different from similar variations in, for example, aphids. Like aphids, however, many psocids are parthenogenic, and the presence of males may even vary between different races of the same species.

Psocids lay their eggs in minute crevices or on foliage, although a few species are known to be viviparous. The young are born as miniature, wingless versions of the adult. These nymphs typically muolt six times before reaching full adulthood. The total lifespan of a psocid is rarely more than a few months.

Booklice range from approximately 1-2 mm in length. Some species are wingless and they are easily mistaken for bedbug nymphs and vice versa. Booklouse eggs take 2 to 4 weeks to hatch and can reach adulthood approximately 2 months later. Adult booklice can live for six months. Besides damaging books, they also sometimes infest food storage areas, where they feed on dry, starchy materials. Although some psocids feed on starchy household products, the majority of psocids are woodland insects with little to no contact with humans, therefore they are of little economic importance. They are scavengers and do not bite humans.

Psocids can affect the ecosystems in which they reside. Many psocids can affect decomposition by feeding on detritus, especially in environments with lower densities of predacious micro arthropods that may eat psocids. The nymph of a psocid species, *Psilopsocus mimulus*, is the first known wood-boring psocopteran. These nymphs make their own burrows in woody material, rather than inhabiting vacated, existing burrows. This boring activity can create habitats that other organisms may use.

The Order Psocoptera is divided into three suborders:

(1) Trogiomorpha have antenna with many segments (22–50 antennomeres) and always three-segmented tarsi. Trogiomorpha is the smallest suborder of the Psocoptera *sensu stricto* (i.e., excluding Pthiraptera), with about 340 species in 7 families, ranging from the monospecific fossil family Archaeotropidae to the speciose Lepidopsocidae (over 200 species). Trogiomorpha comprises infraorder Atropetae (extant families Lepidopsocidae, Psoquillidae and Trogiilidae, and fossil families Archaeotropidae and Empheriidae) and infraorder Psocathropetae (families Psyllipsocidae and Prionoglarididae),

(2) Troctomorpha have antennae with 15–17 segments and two-segmented tarsi. Troctomorpha comprises the infraorder Amphientometae (families Amphientomidae, Compsocidae, Electrentomidae, Musapsocidae, Protoctopsocidae and Troctopsocidae) and infraorder Nanopsocetae (families Liposcelididae, Pachytroctidae and Sphaeropsocidae). Troctomorpha are now known to also contain the order Pthiraptera (lice), and are therefore paraphyletic, as are Psocoptera as a whole.

(3) Psocomorpha are notable for having antennae with 13 segments. They have two- or three-segmented tarsi, this condition being constant (e.g., Psocidae) or variable (e.g., Pseudocaeciliidae) within families. Their wing venation is variable, the most common type being that found in the genus *Caecilius* (rounded, free areola postica, thickened, free pterostigma, r+s two-branched, m three-branched). Additional veins are found in some families and genera (*Dicropsocus* and *Goja* in Epipsocidae, many Calopsocidae, etc.)

Psocomorpha is the largest suborder of the Psocoptera *sensu stricto* (i.e., excluding Phthiraptera), with about 3,600 species in 24 families, ranging from the species-poor Bryopsocidae (2 spp.) to the speciose Psocidae (about 900 spp.). Psocomorpha comprises infraorder Epipsocetae (families Cladiopsocidae, Dolabellopsocidae, Epipsocidae, Neurostigmatidae, and Ptiloneuridae).

29. Order Mallaphaga

Mallophaga is a suborder of biting lice, mainly the ectoparasites of birds and but less frequently of mammals. They are known as chewing lice, biting lice or bird lice, containing more than 3000 species organized into three suborders and distributed in all faunal regions. They infest both domestic and wild animals and birds and cause considerable irritation to their host. They have paurometabolis or incomplete metamorphosis.

Species of Mallophaga are easily identifiable by their head, which is wider than the prothorax. The species that feed on birds typically have two claws on the end of each tarsus while those that feed on mammals typically have just a single claw. Mallophaga have mandibulate mouthparts which are located on the ventral side of their head. They use these mouthparts to feed on feathers, hair and epidermal skin scales. Some species also use these mouthparts to feed on blood which they obtain by piercing the pulp of young feathers or by gnawing through the skin (Fig. 62).

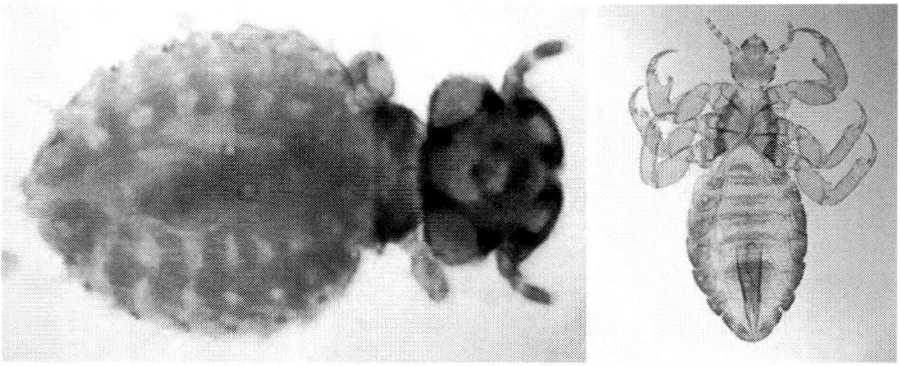

Fig. 62. *Mallophaga* spp. (Source: Wikipedia, the free encyclopedia).

Mallophaga develop by gradual metamorphosis. Females will typically lay 150–300 eggs over an interval of 2–3 weeks. The eggs, commonly known as nits, are oblong and approximately 1 mm long. The eggs are glued to the hairs or feathers of the host with a secretion from the female accessory glands. The eggs typically hatch several days or up to three weeks from the time they are laid. The nymphs that hatch from the eggs resemble the adults except for their smaller size and lighter color. These nymphs go through three nymphal instars during a 2–3-week period. After these three instars, they are considered adults. Most adult species are light tan to brown in colour and are usually 1–4 mm in length although some livestock species can grow to be 5–7 mm and some wild bird species can even get to 10 mm.

The biting lice are often adapted to live on a specific part of their host and typically spend their entire lives on a single host. They can only survive for about three days after their host has died and they typically use phoresis, which is hitching a ride from a fly, as an attempt to reach a new host. Mallophaga may also use phoresis in order to spread to a new host even if the present one is still alive.

30. Order Siphunculata

About 500 species of sucking lice (Anoplura, formerly known as Siphunculata) are currently divided into six families and represent the smaller of the two traditional suborders of lice. As opposed to the paraphyletic chewing lice, which are now divided among three suborders, the sucking lice are monophyletic. The Anoplura are all blood-feeding ectoparasites of mammals. They only occur on about 20% of all placentalian mammal species, and are unknown from several orders of mammals (Monotremata, Edentata, Pholidota, Chiroptera, Cetacea, Sirenia and Proboscidea). They can cause localized skin irritations and are vectors of several blood-borne diseases.

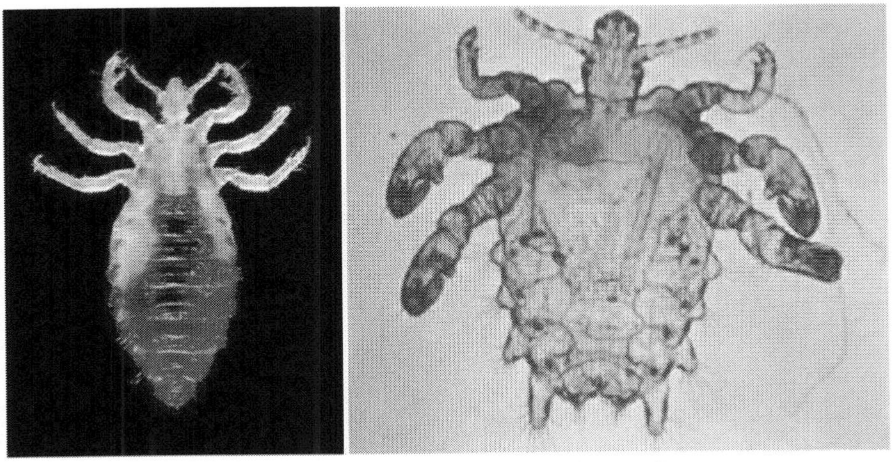

Fig. 63. Anoplurans: (a) *P. humanus,* (b) *P. pubis*

At least three species or subspecies of Anoplura are parasites of humans; the human condition of being infested with sucking lice is called pediculosis. *Pediculus humanus* is divided into two subspecies, *Pediculus humanus humanus*, or the body lice, sometimes nicknamed "the seam squirrel" for its habit of laying of eggs in the seams of clothing, and *Pediculus humanus capitis*, or the head louse. *Pthirus pubis* (the crab louse) is the cause of the condition known as crabs (Piotrowski, 1992) (Fig. 63).

Lice have been infamously known as a pest or plague to man since yore, and have caused devastating miseries as vectors of some of the worst fast spreading epidemic diseases and consequent deaths of many humans. No doubt, therefore, that not a long time ago these were placed as killers beside mosquitoes and fleas. Lice are wingless, dorso-ventrally compressed insects. They have retractable mouthparts which are piercing type. Lice engorge on blood by entering their mouthparts into skin. They project their mouthparts outside only when feeding is desired. Their legs have special claws at the end with which they can grasp hairs. Legs are short and stout and may have the claw on one or on all the three pairs of legs.

Lice, the well known ectoparasites of man, have a long history to live in association with the human-being. Lice are prevalent throughout the world, both in tropical and temperate countries, and in developing and developed nations. There are several species which live as ectoparasites, or nasty vermin, on man but three species, viz., the human body louse (*Pediculus humanus*), the head louse (*P. capitis*) and the crab or pubic louse (*Phthirus pubis*), are encountered most commonly. They all can live synchronously on the same person, without encroaching upon each other's territory. Lice carry enormous public health importance, too, and, in particular, *P. humanus* is a serious vector of two major infections, typhus and relapsing fever, both of which may end up in death of many people if not treated beforehand. These are particularly fearsome in times of war and sudden ecological disasters or catastrophes like earthquakes, which force large masses of people to live in congested congregations with very low hygienic conditions, and where disease epidemics are most possibly predictable. The transmission of disease takes place when an infected person with lice infestation comes in contact with another person and the infection passes over when the infected lice bite him as well. As a principle, all lice infestation on man come either directly or indirectly from another man. Presently, though one of the greatest discomforts and annoyances of human living, they are notwithstanding given sorrowfully but little attention and care to be controlled.

Lice have three stages in their life cycle, namely, egg (called nit), nymph and adult. The human louse is host-specific and does not occur on animals. They are all ecto-parasites and survive by feeding on blood. The development is slow and only three moults occur in the nymphal stages before forming the adult. The development of life cycle stages is optimal at 32°C. Lice discard a dead body at once and move on to another person nearby. Various characteristics of the life history of lice are briefly described below.

(i) *The Egg:* It is the most interesting structure as it is seen only as a cemented body with the hair base close to the scalp (*P. capitis, P. pubis*) or as a body fixed with fine threads on clothes (*P. humanus*). It is just about the size of a pin head (1-1.5 mm). Temperature seems to play a key role in hatching of the eggs of each of the three species in their specific habitat. Eggs of the head louse hatch maximally at 31°C whereupon they emerge in one week's time. At a temperature below than 22°C eggs fail to hatch. Eggs which are found glued to hair more than 6 mm above the scalp have already hatched. The freshly laid eggs are shining white, but as the hair grows they become discernible.

(ii) *The Nymph:* It is a young, creamy white louse which moults thrice before becoming a sexually mature adult. Total duration taken by nymphs to convert into adults range between 7 and 13, largely depending upon temperature. Except for their size and colour, the nymphs resemble adults very much.

(iii) *The Adult:* It has three distinct body parts; a head, a fused chest and a segmented abdomen (or stomach, which measures three-fifths of the body length). The female oversizes the male in length. Males and females mate randomly and several times during adult life. Oviposition follows after nearly 24-48 hours of mating. A female of body louse can lay about 10 eggs per day, but a total of 200-300 eggs in its whole lifetime. In contrast, a female of head louse lays lesser eggs per day totaling around 50 eggs in a lifetime.

Both male and female lice feed on human blood, but hardly to the point of engorgement. The head louse feeds about 5 times a day whereas the body louse feeds about four times per day. Blood is the only nourishment for lice, and they leave a dead body immediately to search for other hosts. They are, therefore, not found away from the human host, but are diversely dispersed through different man-made means such as body contact, change of infested clothes, and hair combs and brushes.

The body louse is commonly found where the clothing is in direct contact with the body including the underwear, crotch or fork of trousers, armpits, wasteline, neck and shoulders. If the clothing is removed or disposed off, the body louse falls in peril of death. *Pediculus humanus* is 10-20% (i.e., 1/10th to 1/5th) larger than *P. capitis* and can live four weeks or more on its host, but can survive for 8-10 days away from the host. The body lice are generally light-coloured than the head lice. The head louse, *P. capitis*, which generally infests children, on the other hand, smartly tries to match the colour of the host's hair. In the head region, their distribution is mainly in areas like back of the head and behind the ears where commensurate temperature and dampness exist to enable profuse hatching of the eggs. The empty eggs, left behind following hatching of young nymphs, remain fastened to the base of the hair while the hair is on the growth, and are identifiable by their shiny papery appearance - a good indicator for comprehending the duration of an infestation. Since human hair grows at the rate of about 1 cm a month, therefore, if egg shells are found 4 cm from the scalp, it is estimated that the infestation should have been nearly four months old. The crab louse (*P. pubis*) differs from both the body louse and the head louse both in its appearance and colour. Although, behaviourally they restrict to mostly pubic and perineal regions, but in the event of heavy infection they may be found over other parts of the body as well, such as chest, thighs, armpits, eyebrows, eyelashes and beard. Young sexually active adults are more prone to *P. pubis* infection, but could be found in children, too. This louse breeds at a much less rate than the other two species, being around 3 eggs a day for a total of about 30 eggs.

Lice may be spread more rapidly and copiously by any means of physical or material (clothing etc.) contact under crowded conditions, although certain human behaviours (like combing etc.) also tend to disseminate its distribution in a small way. One likelihood of spread of infection is the seats of ill-kept public water closets and repeated contact with fomites.

Man is the only reservoir or host of louse-borne diseases. Of all the three forms of human lice, only body lice (*P. humanus*) transmit the disease. For the spread of any louse-borne disease, two factors are important; one, there has to be a disease focus in the area, and secondly, lice have got to be there in high density in association with man. Two diseases, namely, louse-borne typhus (LBT) and louse-borne relapsing fever (LBRF) are spread by lice, although the latter infection now mainly occurs in Ethiopia and Sudan, besides some foci in past in West Africa as well as in South and Central America

31. Order Homoptera

There is no agreement to the classification of some 40,000 known homopteran species (aphids, scale insects, cicadas and leafhoppers, which all have sucking mouthparts) and various higher taxa are assigned different taxonomic levels by different workers (Fig. 64). Homoptera has for long been considered a suborder of order Hemiptera that is considered by some taxonomists to be paraphyletic, and therefore deprecated (obsolete). It was therefore organized into the suborders, Coleorrhyncha, Auchenorrhyncha and Sternorrhyncha. The earlier work was based on nuclear DNA, but more recent phylogenetic analysis using mitochondrial DNA suggest that Homoptera may be a monophyletic group after all, a sister group of Heteroptera. The cause of the disparity in the analyses is suggested to be the long branch attraction effect in phylogenetic analysis, due to rapidly evolving DNA regions. They are believed to have appeared during Permian-time, and are therefore, quite ancient in origin. They are also hemipteroid, meaning they are closely related to Thysanoptera and Psocoptera.

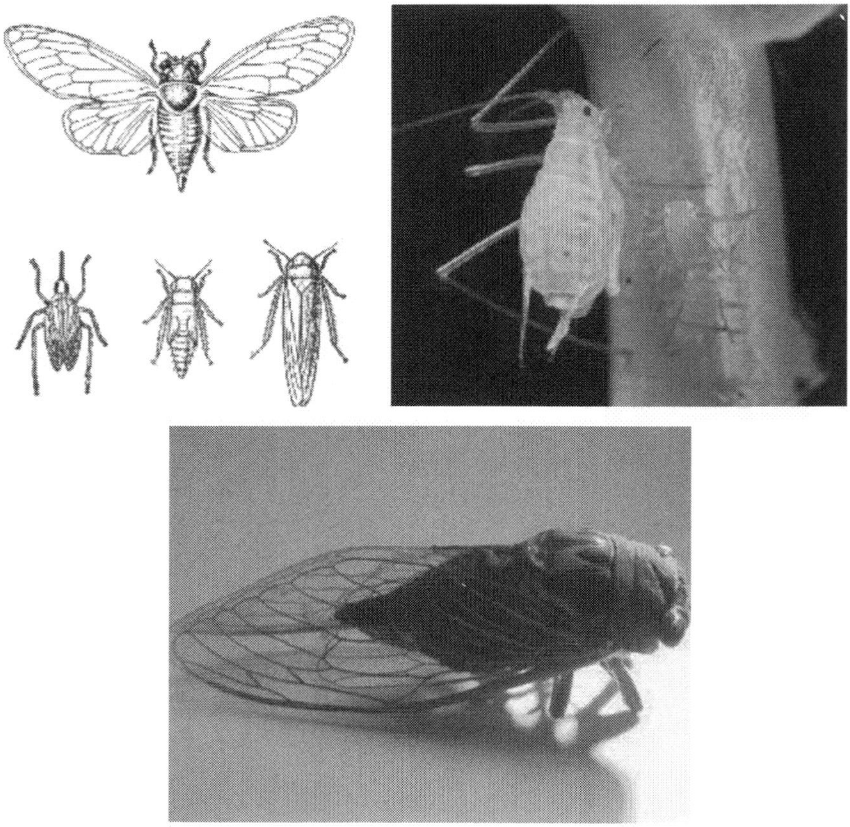

Fig. 64. (a) Adult and nymphs of Homoptera, (b) An aphid, and (c) A Cicada (Source: Wikipedia, the free encyclopedia).

Homoperans are abundant worldwide. While Coleorrhyncha includes only a few species with a scattered distribution in the Southern Hemisphere (South America, New Zealand), the other two occur in all faunal regions. Some species are found high in the alpine zone, particularly those transferred by air currents (e.g., Aphidae). Under Auchenorrhyncha are classified the well-known families of Cicadidae (cicadas), Cercopidae (froghoppers and cocko-spit insects,

Membracidae (treehoppers), and Jassidae (leafhoppers) (all referable to the superfamily Cicadoidea), and the superfamily Fulgoridea. The Sternorrhyncha include the superfamilies Psylloidea (Jumping plant lice), Aleyrodoidea (white flies), Aphidoidea (Plant lice) and Coccoidea (scale insects).

They are hemimetabolous, i.e., their development possesses egg, nymph and adult.

All members of the suborder Homoptera have piercing/sucking mouthparts and feed by withdrawing sap from vascular plants. The proboscis is shorter than that found in true bugs (suborder Heteroptera), and it emerges near the ventral posterior margin of the head capsule (opistognathous). Although some Homoptera are secondarily wingless, the majority have membranous or uniformly textured wings that fold tent-like over the body at rest.

It is difficult to generalize about the biology of these insects. Cicadas are the largest members of the suborder. As nymphs, they live underground and feed on the roots of trees and shrubs. Some species complete development in as little as four years, but others have a 13- or 17-year life cycle. In contrast, the aphids are tiny, soft-bodied insects with multiple generations per year. Many species have complex life cycles involving more than one host plant. Winged and wingless forms of the same species may develop at different times of the year. Asexual reproduction (parthenogenesis) is common and males are unknown in some species. The scale insects are even more specialized. During much of their life cycle, they remain immobile, living beneath an impervious cover of wax or cuticle that they secrete over themselves. Legs and antennae often disappear after the first moult. Only newly hatched nymphs and adult males bear any resemblance to other insects. Females grow to sexual maturity, mate, produce offspring, and die without ever leaving their protective cover.

In most of the Homoptera, a portion of the digestive system is modified into a filter chamber. This structure allows the insects to ingest and process large volumes of plant sap. Excess water, sugars, and certain amino acids bypass most of the midgut and are shunted directly into the hindgut for excretion as honeydew. Only a small volume of filtered plant sap passes through the midgut for digestion and absorption. Many species of ants are attracted by the honeydew and provide care and protection for the homopterans in exchange for the honeydew they excrete.

Adults have antennae which are slender or bristle-like; proboscis is short, arising near lower back margin of head. Front wings, when present, are uniform in texture; at rest, wings fold tent-like over the abdomen. Tarsi are 1- 3-segmented. Immatures are structurally similar to adults and always lacking wings.

Homoptera are among the most abundant herbivores found in terrestrial habitats. Many species are pests of cultivated plants. Aphids and leafhoppers are important carriers of plant diseases. The various families with characteristics include:

- Cicadidae (Cicadas): Nymphs live underground where they feed on the roots of trees and shrubs. Adults are the largest members of the Homoptera. Males produce loud songs to attract a mate.

- Cicadellidae (Leafhoppers): This is the largest family of Homoptera and includes many pests of cultivated plants. Leafhoppers are important carriers of plant diseases, especially mycoplasmas.
- Membracidae (Treehoppers): Ecologically similar to leafhoppers, these insects have a large pronotum that extends over most of the body. They often resemble thorns or small twigs.
- Cercopidae (Spittlebugs or Froghoppers): Nymphs live on plant stems and produce a frothy defensive secretion around themselves. Adults are similar to leafhoppers in size and general appearance.
- Fulgoridae (Planthoppers): This is one of eleven families classified as planthoppers (superfamily Fulgoroidea). These insects are ecologically similar to leafhoppers and treehoppers. Many species are oddly shaped and cryptically coloured.
- Psyllidae (Psyllids or Jumping Plant Lice): Small, aphid-like insects with 3-segmented beaks and 10-segmented antennae. Many species are covered with a woolly layer of wax.
- Aleyrodidae (Whiteflies): Body and wings of adults are covered with a white powdery wax. Nymphs attach to the undersides of leaves and become immobile, resembling scale insects.
- Aphididae (Aphids, Plantlice): Second largest family in the suborder Homptera. Many of these insects are pests of cultivated plants. Aphids are considered the most important carriers of viral plant diseases.
- Coccidae (Soft Scale insects): This is one of 17 families that make up the superfamily Coccoidea (scale insects and mealybugs). Most species are sedentary during most of their life cycle and secrete a protective covering over their bodies. These insects are among the most common pests of cultivated plants.

As homopterous species are highly important economically few of these features are mentioned below: (i) A scale insect, *Laccifer lacca*, is the source of natural shellac. The insect lives on various fig trees in the tropics, (ii) *Dactylopius coccus*, the cochineal insect, is the source of a bright red dye formerly used in the textile industry. It is a scale insect that lives on prickly pear cacti, (iii) Aphids in the subfamily Pemphiginae are gall-makers. The galls are usually open at one end so the insects can come and go freely, (iv) The ground pearls (family Margarodidae) are a group of scale insects that live on plant roots. In some tropical species, the females form large wax cysts, often bronze or gold in colour, that people collect and use as beads, (v) Honeydew, an excretory product that is rich in sugars and amino acids, is produced by many species of Homoptera. Other animals use honeydew as a source of food, and (vi) Honeydew from a mealybug, *Trabutina mannipara*, is regarded as the probable source of Biblical manna.

32. Order Heteroptera

There are well over 40,000 known species of bugs, classified into two suborders, viz., the terrestrial Gymnocerata and the aquatic Cryptocerata. The majority of them occurs in tropical and subtropical regions, but some, particularly the cryptocerates, penetrated high into Arctic. In the high alpine ecosystems bugs represent but a minor faunal element, though some species are found, in Mt. Everest region, up to the 5000 m mark. The name Heteroptera, derived from the Greek "hetero-" meaning different and "ptera" meaning wings, refers to the fact that the texture of the front wings is different near the base (leathery) than at the apex (membranous). Most species have forewings with both membranous and hardened portions (called hemelytra); members of the primitive Enicocephalomorpha have completely membranous wings.

The blackspined soldier bug (Podisus nigrispinus) is a predatory stink bug (family Pentatomidae). This stamp, and three others that depict benefical insects, was issued by Argentina in 1990.

Heteropterans are hemimetabolous, i.e., they go through incomplete development (egg, nymph, adult). They are orthopteroid meaning that they are closely related to orders Thysanoptera and Psocoptera. Heteropterous taxa are abundant worldwide being found in most terrestrial and freshwater habitats.

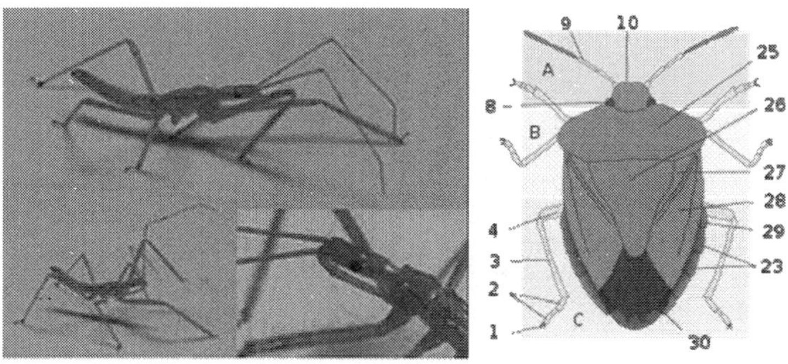

Fig. 65: (a) An assassin nymph (Zelus sp.), and (b) Generalized morphology of a shield bug Pantatomoidea (Source: Wikipedia, the free encyclopedia): (A: head; B: thorax; C: abdomen. 1: claws; 2: tarsus; 3: tibia; 4: femur; 8: compound eye; 9: antenna; 10: clypeus; 23: laterotergites; 25: pronotum; 26: scutellum; 27: clavus; 28: corium; 29: embolium; 30: membrane).

Members of the suborder Heteroptera are known as "true bugs" (Fig. 65). They have very distinctive front wings, called hemelytra, in which the basal half is leathery and the apical half is membranous. At rest, these wings cross over one another to lie flat along the insect's back. These insects also have elongate, piercing-sucking mouthparts which arise from the ventral (hypognathous) or anterior (prognathous) part of the head capsule. The mandibles and maxillae are long and thread-like, interlocking with one another to form a flexible feeding tube (proboscis) that is no more than 0.1 mm in diameter yet contains both a food channel and a salivary channel. These stylets are enclosed within a protective sheath (the labium) that shortens or retracts during feeding. Adults have antennae slender with 4-5 segments; proboscis 3-4 segmented, arising from front of head and curving below body when not in use; pronotum usually large, trapezoidal or rounded; triangular scutellum present behind pronotum; front wings with basal half leathery and apical half membranous (hemelytra). Wings lie flat on the back at rest, forming an "X". Tarsi 2- or 3-segmented. Immatures are structurally similar to adults, always lacking wings though.

The Heteroptera include a diverse assemblage of insects that have become adapted to a broad range of habitats; terrestrial, aquatic and semi-aquatic. Terrestrial species are often associated with plants. They feed in vascular tissues or on the nutrients stored within seeds. Other species live as scavengers in the soil or underground in caves or ant nests. Still others are predators on a variety of small arthropods. A few species even feed on the blood of vertebrates. Bed bugs, and other members of the family Cimicidae, live exclusively as ectoparasites on birds and mammals (including humans). Aquatic Heteroptera can be found on the surface of both fresh and salt water, near shorelines, or beneath the water surface in nearly all freshwater habitats. With only a few exceptions, these insects are predators of other aquatic organisms.

Plant feeding bugs are important pests of many crop plants. They may cause localized injury to plant tissues, they may weaken plants by removing sap, and they may also transmit plant pathogens. Predatory species of Heteroptera are generally regarded as beneficial insects, but those that feed on blood may transmit human diseases (Fig. 66). Chagas disease, for example, is transmitted to humans by conenose bugs (genus Triatoma, family Reduviidae). Although bed bugs (family Cimicidae) can inflict annoying bites, there is little evidence that they regularly transmit any human or animal pathogen. Two families of Heteroptera are ectoparasites. The Cimicidae (bed bugs) live on birds and mammals (including humans). The Polyctenidae (bat bugs) live on bats. Water striders in the genus Halobates (family Gerridae) are the only insects that are truly marine. They live on the surface of the Pacific Ocean. Unlike other insects, male bedbugs do not place their sperm directly in the female's reproductive tract. Instead, they puncture her abdomen and inject the sperm into her body cavity. The sperm swim to the ovaries where they fertilize the eggs. This unusual type of reproductive behavior is appropriately known as "traumatic insemination". Some members of the family Largidae resemble ants. They live as social parasites in ant nests, mimicking the ants' behavior to get food.

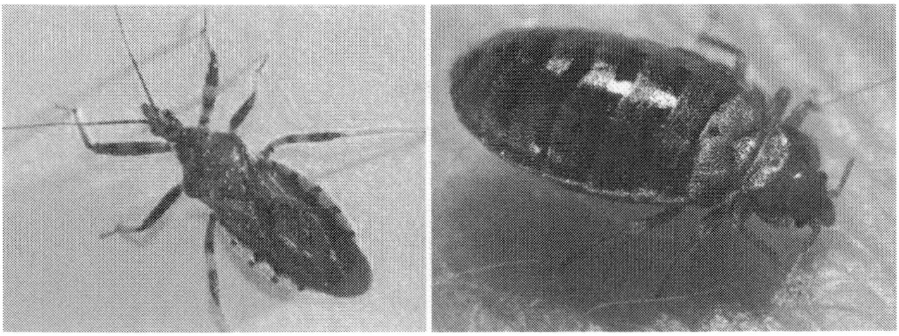

Fig. 66. Heteroptera (True bugs): (a) An assassin bug, and (b) A bed bug.

REFERENCES CITED

Adis, J., O. Zompro, E. Moombolah-Goagoses and E. Marais, 2002. Gladiators: A new order of insect. Scientific American 287:60-65.

Aguiar, A.P., A.R. Deans, M.S. Engel, M. Forshage, J.T. Huber, J.T. Jennings, N.F. Johnson, A.S. Lelej, J.T. Longino, V. Lohrmann, I. Mikó, M. Ohl, C. Rasmussen, A. Taeger & D.S.K. Yu 2013, Order Hymenoptera Linnaeus, 1758. In: Zhang, Z.-Q. (Ed.) Animal Biodiversity: An Outline of Higher-level Classification and Survey of Taxonomic Richness (Addenda 2013). Zootaxa, 3703, 1–82.

Ananthakrishnan, T.N., 1993. Bionomics of Thrips. Annual Review of Entomology 38: 71–92.

Ananthakrishnan, T.N., 2000. Bionomics of Thrips. Annual Review of Entomology 38: 71-92.

Arnett, Ross H. Jr., 2000. American Insects. A Handbook of the Insects of America North of Mexico: A Handbook of the Insects of America North of Mexico, Second Edition. CRC Press.

Baijal, H. N., 1958. Entomological Survey of Himalaya. Part XXVIII, Nival Collembola from the North–West Himalaya. Section B, Part V. Proceedings of the National Academy of Sciences. India, Allahabad 28B(5): 349–360.

Bechly, G. 2002: Phylogenetic Systematics of Odonata. in Schorr, M. & Lindeboom, M., eds, (2003): Dragonfly Research 1.2003. Zerf – Tübingen.

Beeson, C.F.C., 1941. The Ecology and Control of the Forest Insects of India and the Neighbouring Countries. Bishen Singh Mahendra Pal Singh, Dehra Dun, pp. viii + 1007, incl. 203 Figures.

Beier, M., 1968. Ordnung Mantodea (Fangheuschrecken). Handbuch der Zoologie 4(2): 3–12.

Bhatti, J.S., 1994. Phylogenetic relationships among Thysanoptera (Insecta) with particular reference to the families of the Order Tubulifera. Zoology (Journal of Pure and Applied Zoology) 4 (1993): 93-130.

Bhatti, J.S., 1998. New structural features in the Order Tubulifera (Insecta). 1. Amalgamation of labro-maxillary complex with cranium and other cephalic structures. Zoology (Journal of Pure and Applied Zoology) 5: 147-176.

Bhatti, J.S., 2006. The classification of Terebrantia (Insecta) into families. Oriental Insects 40: 339-375

Campbell, K.G., 1959. The importance of research into forest insect problems in New South Wales. Australian Forestry 23 (1): 19–23.

Carapelli, F. Nardi, R. Dallai & F. Frati, 2006. A review of molecular data for the phylogeny of basal hexapods. Pedobiologia 50(2): 191-204.

Chauhan Neena & Vijay Veer, 2008. Notes on diagnosis and bioecology of thrips pest of floral, fruit, vegetable, agronomic crops and forestry plants in the Dehra Dun Valley (Uttarakhand, India). [In: Pests of Forest Importance and their Management. Eds. B.K. Tyagi, Vijay Veer and Shri Prakash]. Scientific Publishers (Jodhpur), pp. 195-228.

Chhotani, O.B., 1997. Fauna of India (Isoptera) - Volume II. Zoological Survey of India, Calcutta.

Conde, B. & J. Thomas 1957. Contribution a la faune des campodeides de Califomie (Insectes Diploures). Bull. Soc. Linn. Lyon, 26: 81-96; 118-127; 142-155.

Corbet, P.S., 1980. Biology of Odonata. Annual Review of Entomology, 25: 189-217.

Corbet, Philip S. & Stephen J. Brook (2008). Dragonflies. London: Collins.

Dakshini, K.M.M. 1974. The Flora of Mothronwala Swamp. Jour. Bomb. Nat. Hist. Soc. 71(2): 235-243.

Enrique Baquero1, Rafael Jordana and Gurupada Mandal, 2013. Redescription of Baijal's Entomobrya species from Indian Himalayan Region (Collembola, Entomobryidae). 85 (3): 171–180.

Faisal, M., Singh, S., Yousuf, M., 2014. Cleridae (Insecta: Coleoptera) type collection at National Forest Insect Collection (NFIC), Forest Research Institute, Dehradun (India). Zootaxa 3846(1):105-18.

Fraser, F.C., 1933. The Fauna of British India, including Ceylon and Burma. Odonata. Vol. I, xiii+423 pp. Taylor and Francis Ltd., London.

Fraser, F.C.,1934. The Fauna of British India including Ceylon and Burma. Odonata. Vol. II, xxiii+398 pp., 4 col. Pls. Taylor and Francis Ltd., London.

Fraser, F.C., 1936. The Fauna of British India including Ceylon and Burma, Odonata. Vol. III, xi+461 pp., 2 pls. Taylor and Francis Ltd., London.

Fraser, F.C. & R.J. Tillyard, 1957. A reclassification of the order Odonata. Royal Zoological Society of New South Wales. Australian zoological handbook. Sydney: Royal Zoological Society of New South Wales.

Ghildiyal, J.C and Srivastava, M.M. 1989. The vegetation of Manu Swamp: A Tropical fresh - water swamp forest. Indian Forester 115(3): 183-191.

Ghosh, A.K. 1996. Insect biodiversity in India. Oriental Insects 30: 1-10.

Gillot, C., 1995. Butterflies and moths. Entomology (2 ed.), pp. 246–266.

Grimaldi, D. and Engel, M.S. 2005. Evolution of the Insects. Cambridge University Press. ISBN 0-521-82149-5.]

Gordon Aalexander, 1962. Ecology of high altitude insects. Ecology 43(4): 780-781.

Gullan, P. J. and Cranston, P. S. 2014. The Insects: An Outline of Entomology (Fifth ed.). Wiley Blackwell.

Gupta, Nutan, Ashish Anthwal and Abhay Bahuguna, 2006. Biodiversity of Mothronwala Swamp, Doon Valley, Uttaranchal. The Journal of American Science, 2(3):33-40.

Hamalalnen, M., 1989. Odonata from the Dehra Dun Valley (Uttar Pradesh, India) with notes on synonymy of some west Himalayan species. Odonatologica 18(1): 13-20.

Hazra, A. K., 2015. A Review on Paronellinae (Collembola: Paronellidae) of India. Records of the Zoological Survey of India. Occasional Paper No. 368, pp 63.

Hoell, H.V., Doyen, J.T. & Purcell, A.H., 1998. Introduction to Insect Biology and Diversity, 2nd ed. Oxford University Press. pp. 320.

Hughes, D. P.; Beani, L.; Turillazzi, S.; Kathirithamby, J., 2003. Prevalence of the parasite Strepsiptera in Polistes as detected by dissection of immatures. Insectes Sociaux 50 (1): 62–68.

James V., Ward & Peter E. Ward, 1992. Aquatic Insect Ecology, Biology and Habitat. John Wiley & Sons.

James G. Needham, Minter J. Jr. Westfall, and Michael L. May, 2000. Dragonflies of North America. Revised Edition. Scientific Publisher, Gainesville, FL, USA, pp. 939.

Janssens, F., 2012. Checklist of Collembola of the World. www.collembola.org.

Joshi, P.C., R.K. Negi & T. Negi, 2007. Seasonal variation in benthic macro-invertebrates and their correlation with the environmental variables in a freshwater stream in Garhwal region (India). Life Sci. J., 4(4): 85-89.

Kailash Chandra, 2011. Insect fauna of states and union territories in India. (In: *Arthropods and their Conservation in India (Insects & Spiders)*. ENVIS Bulletin 14 (1): 189-218.

Kanjilal,U. 1901.Swamp forest in Dehradun,N.W Province.Indian Forester 27: 228-230.

Kiauta, B., 1974. Introduction to insect cytotaxonomy. Vol. I. Lectures delivered at the Tribhuwan University, Kathmandu. Nepal Research Centre, Kathmandu, pp. 1-81.

Klass, K.Dieter, 1997. The external male genitalia and phylogeny of Blattaria and Mantodea. Zoologisches Forschungsinstitut. ISBN 978-3-925382-45-1.

Klass, K.D., O. Zompro, N.P. Kristensen, J. Adis, 2002. Mantophasmatodea: a new insect order with extant members in the afrotropics Science 296: 1456-1459.

Kulshrestha, S.K. & Vijay Veer, 1984. Two new species of Thysanoptera (Insecta) from India. Bull. Ent. 25 (1): 33-37.

Kulshrestha, S.K. & Vijay Veer, 1990. A new species of Liothrips Uzel from Dehra Dun (India) (Insecta: Thysanoptera; Phlaeothripidae). Ann. Entomol. 8 (1): 55-60.

Kumar, A. and Khanna, V., 1983. A Review of the Taxonomy and Ecology of Odonata larvae from India. Oriental Insects, 17: 127-157.

Kumar, A. & A. Mitra, 1998. Odonata diversity at Sahastradhara (Sulphur Springs) Dehra Dun, India with a note on their habitat ecology. Fraseria (NS) 5(1/2): 23-46.

Kumar, A. and Prasad, M., 1978. On a new species of Agriocnemis Selys, 1869 with description of its larva from Dehra Dun valley, India. J. Bombay Nat. Hist. Soc., 75: 174-179.

Kumar, A. and Prasad, M., 1981. Field ecology, zoogeography, and taxonomy of the Odonata of western Himalaya, India. Zoological Survey of India, Calcutta, pp. 118.

Lohmann, H., 1996. Das phylogenetische system der Anisoptera (Odonata). Entomologische Zeitschrift, 106: 209-252.

Mandal, G. P., 2009. Checklist of Collembola of India. http://zsi.gov.in/checklist/collembola.pdf Zoological Survey of India, M-Block, New Alipore, Kolkata.

Mandal, G. P., 2010b. http://zsi.gov.in/checklist/Diplura%20(Insecta).pdf Zoological Survey of India, M-Block, New Alipore, Kolkata.

Mandal, G. P., 2010c. http://zsi.gov.in/checklist/Thysanura%20(Insecta).pdf Zoological Survey of India, M-Block, New Alipore, Kolkata.

Mandes, L.F. 1990. An annotated list of generic and specific names of Machilidae (Microcorphia) with identification keys for the genera and geographical notes: Estudoes, Ensaios e Documentos, 11CT/Lisboa, 155: 1-127.

Mani, M.S., 1962. Introduction to high altitude entomology. Insect life above the timber-line in the Northwest Himalaya. Methuen and Co., Ltd., London. xix + 302 pp.

Maxwell-Lefroy, H. and Howlett, 1909. Indian Insect life. A Manual of the Insects of the plains (Tropical India). Thacker & Spink, Calcutta, pp. xii + 786 + 84 plates.

Menon, M.G.R., 1965. Systematics of Indian insects. In: Entomologiy in India (Supplement). Entomological Society of India, Delhi, pp. 146-157.

Mendes, L.F. 1990. An annotated list of generic and specific names of Machilidae (Microcophia) with identification keys for the genera and geographical notes. – Estudos, Ensaios e Documentos, IICT/Lisboa, 155:1-127.

Meyer-Rochow, V. Benno and Essi Keskinen, 2003. Post-embryonic photoreceptor development and dark/light adaptation in the stick insect *Carausius morosus* (Phasmida, Phasmatidae). Applied Entomology and Zoology 38: 281–291.

Mitra, T.R., 1999. Biology and ecology of dragonflied (Insecta: Odonata) with note on their adaptations in different ecosystems in India. Rec. zool. Surv. India 97 (2): 173-188.

Nagpal B. N. & V. P. Sharma, 1995. Indian anophelines. Science Publishers, pp. 416.

Nautiyal, P., 2001. Ecology of mountain streams and rivers in Garhwal Himalaya. In: Ecology and Conservation of Lake, Reservation and Rivers Edited by A. Kumar, A. B. D. Publishers Jaipur, pp. 393-417.

Nautiyal, P., 1996. The River Valley Project in Garhwal region: Impact on the Population Dynamics of Endangered Himalayan Mahseer with Emphasis on Bioconservation. Final technical report submitted to Ganga Project Directorate, Ministry of Environment and Forests Government of India, New Delhi (J 11013/8/92-GDP/24.09.92), pp. 96.

Nautiyal, P. & V.P. Semwal, 2006. Benthic macroinvertebrate community in the mountain streams: Longitudinal patterns of distribution in west Himalaya (Gangetic drainage – Mandakini basin). J. Mount. Res. 1: 39-49.

Nautiyal, P., Asheesh Shivam Mishra and Krishna Raj Singh, 2013. Benthic macro-invertebrate fauna in the headwaters of Doon Valley streams, India. Journal of the Kalash Science 1(2):1-10.

Needham, J.G., M.J. Westfall, Jr., and M.L. May 2000. Dragonflies of North America. Scientific Publishers, Gainesville, FL.

Nelson, C. Riley, 1996. Plecoptera. Stoneflies. Tree of Life Web Project.

Nelson C.R. & Baumann R.W. 1987. Gynandromorphism in the winter stonefly genus Capnia (Plecoptera: Capniidae). Ent. News 98:224 229.

Oldroyd R. J., 1963. Collecting, preserving and studying insects. Hutchinson, Scientific and Technical, London.

Pahari, Dhrubajyoti; Mandal, Gurupada; Hazra, Ashis Kumar, 2014. A Study of Diversity of Collembola (Insecta) Fauna of Darjeeling Himalayas, West Bengal, India. Records of The Zoological Survey of India. Occasional Paper No. 364, pp 144. |

Piotrowski, F., 1992. Anoplura (echte Läuse). de Gruiter, pp. 61.

Pemola Devi, N. and R.K. Jauhari, 2008. Reappraisal on Anopheline mosquitoes of Garhwal region, Uttarakhand. J. Vector Borne Dis. 45: 112-123.

Prabhu, N. R., 1986. Protura of south India: biogeographic considerations. In: *Proc. 2nd International Seminar on Apterygota*. Ed. R. Dallai, University of Siena, Italy, pp. 43-50.

Prasad, M., 1988. Introduction to the external morphology of the Odonate male accessary genitalia with descriptions of sixty three cases in North-West Indian species. Indian Odonalol., 1 : 45-88.

Prasad, M. & Singh, A., 1976. Odonata of Doon Valley. II. Zygoptera. Ree. Zool. Surv. India, 70: 121-131.

Prasad, M. and Varshney, R.K., 1995. A Check-list of the Odonata of India including data on larval studies. Oriental Insects 29: 385-428.

Pruthi, H.S., 1958. 'AK' grasshopper –*Poecilocerus pictus* (Fabr.) (Acrididae). Sucking citrus in Delhi. 16(2): 196.

Ramamoorthy, V.V. 2009. Annotated checklist of the Ephemeroptera of the Indian subregion. Oriental Insects 43: 315-339.

Richards, O.W. & R.G. Davies, (Eds.), 1977. Imms' General Textbook of Entomology. Vol. 1: Structure, Physiology and Development. Springer Netherlands, pp. 418.

Richards, O.W. & R.G. Davies, (Eds.), 1997. Imms' General Textbook of Entomology. Vol. 2: Classification and Biology. Springer Netherlands, pp. 388.

Roonwal, M.L., 1936. Studies on the embryology of the African migratory locust *Locusta migratoria migratoria* .1. Early development with a new theory of multiphased gastrulation among insects. *Phil. Trans. Roy. Soc.* B226: 391-421.

Roonwal, M.L., 1936. On the existence of two different types of the striped eyes solitary type specimens of the desert locust, *Schistocerca gregaria* Forsk. Current Sci. 5(1): 24.

Roonwal, M.L., 1937. Development of the eye-stripes in the Desert Locust *Schistocerca gregaria* (Forskal). In: Y.R. Rao's *Rept. Work Res. Staff under Locust Res. Entom. To Inm. Counc. Agric. Res., Karachi*, during 1936. Simla (Govt. of India Pr.), 25-26 and 150-151.

Roonwal, M.L., 1938. Some reversal changes among locusts and other Acrididae, and the probable importance of the third instar. Proc. 25th Indian Sci. Congr., Pt. 3. Abstracts, Calcutta, 173-174.

Roonwal, M.L., 1940. Preliminary note on some directional changes among locusts and other Acrididae, and the importance of the third instar. Indian J. Ent. 2(2): 137-144.

Roonwal, M.L. & R.C. Sharma, 1988. Biography and scientific contributions of Professor M.L. Roonwal, zoologist, entomologist, and educationist : 80th birthday souvenir volume. Entomological Society of India, Zoological Survey of India, and Dept. of Zoology, University of Jodhpur, pp. 95.

Roonwal, M.L. & O.B. Chhotani, 1989. Fauna of India (Isoptera) - Volume I. Zoological Survey of India, Calcutta.

Roonwal, M. L., 1951a. Introduction. Part-1. 1-5 pp. In A systematic catalogue of the main identified Entomological collection at the Forest Research Institute, Dehra Dun. Parts 1-3. Introduction; Sub class Apterygota; sub-class Pterygota, order Orthoptera (in part), Family Blattidae. The Indian Forester 76(11): 498-505.

Roonwal, M. L., Bhasin, G. D. and Pant, G. D., 1951b. A systematic catalogue of the main identified entomological collection at the Forest Research Institute, Dehra Dun. Parts 1-3. Introduction; Sub class Apterygota; sub-class Pterygota, order Orthoptera (in part), Family Blattidae. The Indian Forester 76(11): 498-505.

Roonwal, M. L., Bhasin, G. D., Pant G. D. and Misra, S. D. 1951c. A systematic catalogue of the main identified entomological collection at the Forest Research Institute, Dehra Dun. Parts 4-8. Order Orthoptera (concluded), Dermaptera and Plecoptera. The Indian Forester 75(5): 313-342.

Roonwal, M. L., Mathur, R. N., Bhasin, G. D., Balwant Singh, Pant, G. D. Avinash, Chandra, Katiyar, K. K. and Mathur, R. K. 1953. A systematic catalogue of the main identified entomological collection at the Forest Research Institute, Dehra Dun. Parts 9-21. Orders Isoptera, Embioptera, Psocoptera, Anoplura, Ephemeroptera, Odonata, Thysanoptera and Hemiptera. Indian Forest Leaflet No. 121 (Part 3) (Entomology): 39-187.

Roonwal, M. L., Mathur, R. N., Bhasin, G. D., Chatterjee, P. N., Sen-Sarma, P. K., Balwant Singh, Avinash Chandra, Thapa, R. S., and Kumar Krishana. 1963. A systematic catalogue of the main identified entomological collection at the Forest Research

Institute, Dehra Dun. Parts 22-38. Orders Neuroptera, Mecoptera, Trichoptera and Lepidoptera. Indian Forest Leaflet No. 121 (Part 4) (Entomology): 189-540.

Sen-Sarma, P.K, 1974. Ecology and biogeography of the termites of India: 421-472. In: Ecology and biogeography in India (Ed. M.S. Mani), Monographiae Biologicae, vol. 23, 773 pp+xv. Published by Dr W. Junk, The Hague.

Singh, J., M. Raghvraman, R.S. Yadav and S. Prasad 2008. First report of dipterans (Insecta : Diptera) from Uttar Pradesh, India. *Indian Journal of Entomology* 70(4): 337-340.

Singh, A. & Prasad, M. 1976. Odonata of Doon Valley - I. Anisoptera. Rec. zool. Surv. India, 70: 21-38.

Singh, S. & Singh, K.P., 2012. Forest Entomology and National Forest Insect Collection (NFIC). [In: *Forestry in Service of Nation: ICFRE Technologies*. Bhauguna, V.K., Kumar, R., Singh, R.P. & Misra, R.K. (Eds.)], ICFRE Publication No 137 ICFRE Bk 81/2011, Dehradun, pp. 368–398.

Sivaramakrishnan, K.G., K.A. Subramanian, V.Y. Ramamoorthy, R.M. Sharma and Kailash Chandra 2009. Check-list of Ephemeroptera of India. E-pubn. ZSI, Calcutta.

Subramanian, K.A. 2014. Checklist of Odonata of India. Zoological Survey of India. www.zsi.gov.in.

Stebbing, E.P., 1923. The Forests of India, Vol. 2. J. Lane, The University of California.

Tillyard, R.J., 1917. The biology of dragonflies (Odonata or Paraneuroptera). Cambridge [Eng.] : University Press. Pp. 436.

Tyagi, B.K., 1977. A note on the karyotypes of Burmagomphus pyramidalis Laidlaw and Onychogomphus saundersi duaricus Fraser (Anisoptera: Gomphidae). Odonatologica (Holland) 6: (4): 277-282.

Tyagi, B.K., 1978a. The chromosome numbers and sex determining mechanisms newly recorded in thirteen Indian dragonflies (Odonata). Chromosome Information Service (Japan) 25: 5-7.

Tyagi, B.K., 1978b. New dragonfly records from India and the Dehradun Valley (Northern India). Current Science 47 (18): 698.

Tyagi, B.K., 1978c. Studies on the chromosomes of Odonata of Dun Valley (Dehra Dun, India). Ph.D. Thesis, Garhwal Univ., Srinagar, Garhwal. VIII+285 pp. (17 Tabs, 335 Figs.excl.).

Tyagi, B.K., 1982. Notes on some dragonflies from the Dehra Dun Valley (Uttar Pradesh, India). Fraseria 3: 9-10.

Tyagi, B. K., 1985. Dragonfly conservation in India: present status and future prospects. Rep. Odon. Specialist Group Int. Un. Conserv., No.6: 1-5.

Tyagi, B.K., 1986. Cytogenetics, karyosystematics and cytophylogeny of the Indian odonata. Indian Review of Life Sciences 6: 215-219.

Tyagi, B.K., 1987. Review of the cytotaxonomy of Isoptera (Insecta), with a description of the male germ cell chromosomes of Microcerotermes beesoni Snyder from the Dehradun Valley, India. Indian Review of Life Sciences 7: 263-272.

Tyagi, B.K., 1997. Dragonfly wealth of India: extant species – a preliminary systematic list. Zoos' Print 12: 3-10.

Tyagi, B.K., 2002. Malaria in the Thar Desert: Facts, Figures and Future. Agrobios (India) Jodhpur, pp. 162.

Tyagi, B.K., 2007 (Ed.). Odonata: Biology of dragonflies. Scientific Publishers (India), Jodhpur, pp. 323-366.

Tyagi, B.K. and Ajita Tyagi, 1981. A note on the asymmetrical antennae in some Collembola. Biologia 27 (1): 15-17.

Tyagi, B.K., A. Tyagi and S.K. Sangal, 1986. An up-to-date catalogue and bibliography of dragonfly fauna of the Dehradun valley (India) Occ. Publ. SIO Natl. Off. India 2:1-14.

Tyagi, B.K., P.C. Pandey, P.S. Rehill and P.K. Sen-Sarma, 1984. Termite fungi interactions. II. Laboratory testing of the decayed wood blocks to the wood carton nest building termite, Microcerotermes beesoni Snyder. Material und Organismen (Germany) 19: 69-75.

Tyagi, B.K., A.Munirathinam, and A. Venkatesh, 2015. A catalogue of Indian mosquitoes. Intl. J. Mosq. Res. 2(2): 50-97.

Varshney, R.K. 1998. Faunal diversity in India: Insecta. Edited by J.R.B. Alfred *et al.,* Zoological Survey of India, pp. 146-157.

Venkatesh, A. and Tyagi, B.K., 2013. Predatory potential of bradinopyga geminata and ceriagrion coromandelianum larvae on dengue vector aedes aegypti under controlled conditions (Anisoptera; libellulidae; zygpotera: coenagrionidae; diptera: culicidae). Odonatologica 42: 139.

Vijay Veer & B.K. Tyagi, 2008. Prof. P.K. Sen-Sarma, PhD, FNASc, FASc, FESI, FISWSc, FNAAS – A Termitologist and Forest Entomologist: Biography and contributions (1929-2002). [In: *Pests of Forest Importance and their Management.* Eds. B.K. Tyagi, Vijay Veer and Shri Prakash]. Scientific Publishers (Jodhpur), pp. 1-38.

Vijay Veer & Chauhan N. 1990. New species of genera Azaleothrips Ananthakrishnan and Gigantothrips Zimmermann (Thysanoptera: Tubulifera: Phlaeothripidae) from India. Annals of Entomology, 8: 65-70, 1990.

Vijay Veer & Chauhan Neena, 1992. A new species of Pseudodendrothrips Schmutz (Insecta: Terebrantia: Thirpidae) from Dehradun (India). Zoology (Journal of Pure and Applied Zoology) 2(3): 169–172.

Westfall, M.J. and M.L. May, 1996. Damselflies of North America. Scientific Publishers, Gainesville, FL.

Williams, C. B., 1927. A study of butterfly migration in south India and Ceylon, based largely on records by Messrs. G. Evershed, E. E. Green, J. C. F. Fryer and W. Ormiston. Transactions of the Royal Entomological Society of London 75 (1): 1–33.

Zhang, Z.Q. 2011. Animal biodiversity: An introduction to higher level classification and taxonomic richness. Zootaxa 3148:7-12.

ZSI, 2010. Fauna of Uttarakhand. Pt. 2. Insects. Zoological Surcey of India, State Fauna Series, 18, Kolkata.

Annex 1

CATALOGUES OF EXTANT TAXA OF SELECTED INSECT ORDERS FROM THE DOON VALLEY

1. ODONATA

Professor B. Kiauta (Department of Cytotaxonomy & Cytogenetics, Institute of Genetics, Utrecht University, Utrecht, The Netherlands and the Executive Editor of ODONATOLOGICA), in his inspirational communication in 1975 to Dr. B.K. Tyagi, had referred to the Doon Valley as "*...developing into a stronghold in odonatological research...* ." In the same spirit Dr. M.A. Lieftinck (Rhenen, The Netherlands), and one of the world's most celebrated odonate taxonomists, wrote dated the 20th August, 1977, ".... *The Dehra Dun is a region of great interest to a student of Odonata taxonomy and systematics.*' Dr. Lieftinck had later paid a visit to the Doon Valley's Sahastradhara (the Sulphur Springs) in early 1980s. Two other internationally acclaimed odonatologists, namely, Prof. J.M. van Brink (Editor, GENETICA) and Dr. Matti Hamalainen (Finland) visited the Doon Valley in 1986 in connection with participating in the 2nd Indian Symposium of Odonatology, held at the Zoology Department, DAV (PG) College, Dehra Dun (*Organizing Secretary*: Prof. S.K. Sangal; *Conference Coordinator*: Dr. B.K Tyagi, National Representative in-Charge, SIO National Office in India).

So far over seven dozen species and subspecies have been reported from the Doon Valley (which is about 18% of the total species occurring in India; *cf.* Prasad & Varshney, 1995, Tyagi, 1997). Many of these taxa are either rare or endemic to the Valley and are of great academic interest, such as *Onychogomphus garhwalicus* Singh & Baijal, 1954, *Agriocnemis corbeti* Kumar & Prasad, 1981 and *Calicnemia doonensis* Sangal & Tyagi, 1984 which have been described from the Doon Valley as new species. This inventory of the dragonfly species and subspecies has been principally adapted from Tyagi et al. (1986), which was an outcome of thorough literature review and consultation of different forms of "Types" deposited in the museums of FRI and ZSI, in particular, besides expert opinion in case of certain ambiguous taxa. Incidentally Tyagi et al. (1986) also compiled the entire odonatological literature from the Doon Valley, hence served a good reference for the beginners in the order Odonata.

ORDER ODONATA

Suborder Zygoptera

FAMILY PLATYSTICTIDAE

- *Drepanosticta* spec. indet. (not *charmichaeli*)

FAMILY PROTONEURIDAE

- *Caconeura autumnalis* (Fraser)

FAMILY PLATYCNEMIDIDAE

- *Copera marginipes* (Rambur)
- *C. vittatus* Selys
- *Calicnemia doonensis* Sangal & Tyagi
- *C. eximia* (Selys) (*cf.* Hamalainen, 1989)

FAMILY COENAGRIONIDAE

- *Agriocnemis clauseni* (Fraser)
- *A. corbeti* Kumar & Prasad
- *A. pygmaea* (Rambur)
- *Ceriagrion cerenorubellum* (Brauer)
- *C. coromandellianum* (Fabricius)
- *Enallagma parvum* Selys
- *Ischnura a. arora* Brauer
- *I. forcipata* Morton
- *I. senegalensis* (Rambur)
- *Pseudagrion decorum* Rambur
- *P. laidlawi* Fraser
- *P. rubriceps* Selys
- *P. spencei* Fraser

FAMILY LESTIDAE

- *Lestes p. praemrosa* Selys
- *L. thoracica* Laidlaw
- *L. viridulus* Rambur

FAMILY CHLOROLESTIDAE

- *Megalestes major* Selys

FAMILY CHLOROCYPHIDAE

- *Rhinocypha bifasciata beesoni* Fraser
- *R. quadrimaculata* Selys
- *R. trifasciata* Selys
- *R. unimaculata* Selys

FAMILY EUPHEIDAE

- *Anisopleura comes* Hagen
- *A. lestoides* Selys
- *Bayadera indica* Selys

FAMILY CALOPTERYGIDAE

- *Neurobasis c. chinensis* (Linnaeus)

Suborder Anisoptera

FAMILY GOMPHIDAE

- *Anisogomphus occipitalis* Selys
- *Anormogomphus heteroptera* Selys
- *A. kritshenkoi* Bartenef
- *Burmagomphus basimaricus* Fraser
- *B. pyramidalis* Laidlaw
- *B. siwalikensis* Laidlaw
- *Burmagomphus* cf. *arboreus* Lieftinck
- *Ictinogomphus rapax* Rambur
- *Onychogomphus bistrigatus* (Selys)
- *O. cerastis* Selys
- *O. garhwalicus* Singh & Baijal (According to van Tol J., *cf.* GBIF, this name was made a synonym of *Onychogomphus bistrigatus* Hagen in Selys, 1854)
- *O. m-flavus* Selys
- *O. modestus* Selys
- *O. risi* Fraser
- *O. saundersi duaricus* Fraser
- *O. schmidti* Fraser
- *Paragomphus l. lineatus* (Selys)

FAMILY AESHNIDAE

- *Anax guttatus* Burmeister
- *A. immaculifrons* Rambur
- *A. imperator* Leach
- *Gynacantha hyalina* Fraser
- *Gynacanthaeschna sikkima* (Karsch)

FAMILY CORDULEGASTERIDAE

- *Cordulegaster brevistigma* Selys

FAMILY CORDULIIDAE

- *Macromia m. moorei* Selys

FAMILY LIBELLULIDAE

- *Acisoma p. panorpoides* Rambur
- *Brachythemis contaminata* Fabricius
- *Cratilla l. lineata* Brauer
- *Crocothemis erythraea* Brulle
- *C. s. servillia* Drury
- *Crocothemis servillia* subspec. indet.

- *Diplacodes nebulosa* Fabricius
- *D. trivialis* (Rambur)
- *Neurothemis fulvia* Drury
- *N. i. intermedia* Rambur
- *N. t. tullia* (Drury)
- *Palpopleura s. sexmaculata* (Fabricius)
- *Pantala flavescens* (Fabricius)
- *Potamarcha obscura* Rambur
- *Rhyothemis v. variegata* Linnaeus
- *Sympetrum commixtum* (Selys)
- *Tetrathemis* spec. indet. (not *platyptera*)
- *Tholymis tillarga* Fabricius
- *Tramea basilaris burmeisteri* Kirby
- *T. virginea* Rambur
- *Trithemis aurora* (Burmeister)
- *T. festiva* (Rambur)
- *T. k. kirbyi* Selys
- *T. pallidinervis* Kirby
- *Zygonyx torrida* subspec. indet. (not *isis*)
- *Zyxomma petiolatum* Rambur
- *Orthetrum brunneum brunneum* Fonscolombe
- *Orthetrum chandrabali* Mehrotra (synonymised with *O. triangulare* (Selys, 1878). Ref: Mitra, 2002).
- *Orthetrum ganeshii* Mehrotra (synonymised with *O. triangulare* (Selys, 1878). Ref: Mitra, 2002).
- *Orthetrum garhwalicum* (synonymized with *Orthetrum taeniolatum*; *Ref.*- Mitra, A. 2010. *Orthetrum taeniolatum*. In: IUCN 2012. IUCN Red List of Threatened Species. Version 2012.2. [1].
- *Orthetrum glaucum* (Brauer)
- *Orthetrum guptai* Baijal & Singh (Synonymised with *Orthetrum cancellatum* (Linnaeus, 1758). Ref: www.odonata.info)
- *Orthetrum japonicum internum* McLachlan
- *O. luzonicum* (Brauer)
- *Orthetrum pruinosum neglectum* (Rambur)
- *Orthetrum s. sabina* (Drury)
- *Orthetrum taeniolatum* (Schneider)
- *Orthetrum t. triangulare* (Selys)

2. ORDER THYSANOPTERA

Annotated checklist of Thrips or Thysanoptera (Insecta) from Dehra Dun Valley, Uttarakhand (India), with description of new genus *Kulshresthothrips*

NEENA CHAUHAN and VIJAY VEER*

Forest Entomology Division, Forest Research Institute, Dehra Dun -248006, India.
E-mail: Neenachauhan2015@gmail.com
*Defence Research Laboratory, Tezpur-784001, Assam E-mail: vijayveer50@yahoo.com

INTRODUCTION

The thrips or thysanopterans are minute insects of 0.5 to 2.0 mm in body length. Both adults and larvae thrips are characterized by having asymmetrical piercing and sucking mouth parts where right mandible is vestigial, an eversible sac like 'arolium' at the end of each leg and adults with fringed wings when present. Currently more than 5822 species are known worldwide in two suborders, nine families and 779 genera. (anic,ento,csiro.au/thrips accessed on 1.4.2016). They exhibit wide range of biologies with a life history that is intermediate between those of hemi- and the holometabola, and a haplodiploid sex control system. Many species exhibit complex behavioural pattern including male swarming, leking, fighting, predatory, eusociality and kleptoparasitism.

Thrips being minute and having conservative association with their niches usually escape the notice of the general collector. Their common sites of occurrence are tender and succulent plant parts like buds, profusion of flowers and leaves, leaf-sheaths, grass tussocks, under the bark of living and dead trees, leaf litter, plant galls, etc. Majority of them are phytophagus and saprophagus. Some are predatory on mites, thrips, etc. and one each species is aquatic and ecotparasitic. Saprophagus thrips representing about 40% of thrips species; they feed mostly on hyphae but a few on spores of fungi growing on dead branches, leaf litter, etc. Phytophagus thrips feed in flowers and on leaves and some of those induce galls on leaves and stem. More than 50 species are serious pests of agricultural, horticultural and forestry plants causing feeding damage and vectoring tospoviruses, bacterial and fungal pathogens and inflicting huge loss in crop production. Bioecological studies on some important thrips species in Doon valley have been made by Vijay Veer (1980, 1986) and Srivastava *et al.* (1982, 1986). Notes on pest species of thrips in Dehradun valley is given by Chauhan & Vijay Veer (2007).

The first species of thrips recorded from the valley was a tubuliferan, *Leeuwenia coriacea* by Bagnall in 1912 and in 1919 he described another thrips species, *Aeolothrips fulvicollis*. Later Bagnall, Moulton and Ramakishna Ayyar have added 19 species of thrips from this region (Bagnall, 1921, 1923, 1924, 1926; Moulton, 1927, 1928, 1929, 1933; Ramakishna, 1928). However, Ramakrishna and Margabandhu (1940) in their catalogue on Indian Thysanoptera have included only 15 species instead of 20 known species at that time from the valley. Those five tubuliferan species not included inadvertently are *Liothrips flavitibia* Moulton, 1929, *Liothrips litseae* Moulton, 1933, *Liothrips gardneri* Moulton, 1933 (now known as *Gynaikothrips gardneri* (Moulton)), *Androthrips flavitibia* Moulton, 1933 and *Gynaikothrips caseariae* Moulton, 1933 (now known as *G. flaviantennatus* Moulton, 1929).

Shumsher Singh (1946) has described a new species, *Taeniothrips rhopalantennatis* which was later synonymised with *Thrips flavus* Schrank by Bhatti (1970). Ananthakrishnan (1960, 1962, 1972), Ananthakrishnan and Jagadish (1968), Bhatti (1968, 1970, 1972, 1973, 1978, 1980) and Kulshrestha and Vijay Veer (1980) have recorded 15 species more including several new species. Thus by the end of 1980, 35 species of thrips were known form the valley. However Anathakrishnan and Sen (1980) have included 23 species from this region in their book on Indian Thysanoptera. Subsequently 24 species includes some new species were added to the thrips fauna of the valley by Vijay Veer and Srivastava (1981), Kulshrestha *et al.* (1982, Kulshrestha and Vijay Veer (1983, 1984, 1990, 2010), Bhatti *et al.* (1993, 1998), Chauhan and Vijay Veer (1991), Chauhan (2010).

The Dehra Dun Valley bordered by the hill ranges of Himalaya in the north and by the Siwalik Hill in the south, provides an interesting ground for study of thrips as it enjoys varied climates ranging from temperate to tropical, and has rich and diverse vegetation, both native and exotic plants. About 1230 species of plants belonging to 115 families are known from the Dun valley including 239 exotic plant species (Babu, 1977). Therefore thrips collecting was carried out by the present authors during 1977-1990 and added 75 species of thrips to the thrips fauna of the valley and now total number of thrips raises to 169 (Terebrantia-117 spp. and Tubulifera-52 spp.). The all species reported in the list below are based on mainly collection done by the present authors, except 20 species which we could not become able to collect. But thrips fauna of Dehra Dun Valley is undoubtedly far more rich than reported in this list and therefore systematic collecting should be done so that the biodiversity of thrips in the valley may be more fully understood.

Kulshresthothrips Chauhan and Vijay Veer gen. nov.

Aeolothrips moundi Kulshrestha & Vijay Veer, 1984. *Bull. Entomol.*, 25(11): 33.

Type species *Aeolothrips moundi* Kulshrestha & Vijay Veer.

Habitus of *Coleothrps*. Head broader than long, ventrally strongly asymmetrical; dorsum of head with numerous setae; one pair of interocellar setae little longer. Maxillary palpi 3-segmented; labial palpi 4-segmented. Antennae 9-segmented; III and IV with short, linear sensory areas, on IV sensory area claviform and curving around the apex of the segment; V to IX segments connate and forming a distinct compact unit

Pronotum longer than broad with numerous recurved small setae, posteromarginal setae slightly longer; mesonotum transversely striate with well developed setae, median pair of setae placed far ahead of posterior margin. Metanotum reticulate, reticules without internal markings, median pair of setae inserted near to the anterior margin. Ferna entire and broad; spinula present on both pterygosterna.

Forewing broad, parallel sided with slight constriction at submedian region, rounded apicallly, with two grey cross bands and distal tip shaded grey, first band at medially and smaller in size, 4 cross veins. Scale with 12-17 venal setae. Fore tarsi with well developed recurved tooth and hook on inner margin of distal segment.

Abdomen vespiform in female, sternites II to VI without accessory setae, VII with two pairs of accessory setae and a pair of lateral accessory setae. Posteromarginal setae on sternite II with two pairs, III to VIII with 3 pairs. Abdominal tergite I in male long and bearing two longitudinal ridges, tergites II to VII without tubercles or appendices, sternites without glandular areas, terminal abdominal segments without claspers.

Remarks: *Kulshresthothrips* **gen. nov**. comes close to the Australian genus *Desmothrips* Hood in having lateral accessory setae on abdominal sternite VII; tips of the fore wing shaded; metanotum reticulated. But it is readily separated from *Desmothrips* by the presence of 12-17 venal setae on scale of fore wing. It also comes close to the *Gelothrips* Bhatti known from India and Australia which has shaded tip of fore wing and acessoroy setae on the sternite VII and males without claspers. However it is readily separated from *Gelothrips* on basis of not having arcute sculture on metanotum; microtrichia along striae on dorsum of head and by having antennal segments V to IX connate.

It seems that *Kulshresthothrips* gen. nov. is more primitive than the genus *Coleothrips* and it appears that it has evolved from *Desmothrips* alongwith *Andrewarthia* and *Gelothrips*.

1. Suborder TEREBRANTIA

1. Family AEOLOTHRIPIDAE

   1. *Aduncothrips asiaticus* (Ramakrishna Ayyar & Margabandhu, 1931)
   2. *Aeolothrips collaris* (Priesner, 1919)
   3. *Aeolothrips distinctus* Bhatti, 1971
   4. *Aeolothrips meridionalis* (Priesner, 1948)
   5. *Franklinothrips megalops* (Trybom, 1912)
   6. *Franklinothrips uttarakhandiensis* Vijay Veer, 2010
   7. *Indothrips bhusani* Bhatti, 1967
   8. *Kulshresthothrips moundi* (Kulshrestha & Vijay Veer, 1984) **n. com.**
   9. *Streothrips arorai* Bhatti, 1967
   10. *Streothrips anshumani* Chauhan, 2010

2. Family STENUROTHRIPIDAE

   11. *Holarthrothrips indicus* Bhatti & Ananthakrishnan, 1978

3. Family THRIPIDAE

Sub family Panchaetopripinae

12. *Astrothrips globiceps* (Karny, 1913)
13. *Astrothrips stannardi* Bhatti, 1967
14. *Astrothrips tumiceps* Karny, 1923
15. *Caliothrips indicus* (Bagnall, 1913)
16. *Helionothrips kadaliphilus* (Ramakrishna & Margabandhu, 1931)
17. *Helionothrips parvus* Bhatti, 1968
18. *Monilothrips kempi* Moulton, 1929
19. *Panchaetothrips indicus* Bagnall, 1912.
20. *Phibalothrips peringueyi* (Faure, 1925)
21. *Rhipiphorothrips cruentatus* Hood, 1919
22. *Selenothrips rubrocinctus* (Giard, 1901)
23. *Zaniothrips recini* Bhatti, 1967

2. Subfamily Thripininae

1. Tribe Dendrothripini

24. *Asprothrips indicus* (Bagnall, 1919)
25. *Dendrothrips cameroni* Priesner, 1965
26. *Dendrothrips sexmaculatus* Bagnall, 1916
27. *Dendrothrips strasseni* Bhatti, 1971
28. *Parsiothrips fuscus* Bhatti, 1970
29. *Pseudodendrothrips kulshresthai* Chauhan & Vijay Veer, 1992

2. Tribe Sericothripina

30. *Hydatothrips boerhaaviae* (Seshadri & Ananthakrishnan, 1954)
31. *Hydatothrips proximus* Bhatti, 1973
32. *Hydatothrips ramaswamiahi* Priesner, 1926
33. *Neohydatothrips smayunkur* (Kudo, 1995)

3. Tribe Scirtothripini

1. Subtribe Scirtothripina

34. *Ajothrips gara* Bhatti, 1967
35. *Ajothrips karma* Bhatti, 1967
36. *Scirtothrips bispinosus* (Bagnall, 1924)
37. *Scirtothrips dorsalis* Hood, 1919
38. *Scirtothrips oligochaetus* (Karny, 1926)

4. Tribe Chirothripini

39. *Arorathrips mexicanus* Craford, 1909 (syn. *Chirothrips mexicanus* Craford, 1909
40. *Bregmatothrips binervis* (Kobus, 1893)
41. *Chirothrips meridionalis* Bagnall, 1927
42. *Sciothrips cardamomi* (Ramakrishna, 1935)
43. *Sorghothrips jonnaphilus* (Ramakrishna, 1928)

5. Tribe Thripini

1. Subtribe Anaphothripina

44. *Anaphothrips doonensis* Chauhan & Vijay Veer, 1992
45. *Anaphothrips latis* Bhatti, 1967
46. *Anaphothrips obscurus* (Muller, 1776)
47. *Anaphothrips sudanensis* Trybom, 1911
48. *Aptinothrips rufus* (Haliday, 1836)
49. *Aptinothrips stylifer* Trybom, 1894
50. *Dendrothripoides innoxius* (Karny, 1914)
51. *Exothrips hemavarna* (Ramakrishna & Margabandhu, 1931)
52. *Rhamphothrips aureus* (Ananthakrishnan, 1954)
53. *Rhamphothrips parviceps* (Hood, 1919)
54. *Rhamphothrips santokhi* Kulshrestha & Vijay Veer, 1984
55. *Salpingothrips hoodi* Ananthakrishnan, 1969

2. Subtribe Thripina

56. *Ayyaria chaetophora* Karny, 1926
57. *Bathrips jasmine* Ananthakrishnan, 1968
58. *Bathrips melanicornis* (Shumsher, 1946)
59. *Bolacothrips indicus* (Ananthakrishnan, 1966)
60. *Craspedothrips minor* (Bagnall, 1921)
61. *Ctenothrips smilax* Bhatti, 1976
62. *Dichromothrips indicus* Mound, 1976
63. *Dichromothrips phalaenopsidis* Sakimura, 1955
64. *Dichromothrips smithi* (Zimmermann, 1900)
65. *Doonthrips sector* Bhatti, Vijay Veer & Chauhan, 1994
66. *Dorcadothrips albus* Bhatti, 1978
67. *Ernothrips lobatus* Bhatti, 1967
68. *Euphysothrips subramanii* (Ramakrishna & Margabandhu, 1939)
69. *Florithrips traegardhi* (Trybom, 1911)
70. *Frankliniella shultzei* (Trybom, 1910)
71. *Frankliniella intonsa* (Trybom, 1895)
72. *Fulmekiola serrata* Kobus, 1893
73. *Laplothrips bicolor* Bhatti, 1972
74. *Lefroyothrips lefroyi* (Bagnall, 1913)
75. *Megalurothrips distalis* (Karny, 1913)
76. *Megalurothrips peculiaris* (Bagnall, 1918)
77. *Megalurothrips usitatus* (Bagnall, 1913)
78. *Microcephalothrips abdominalis* (Crawford DL, 1910)
79. *Mycterothrips nilgiriensis* (Ananthakrishnan, 1960)
80. *Mycterothrips ricini* (Shumsher, 1946)
81. *Mycterothrips setiventris* (Bagnall, 1918)

82. *Organothrips indicus* Bhatti, 1974
83. *Oxythrips indicus* Bhatti, 1967
84. *Plutonomothrips cus* (Bhatti, 1967)
85. *Scolothrips rhagenianus* Priesner, 1950 (syn. S. *indicus* Priesner, 1950)
86. *Stenchaetothrips biformis* (Bagnall, 1913)
87. *Stenchaetothrips dissidens* (Ananthakrishnan & Jagadish, 1967)
88. *Stenchaetothrips divisae* Bhatti,1982
89. *Stenchaetothrips faurei* (Bhatti, 1962)
90. *Stenchaetothrips indicus* (Ramakirshna & Margabandhu, 1931)
91. *Stenchaetothrips. garhwalensis* sp nov
92. *Stenchaetothrips minutus* (Deventer, 1906)
93. *Taeniothrips major* Bagnall,1916
94. *Taeniothrips jitendraveeri* sp nov
95. *Taeniothrips orionis* Treherne,1924
96. *Trichromothrips nilgiricus* (Ramakrishna & Margabandhu, 1939)
97. *Thrips alatus* Bhatti, 1980
98. *Thrips andrewsi* (Bagnall, 1921)
99. *Thrips apicatus* Priesner, 1934
100. *Thrips apicatus* Priesner, 1934
101. *Thrips arorai* Bhatti, 1980
102. *Thrips carthami* Shumsher, 1946
103. *Thrips coloratus* coloratus Schmutz, 1913
104. *Thrips flavidulus* (Bagnall, 1923)
105. *Thrips flavus* Scharank, 1776
106. *Thrips florum* Schmutz, 1913
107. *Thrips formosanus* Priesner, 1934
108. *Thrips hawaiiensis* (Morgan, 1913)
109. *Thrips orientalis* (Bagnall, 1915)
110. *Thrips palmi* Karny, 1925
111. *Thrips subnudula* (Karny, 1926)
112. *Thrips sumatrensis* Priesner, 1934
113. *Thrips tabaci* Lindemann, 1889
114. *Thrips tanicus* Bhatti, 1970
115. *Thrips taurus* Bhatti, 1980
116. *Thrips tectus* zur Strassen, 1996
117. *Tusothrips sumatrensis* (Karny, 1925) (syn, *seudosetiprivus* Ramakrishna & Margabandhu, 1939)

2. Suborder TUBULIFERA

1. Family PHLAEOTHRIPIDAE
1. Sub Family Idolothripinae

118. *Aesthesiothrips jatrophae* Ananthakrishnan, 1961
119. *Allothrips bicolor* Ananthakrishnan, 1964
120. *Dinothrips spinosus* (Schmutz, 1913)
121. *Elaphrothrips procer* (Schmutz, 1913)
122. *Ethirothrips beesoni* (Moulton, 1928) (= *Dichaetothrips beesoni* Moulton)
123. *Ethirothrips indicus* (Baganll, 1921) (=*Mesothrips indicus* Bagnall)
124. *Machatothrips silvaticus* Ananthakrishnan, 1972

2. Subfamily Phlaeothripinae

125. *Andothrips flavitibia* Moulton, 1933
126. *Areothrips longisetis* Ananthakrishnan, 1976
127. *Athlibothrips inquilinus* (Ananthakrishnan & Varadrasan, 1978)
128. *Austrothrips cochinchinensis* Karny, 1922
129. *Azaleothrips amabilis* Ananthakrishnan, 1964
130. *Azaleothrips bhattii* Vijay Veer & Chauhan, 1990
131. *Bamboosiella bicoloripes* Ananthakrishnan, 1957
132. *Dolicothrips indicus* (Hood, 1919)
133. *Ecacanthothrips tibialis* (Ashmead, 1905)
134. *Gigantothrips gardneri* Ananthakrishnan, 1960
135. *Gigantothrips himalayanus* Vijay Veer & Chauhan, 1990
136. *Gigantothrips tibialis* (Bagnall, 1921)
137. *Gynaikothrips flaviantennatus* Moulton, 1929
138. *Gynaikothrips gardneri* (Moulton, 1933)
139. *Haplothrips ceylonicus* Schmutz, 1913
140. *Haplothrips ganglbaueri* Schmutz, 1913
141. *Haplothrips gowdeyi* (Franklin, 1908)
142. *Haplothrips reuteri* (Karny, 1907)
143. *Haplothrips tenuipennis* Bagnall, 1918
144. *Holothrips stannardi* (Ananthakrishnan, 1972)
145. *? Hoplandrothrips natalensis* (Trybom, 1912)
146. *Leeuwenia coriacea* (Bagnall, 1912)
147. *Leeuwenia eugeniae* Bagnall, 1924
148. *Liophlaeothrips ablusus* Ananthakrishnan & Jagadish, 1971
149. *Liothrips associatus* Ananthakrishnan & Jagadish, 1969
150. *Liothrips bosei* Moulton, 1928
151. *Liothrips brevitubus* Karny, 1912
152. *Liothrips flavitibia* Moulton, 1933
153. *Liothrips gracilis* Karny, 1913
154. *Liothrips himalayanus* Ananthakrishnan & Jagadish,1970
155. *Liothrips interlocatus* (Karny, 1926)
156. *Liothrips litseae* Moulton, 1933

157. *Liothrips morulus* Ananthakrishnan & Jagadish, 1970
158. *Liothrips siwalicus* n. sp.
159. *Liothrips sangali* Kulshrestha & Vijay Veer, 1990
160. *Mallothrips indica* Ramakrishna, 1928
161. *Mesothrips malloti* Moulton, 1929
162. *Pegothrips meghalaya* Sen & Muraleedharan, 1976
163. *Phenicothrips eugeniae* Priesner, 1930
164. *Phorinothrips loranthi* Ananthakrishnan, 1968
165. *Plicothrips apicalis* (Bagnall, 1915)
166. *Podothrips bicolor* Seshadri & Ananthakrishnan, 1954
167. *Xylaplothrips flavitibia* Ananthakrishnan & Jagadish, 1969
168. *Xylaplothrips pictipes* (Bagnall, 1919)
169. *Xylaplothrips tener* Ananthakrishnan & Jagadish, 1969

Acknowlegements

The authors express their sincere thanks to Prof. J.S. Bhatti, Delhi University, Delhi for very kindly confirming the identification of several thrips species and for providing valuable information on thrips. The first author is highly thankful to the Director, Forest Research Institute & Colleges, Dehra Dun for facilities.

References

Ananthakrishnan, T.N. 1960 New Tubuliferan Thysanoptera from India. *J. Zool. Soc. India* 12: 250-258.

Ananthakrishnan, T.N. 1962. Some little known Indian Terebrantia. *Proc. R. ent Soc. London* 318 (7-8): 87-91.

Ananthakrishnan, T.N. 1972. Mycophagous Thysanoptera IV. *Oriental Ins.*(4): 425- 437.

Ananthakrishnan, T.N. and Jagadish, A. 1968. Gall inhabiting *Liothrips* Uzel from India. Oriental Ins., 2(2); 205-217.

Ananthakrishnan, T.N., 1979. Biosystematics of Thysanoptera. Annual Review of Entomology 24: 159-183

Ananthakrishnan, T.N. and Sen, S. 1980. Taxonomy of Indian Thysanoptera. Zoological Survey of India, Calcutta. Hand book Series No.,1, 234 pp.

Babu, C.R. Herbaceous flora of Dehradun. Publications & Information Directorate (CSIR), New Delhi. 721 pp.

Bagnall, R.S. 1912 Preliminary description of three new species of Thysanoptera. Ann. Mag. nat. Hist., (8)9:214- 217.

Bagnall, R.S. 1915. Brief descriptions of new Thysanoptera V. Ann. Mag. nat. Hist., (8):15:588-597.

Bagnall, R.S. 1919. Brief descriptions of new Thysanoptera X. Ann. Mag. nat. Hist., (9) 4: 253-277.

Bagnall, R.S. 1921. Brief descriptionsof new Thysanoptera XII. Ann. Mag. nat. Hist., (9)8:393-400.

Bagnall, R.S. 1923. Brief descriptions of new Thysanoptera XIII. Ann. Mag. nat. Hist., (9)12: 624-631.

Bagnall, R.S. 1924. Brief descriptions of new Thysanoptera XIV. Ann. Mag. nat. Hist., (9) 14: 625-640.

Bagnall, R.S. 1926. Brief descriptions of new Thysanoptera XV. Ann. Mag. nat. Hist., (9): 18: 98-114.

Bhatti, J.S. 1968. The genus Helionothrips Bagnall in India (Thysanoptera) Oriental Ins. 2(1): 35-39.

Bhatti, J.S. 1970. Taxonomic studies in some Thripini (Thysanoptera: Thripidae) Oriental Ins. 3(4): 373-382.

Bhatti, J.S. 1970. A new genus Parsiothrips of the Tribe Dendrothripini (Thysanoptera: Thripidae) Oriental Insects. 4(2):205-206.

Bhatti, J.S. 1972. A review of the Indian species of *Caliothrips* with notes on other Heliothripinae. Oriental Insects, 6 (2):217-228.

Bhatti, J. S. 1973. A preliminary revision of *Sericothrips* Haliday sensu let and related genera with a revised concept of the tribe Sericothripini (Thysanoptera:Thripidae). Oriental Insects, 7(3):403-444.

Bhatti, J.S. 1978. Four new species of Thysanoptera from India. Oriental Insects, 12(3): 419-432.

Bhatti, J.S. 1980. Studies on the genus *Thrips* from India (Thysnoptera). Systematic Entomology, 5: 109-166.

Bhatti, J.S. 1990. Catalogue of insects of the order Terebrantia from the Indian Subregion. Zoology, 2(4):205-352.

Bhatti, J.S., Tyagi, K. & Kumar, 2003. Variation in population of *Plicothrips apicalis* on *Cynodon dactylon*: 672-673. Proc. National Symposium on Frontier areas of entomological research, IARI, New Delhi.

Bhatti, J. S., Varatharajan, R., Vareichung, S.C., Kumar, V. & Tyagi, K. 2003. The exotic species *Neohydatothrips samayunkur* (Kudo, 1995) (Terebrantia: Sericothripidae) is now discovered from 6 states in India: 666-667. Proc. National Symposium on Frontier areas of entomological research, IARI, New Delhi.

Bhatti, J.S., Vijay Veer and Negi, B.K. 1998. Discovery of the natural habitat of the only fully aquatic Thysanopteran, *Organothrips indicus* (Terebrantia: Thripidae). Oriental Insects, 32: 259-266.

Bhatti, J.S., Vijay Veer and Chauhan, Neena. 1993. *Doonthrips setor*, a new genus and species of Thripidae (Insecta: Terebrantia) from Dehradun (India). In Advances in Thysanopterology (ed. J.S. Bhatti). Zoology, 4: 177-192.

Chauhan, Neena and Vijay Veer. 1992. A new species of *Pseudodendrothrips* Schmutz (Insecta: Terebrantia: Thripidae) from Dehradun (India). Zoology 3(2):169-172.

Chauhan, Neena and Vijay Veer 1992. *Anaphothrips doonensis*, a new species from the tussock of plant, *Typha elephantiana* (Family Typhaceae) from Dehradun (India) (Thysanoptera: Terebrantia: Thripidae). Annals of Entomology 10 (1): 41-44.

Kulshrestha S.K., Srivastava, A.K. and Vijay Veer. 1982. Some observations on the relative abundance and sex-ratio of the thrips species (Thysanoptera: Insecta) infesting *Oryza sativa*, *Saccharum officinarum* and *Triticum vulagare* in Doon Valley. Indian Journal of Forestry, 6 (3): 211-215.

Kulshrestha, S.K. Srivastava, A.K., Nishi Kapoor and Negi, B.K. 1986. On the relative abundance and interspecific competion among different Thrips species (Thysanoptera: Insecta) infesting flowers of *Tagetes erecta* L. in the Doon valley. Indian J. Forestry, 9 (3): 253-257.

Kulshrestha, S.K. and Vijay Veer. 1980. Studies on the population ecology of *Thrips coloratus* Schmutz (thysanoptera : Insecta) from Doon Valley. Symposium on Ecology and Environment Science, Jaipur, 18-20 Feb, 1980.

Kulshrestha, S.K. ad Vijay Veer. 1983. Systematics of undescribed males of Thysanoptera from Dehradun. J. Bombay nat. Hist. Soc. 80 (1) :240-243.

Kulshrestha, S.K. and Vijay Veer. 1984 Two new species of Thysanoptera from Dehradun. Bull. Entomol., 25 (1):33-37.

Kulshrestha, S.K. and Vijay Veer. 1990. A new species of *Liothrips* Uzel from Dehradun (India) (Insecta: Thysanoptera: Phlaeothripidae). Ann. Entomol., 8(1) 55-60.

Mani, M.S., 1968. Ecology and Biogeography of High Altitude Insects. Series Entomologica IV. Springer Netherlands, pp. xiv+527.

Moulton, D. 1927. Thysanoptera- New species and notes. Bull. Brooklyn ent Soc. 22:181-201.

Moulton, D. 1928. New Thysanoptera from India. Indian forest Records, 13:285.

Moulton, D. 1929. Thysanoptera from India. Rec. Indian Mus., 31:93-100.

Moulton, D. 1933. New Thysanoptera from India. Indian For. Rec., 19 :1-6.

Chauhan, N & Vijay Veer, 1992. Anaphothrips doonensis new species from the tussock of plant, Typha elephantiana (Family- Typhaeceae) from Dehradun (Thysanoptera: Terebrantia: Thripidae). Annals of Entomology 10(1): 41-44, 1992.

Ramakrishna Ayyar, T.V. 1928. A contribution to our knowledge of the Thysanoptera of India. Mem Dept Agric. India (Ent.), 10(7): 217-316.

Ramakrishna Ayyar, T.V. and Margabandhu, V. 1940. Cataloque of Indian Insects: Pt. 25 Thysanoptera. Govt. of India press, New Delhi, 64 pp.

Shumsher Singh. 1946. Studies on the systematics of Indian Thysanoptera: Terebrantia. Indian J. Entomol., 7 :147-188. (1945).

Srivastava, A.K. and Vijay Veer. 1983. Preliminary observation on the life history of Lefroyothrips lefroyi (Bagnall) (Thysanoptera:Thripidae). Indian J. Forestry, 6(1): 84-85.

Vijay Veer. 1978. *Thrips flavus* Schrank as pollinating agent. Indian Journal of Forestry, 1 (4): 278.

Vijay Veer. 1979. A new record of internal parasite of *Gynaikothrips flaviantennatus* Moulton from India. Current Science, 48(12): 558

Vijay Veer, 1980. Studies on the ecology of two species of genus *Thrips* (Thripidae: Terebrantia: Thysanoptera) from Dehra Dun. Garhwal University, Srinagar, Garhwal, 200 pp

Vijay Veer, 1985. Observations on the host preferences and biology of *Thrips flavus* Schrank (Thysanoptera: Thripidae) from Dehra Dun, India. Annals of Entomology, 3(1); 39-48.

Vijay Veer 2010. A new species of *Franklinothrips* (Thysanoptera: Terebrantia: Aeolothripidae) from Uttarakhand, India. Annals of Entomology, 28(2): 7-9.

Vijay Veer and Srivastava, A.K. 1981. *Ajothrips gara* Bhatti damaging leaves of *Mangifera indica* (mango) from Dehradun. Indian Journal of Forestry, 4(1):66.

Vijay Veer and Chauhan, Neena.1983. A study on the genus *Arathrips* Bhatti (Thysanoptera: Thripidae) from Dehradun (India). Entomon, 8(2):131-135.

Vijay Veer and Chauhan, Neena. 1990. New species of genera *Azaleothrips* Ananthakrishnan and *Gigantothrips* Zimmermann (Thysanoptera: Tubulifera : Phlaeothripidae) from India. Ann. Entomol., 8(2):

Chauhan, N. & Vijay Veer. 2007. Notes on diagnosis and bioecology of thrips pests of floral, fruit, vegetable, agronomic crops and forestry plants in the Dehra Dun valley (Uttaranchal), India: Chapter 10: 195-228. In: *Pests of forest importance and their management* (Edit. by Tyagi, B.K, Vijay Veer & Shri Prakash). Scientific Publishers (India), Jodhpur, xiv + 299 pp.

• • •

Annex 2

THE DOON VALLEY'S EMINENT ENTOMOLOGISTS

1. Dr. Edward Percy Stebbing

E.P. Stebbing
(Dec. 1900-1902 and Aug.1906-Dec., 1909)

Mr. Stebbing was the first Head of the Division. He worked as forest zoologist before the establishment of Forest Research Institute and initially served for two years from end of 1900 to 1902. He was appointed Imperial Forest Zoologist in April, 1906, and held the office until December, 1909. He worked on the biology and management of wood borers. Some of his contributions are: Bee-hole borer of teak in Burma (Myanmar), Quetta borer, Chilgoza bark-boring beetles of Zhob, life history of sal borer, duki-fig tree borer of Baluchistan, some insect pests of Assam sal, Blue pine bark borer, Sal bark borer, teak defoliator, teak skeletonizer, larger deodar bark-borer, blue pine bark borer, notes on the lac insects and insect pests of conifers of the Himalaya. Apart from biology and control for insects he also worked on the systematics of Scolytidae of economic importance. He also published two books "Manual of Forest Zoology for India" and "Indian Forest Insects of Economic Importance – Coleoptera" which are very important documents on forest zoology and entomology.

2. Shri V. Subramaniya Iyer

Mr. Iyer, PFS (1899-1913) was appointed temporary Forest Zoologist and held the post from January, 1910 till April, 1911. Dr. S. Maulik was his assistant in

1910 and 1911, who authored the volumes in the Fauna of British India on Chrysomelidae.

3. Dr. Agustus Danial Imms, FRS

A.D. Imms
(Nov. 1911- Feb., 1913)

Dr. Imms was graduate from Cambridge in London and awarded scholarship of an "1851 Research Scholarship" at Corpus Christi College, Cambridge, from 1905 to 1907. He worked on systematics of midge larvae before leaving for India to teach. He was appointed professor of biology in the University of Allahabad. In October, 1911, A.D. Imms was appointed as Forest Zoologist with one assistant, N. C. Chatterjee. Beeson in his famous book writes about Dr. Imms "..During his sixteen months' tenure of office he introduced valuable reforms; extensive additions were made to the library; contacts were made with entomological specialists in many countries and large collections were distributed to them. He designed an insectory and laboratories, which were erected, according to his plans after his departure. Dr. Imm's association with forest entomology at a time when a new Institute was projected and funds for expansion of entomological research had to be secured, was a most fortunate occurrence." It was unfortunate that Dr Imms headed this division only for a short time and due to health reasons (lifelong asthmatic) left India in 1913. After leaving India he accepted a post of reader in Agricultural Entomology under Professor S.J. Hickson at Manchester (1913-18); chief entomologist at Rothamsted (1918-31), FRS (1929), reader in entomology at Cambridge (1931-45), and Fellow of Downing College (1940). In 1924 Dr. Imms published his famous "Text-book of Entomology", and at once he became recognized as the leader of entomology in Great Britain. This excellent book has gone through three editions, and is almost as widely used in the United States as it is in Britain. "Recent Advances in Entomology" followed in 1931, and during the War Dr. Imms produced an admirable "Introduction to Entomology". The influence of these books has been immense, and will long continue. He studied lac cultivation in the Central Provinces and the pests of coniferous forests. During his short time at Forest Entomology Division he acted as an architect and gave direction for creating most needed facilities for forest entomology research in the country. Now the Entomology Library is the richest library in the country replete with old and rare entomological literature.

Knowledge seekers from all over the country visit and gain knowledge from this temple. Similarly the Insect Reference Collection now known as National Forest Insect Collection (NFIC) has become an asset and pride of the nation.

4. Dr. Cyril Frederick Cherrington Beeson, D.Sc.

C.F.C. Beeson
(Mar.1913- Aug., 1941)

Dr. Beeson served in the Indian Forest Service from 1911 - 1942. During the first year of his service he studied entomology under Prof. H. M. Lefroy in London and K. Escherich in Tharandt. He was a forest officer in the Punjab and was transferred from there to take over the post of Forest Zoologist in August, 1913. This post was renamed in 1922 as Forest Entomologist. He headed the division for the longest period, 27 years. During his time more posts were added: (a) three assistants S.N. Chatterjee (in 1917), B.M. Bhatia (in 1918) and G.D. Bhasin (in 1923) were appointed, and (b) three posts of entomologists, viz., Dr. M. Cameron (Dec., 1920-March, 1923), J.C.M. Gardner, who studied entomology under Lefroy and forestry at Cambridge replaced Cameron (in 1923), and D.J. Atkinson were filled. Other assistant entomologist, S.K. Pillai who had been trained in Germany, was appointed in December 1921 [first. Asstt. entomologist being S. Maulik (1910-1911) and author of volumes in the Fauna of British India on Chrysomelidae)].

During his time lot of research work and publications on forest entomological research was done in the country. All the entomologists and other supporting staff were engaged in exploration of forest insect and their management. Huge information on Indian forest insects were produced and published. He studied the life history, ecology and control of various insects like- Construction of calcarious opercula by longicorn larvae of Cerambicinae, life history of *Diapus furtivus*, larvae and life history of Prionine beetles, *Hypsipylla robusta*, *Duomitus ceramicus*, Indian Cleridae. Biology and new species of Xyleborus, *Celosterna scabrator*; Braconidae, Ichneumonidae, Tachinidae, Bostrychidae, Cerambycidae, etc. were described.

He was also involved in entomological investigations on the sandal spike disease (1933). He worked also on the systematics of borers and described new species of Crossotarsus, Coccotrypes and Thamnurgides. In 1941 he published a monumental work in form of a book entitled "The Ecology and Control of the Forest Insects of India" which is studied by all forest entomologists in the country and abroad. He returned to England and was the Director of the Imperial Forestry Bureau from 1945-1947. In his retirement he pursued his interest in horology (science of measuring time or making instruments for measuring time), and wrote two books and numerous papers on this subject. He died in 1975 at an age of 86 years.

5. Dr. J. C. M. Gardner

J.C.M. Gardner
(Sept. 1941- May, 1947)

Dr. Gardner served Indian Forest Service (1921), like Dr. Beeson studied entomology under Lefroy and forestry at Cambridge and became head of division after Beeson from 1941-1947. His field of research was the systematics and morphology of mostly immature stages of beetles, moths and also on pupa of the order Diptera. He also worked on insect borers of Bamboos and their control. His research contributions are: Indian Coleoptera (1926), immature stages of Indian Coleoptera; immature stages of families: Anthribidae, Platypodidae, Bostrichidae, Curculionidae, Scolytidae, Scarabaeidae, Eucnemidae, Brenthidae, Cerambycidae, Carabidae and Cleridae; list of described Indian Coleoptera; immature stages of Indian Lepidoptera of families Lymantridae,

Noctuidae and Hypsidae, Zygaenidae; puparia of some Indian Tachinidae. He also worked on Systematics of Niponiidae and Cerambycidae.

6. Shri A. H. Khan

Shri A.H. Khan was in-charge of the division for a very short period of just four months from June, 1947 to Sept., 1947. After partition of the country he migrated to Pakistan. He worked on insect borers of newly felled timber and their control.

7. Dr. Nibaran Chandra Chatterjee, D.Sc.

N.C. Chatterjee (Mar.1948-Nov., 1949)

Dr. Chatterjee was the first head of the division of independent India and served the division as head for almost one and a half year. His earlier working with the eminent entomologists and heads of the division, especially Dr. Imms under whom he was first appointed, was a guiding experience. He worked extensively on the entomological investigations on sandal spike disease. Some of his contributions on sandal spike disease are as follows: life history and morphology of *Sarima nigraclypeata*, *Euryrbachys tomentosa*, *Petalocephala negrilinea*. During this Entomological investigation on sandal spike disease he worked on the various families of insects associated with sandal and produced a series of publications as follows on - Membracidae and Cercopidae, Cicindelidae, Neuroptera and Elateridae, Coccinellidae, Fulgoridae, Pentatomidae, Lepidoptera, Coccidae, Chrysomelidae, Coreidae and Berytidae, Reduviidae, Dermaptera and Orthoptera, Lygaeidae and Heteroptera. Apart from sandal survey he also worked on insect borers of newly felled timber and their control and explored the possibilities of control of Lantana by indigenous insect pests.

8. Dr. Mithan Lal Roonwal, D. Sc.

M. L. Roonwal (Dec.1949-July, 1956)

Prof. Roonwal was born at Jodhpur in 1908. He had a brilliant academic career, did his M.Sc. (Hons) in 1930 from Lucknow University and Ph. D. from Cambridge University, U.K. in 1935. For his published research work in the field of Zoology, the Cambridge University also conferred upon him the D.Sc. Degree in 1962. Prof Roonwal excelled in his career spanning nearly three decades and held many important positions. He started his career as a research fellow in 1930 at Lucknow University, became research assistant (1931-33), Locust Research Laboratory ICAR, Punjab Agricultural University, Lyallpur; Asstt. Locust Entomologist and officer in-charge, Locust field station, Pasni Baluchanistan (1935-39); and officer in-charge, bird and mammal section, ZSI, Calcutta (1939-49).

During the II-World War he served in the army as a major attached to 15 Punjab Regiment, Southeast Asia command (Assam and Burma) and was awarded Burma Star and War Medals. Reverting back in peace time he served as Forest Entomologist from 1949-56 at FRI, Dehradun. He served at the division for about five and half years and worked on the forest insect pests and started work on termites. His major contributions are: Control of Indian book worm beetle. List of insect pests of forest plants in India and adjacent countries (arranged alphabetically according to plant name; efficacies of insecticides against timber borers, bamboo ghoon, anti-borer protection of Salai and Bijasal timbers, insect borers of felled timbers and their control, anti-termite characteristics of new synthetic fiber cloth "PAN" and festing of Chlorinated hydrocarbons for controlling mound building termites (1961). He also worked on tree poisoning experiments for controlling insects (1960), and water seasoning of Indian timbers for anti-insect protection. He also studied ultrastructures in the wings of termites and used them in identification and classification.

In 1956 he was selected as Director ZSI, Calcutta where he served from 1956-66. From ZSI he joined Department of Zoology Jodhpur University as Prof and head. He also became Vice Chancellor of Jodhpur University and served from 1966-1969. He also worked as emeritus scientist, CSIR from 1969-1971. After retirement, he continued his research at ZSI, Jodhpur till his death. He was awarded the Tata Gold medal of Zoological Society of India in 1956. Dr. Harswarup memorial lecture award in 1984; he was associated with many national and international agencies and organizations such as UNESCO, FAO, etc. He published more than 400 research papers, many books and nearly 102 reviews and popular articles on pure and applied zoology.

9. Dr. Ram Nath Mathur

R.N. Mathur (Aug.1956-Oct., 1961)

Dr. Mathur was appointed as assistant Entomologist in the year 1923 and was fortunate enough to have joined and worked with Drs. Beeson, Gardner, Chatterjee and Roonwal. During his long tenure of service he published 115 papers on the systematics, biology, ecology and control of forest insects. He was associated with listing of insect pests and their host plants and produced many volumes of lists of insect pests of forest plants in India and adjacent countries, arranged alphabetically according to plant name in 1960 and 1961 along with Mr. Balwant Singh. He also worked on clum curing of bamboos and their susceptibility to powderpost beetles and effect of curing. He explored insect pests of flowers, seeds and fruits of forest trees, pests of teak and their control.

His work on Sal heart wood borer and its control by Trap Tree Method is well known and practiced methodology. He worked on efficacies of insecticide against insect borers of freshly felled timbers. He worked the biologies of Mantidae, Noctuidae, and shisham defoliator in Punjab and its parasites. He described puparia of some Indian Tachinidae, immature stages of Indian Coleoptera and Lepidoptera like Curculionidae, Bostrychidae, Pyralidae, Anthribidae, Cerambycidae, Pyralidae etc. He is more known by his publications in systematics of Lyctidae and Dinoderus spp., described new species of Bootanomyia and studied termite imago castes of genera Globitermes, Microcerotermes and Odontotermes. Insect fauna of Nepanagar is a unique work of his time. He revised catalogue of Isoptera in entomological collection at FRI. After his retirement he got a project on Psyllidae and worked in the division with the microscopes though his vision had considerably deteriorated. He completed the assignment and published a book "Psyllidae of the Indian Subcontinent" ICAR publication, New Delhi, 429 pp. (1975) which is a monumental work on Indian Psyllidae.

10. Dr. P.N. Chatterjee

P. N. Chatterjee (Nov.1961-Dec., 1970)

Dr. Chatterjee took over the headship of the division after retirement of Dr. R. N. Mathur in Nov., 1961. He served the division in capacity of Head for about nine years from Dec., 1961 to Dec., 1970. He worked and published on use of DDT and BHC for protection of plywood in storage in goodowns against Lyctus beetles, Anti-termite characteristics of jute board treated with SHIRLAN, described termites imago castes of Euhamitermes, termite control services at FRI, biology and morphology of *Apanteles machaeralis*. He also published many works co-authored with N.C. Chatterjee, viz., Insect borers of newly felled timber and their control; and with Roonwal – Control of Indian book worm beetle, experiments on basal girdling of four timber species in relation to anti-insect protection, bamboo ghoon, efficacies of insecticides against converted timber borers, tree poisoning experiments, water seasoning of timbers, control of mound building and subterranean termites, anti-borer protection of Salai and

Bijasal timbers, insect borers of felled timbers and their control, anti-termite characteristics of new synthetic fibre cloth "PAN", effect of leaching out of water soluble substances in *Bambusa nutans,* He also worked on the systematics of Indian Bostrychidae; revised Indo-Malayan genus Hypotermes, and Indian termite genus Microtermes.

11. Dr. Swaraj Kumar Sangal

Dr. Sangal was a keen entomologist, particularly laying his attention on Odonata, but first he was a teacher – most respected, adorable and resourceful. Dr. Sangal was a senior Professor and Head of the Zoology Department in DAV (PG) College, Dehradun which he had developed into a "stronghold of odonatological research through his personal inflence" (Kiauta, 1975, in a communication). His own doctoral thesis was on an odonatological topic centering on *Crocothemis servilia* Drury. He produced as PhD students two outstanding research scientists who excelled in their respective odonatological disciplines of bioecology (Dr. Arun Kumar) and cytogenetics & cytotaxonomy (Dr. B.K. Tyagi). A fatherly figure, he helped his students and scholars in every possible ways, many a time financially as well. Dr. Sangal will be alsways rembembered by his numerous entomologist students for his kind and generous mentoring and teaching, though his scientific contributions to the science of odonatological and formicid research are also much appreciated and reviewed. Dr. Sangal, along with Dr. B.K. Tyagi, has described a new species of damselfly, *Calicnemia doonensis* Sangal & Tyagi, 1984. He had also enlisted, along with Dr. B.K. Tyagi and Ajita Tyagi, the entire extant taxa of the Doon Vally (Tyagi, et al., 1986).

12. Dr. Parimal Kumar Sen-Sarma, D.Sc.

P. K. Sen-Sarma (June, 1970-May, 1973; & Aug., 1976-Aug., 1982)

Dr. Sen-Sarma was a keen researcher and started his research career from his M.Sc. project work where he s tudied the "Anatomy and affinity of the elephant louse (*Haematomyzus alephantis*)". Results of this research were published in the journal "Nature" in 1951 and other findings in the Journal of Parasitology published from the Cambridge University, London. He joined the Forest Research Instiute and Colleges in the year 1951 under the mentorship of Dr. Roonwal. He made great contribution to the science of forest entomology and published more than 200 research papers and 8 books mostly in the field of taxonomy, biology, biochemistry, ecology phylogeny, physiology, polymorphism, polyethism and control of termites and some other forest insects. Studies on natural resistance of timbers to termites was a great contribution and he published several papers in the years 1963, 1965, 1968, 1970, 1973, 1979, 1982, 1988, etc. He established procedures for laboratory testing of natural resistance in Indian woods. Studies on effect of humidity on longevity, effect and humidity on behaviour, effect of gravity on termites were some of his innovative researches.

Systematic studies of termites: he described 38 new species, 7 new genera and a new family, Indotermitidae of termites. He published 5 monographs co-authored with Roonwal on systematics of Oriental termites. He reviewed following genera: Trinervitermes, Ceylonitermes, grallatotermes, Emersonitermes and Ceylonitermellus. Termites of Kumaon, and numerous other publications are important contributions (at least 15 more) to biodiversity of Indian termites. Systematic catalogue of identified entomological collection at FRI dealing with Lepidoptera, Bombaycoidea and Noctuoidea is another major contribution.

Some of his works on biology and control of forest insect are: Alianthus defoliator (1970), relative toxicity of some insecticides, anti-termite characteristics of some organic fibrous material and insect pests of Dipterocarpaceae. He also contributed towards the investigations on Sandal spike disease and insect vectors of sandal spike disease. He also delivered on strengthening plant quarantine in India.

13. Shri Pratap Singh

Pratap Singh (June, 1973 - August, 1976, and October, 1983 - July, 1989)

Sh. Singh was head of the division for two times from 1973 to 76 and again from 1983 to 1989. He was very experienced field entomologist and teacher. He published about 40 research papers. He was selected as FAO consultant to Bhutan four times for managing forest insect pests of fir. He was awarded a one-year long fellowship to USA by FAO in 1970s on control of forest insect pests. He is credited the most for aerial spraying operations in the country for controlling various insect outbreaks like – chir pine defoliator in Rajauri Forest Division (J&K), against teak defoliator and skeletonizer in young teak plantations of Barnawapara, Raipur and Chhatishgarh; deodar defoliator, *Ectropis deodarae* in Kamraj Forest Division (J&K) and against poplar defoliator *Clostera cuptera* and *C. fulgurita* in Central Tarai Forest Division, Haldwani (Uttarakhand).

He also worked on nematode and microbials against important forest pests. Most of his work on pest and their management was on Eucalyptus, poplars, willows, *Accacia tortolis*, Ailanthus, and bamboos. Some of his works were also on the ecology, life history, natural enemies, larval and adult morphology of Ailanthus defoliator, *Eliogma narcissus*; control of semul shoot borer; epidemic defoliation of poplar and its control in Tarai; effect of neem seed extractives on poplar defoliator; epidemic of sal heart wood borer. He worked on the exotic insect *Hetropsylla cubana* pest psyllid of Leucaena; bionomics of ambara defoliator *Podontia puntata*; external morphology and bionomics of *Pagyda salvalis*, inflorescence and fruit borer of teak; tent caterpillar epidemic in *Betula utillis* forest in Pithoragarh Dist.

After his retirement he joined as Professor in Entomology at University of Horticulture and Forestry at Nauni, Solan in July 1989 till he met with an accident in 1991.

14. Dr. Shailendra Kumar Kulshrestha

Professor Kulshrestha is a highly respected teacher and an outstanding entomologist who has taken during his research scholar days his best lessons in insect research in the famous School of Entomology at St. John's College, Agra, under the tutelage of India's best known entomologist, Prof. M.S. Mani, and collaborating with his equally celebrated desciples, Prof. Santokh Singh and Prof. H.N. Baijal. During his tenure as senior professor and the Head of the Department of Zoology, DAV (PG) College, Dehradun, Dr. Kulshrestha developed a strong research base in entomology with particular emphasis in thrips (Thysanoptera) who are serious pests of agriculture and horticulture. Three of his research students excelled in thrips' research later on (Dr. Vijay Veer, Dr. Neena Chauhan and Dr. A.K. Srivastava), and brought out many publications signifying their role in damaging the crops etc. Dr. Kulshrestha, along with above, has also described many new species of thrips from the Doon Valley. In the current volume two of his distinguished scholars have dedicated in his honour a new genus, *Kulshresthothrips* Vijay Veer & Neena Chauhan, 2016.

Like Dr. CFC Beeson, Dr. Kulshrestha has many interests, a true polymath. Besides entomology, he, as a service to social yeomanship, is also engaged in consumer protection activities, career counselling, and personal enrichment training.

15. Dr. Mohan Lal Thakur

M.L. Thakur (Aug.1989-Aug., 1995)

Dr. M. L. Thakur obtained his Master degree in Zoology with specialization in Entomology (1960) and Ph.D. (1973) from Agra University. Except for a short spell at Zoological Survey of India, Calcutta, the entire period of his professional career, spanning over three and a half decades has been devoted to Forest entomology research at the Forest Research Institute, Dehra Dun. He became the head of the Division of Entomology in Aug. 1989. He has to his credit nearly 150 research publications in reputed Indian and foreign journals/ periodicals including several monographs. He has also authored a book on 'Honey and the Honey bees' published by the ICFRE. His major contributions are in the field of termite, biology, morphology, taxonomy and control. Some of his major publications are: termite fauna of NW Himalaya (1967), imago castes of *Stylotermes bengalensis*, termites of Tripura, revision of genus Heterotermes, termites of Kumaon hills, revision of genus Odntotermes. He described several new species of termites form India. He also worked on other forestry insect pests some of which main works are: Insect pests of Dipterocarpaceae, insect fauna of *Leucaena leucocephala*, pests of nurseries. He retired in 1995 and remained academically active and wrote another book "Forest Entomology: Ecology and Management" (2000).

16-20. Dr. R.S. Bhandari, Dr. Md. Yusuf, Dr. M. Ahmad, Dr. Neena Chauhan and Dr. Sudhir Singh

Forest Entomology Division was in recent years modernized in accordance with the demand in entomological research at present. Though time periods of their application could be slightly apart, nevertheless they all had brought a new look to the whole etomological treasure, both the Library and the Museum, of the Division. They also re-inforced a great degree of systematics research into Hymenoptera (Chalcidae, Encyrtidae), Trichogramma (eggs parasitoids), Thysanoptera, biological control of poplar defoliators, and bioefficacy of insecticides against forestry pests, in particular.

In recent years, entomological explorations in different orders like Thysanoptera (Dr. Neena Chauhan, who during her sojourn at the FRI between 1979 and 2015, extensively researched on bioecology and taxonomy as well as biological control of forest pests from the insect orders like Thysanoptera and Hymenoptera (Icheneumonidae, Braconidae), besides engaging in teaching forest entomology at the FRI Deemed University, Dehradun) and Hymenoptera (Dr. Sudhir Singh, who raised several new species), bore significant implications on conserving rare fauna, on one hand, and prevented damage to forest timber, on the other. Dr. Neena, along with Dr. Vijay Veer, has also described a new thrips genus, *Kulshresthothrips* Vijay Veer & Neena Chauhan, 2016 (see Thysanoptera under Annexure 1).

• • •

Annex 3

BIO-BIBLIOGRAPHIES WITH MOTIVATIONAL ANECDOTES OF SOME ENTOMOLOGISTS FROM THE DOON VALLEY

1. Dr. P.K. Sen-Sarma

'Woods are lovely, dark and deep,
But I have promises to keep,
And miles to go before I sleep,
And miles to go before I sleep"
– *Robert Frost*

The life of Professor Dr. P.K. Sen-Sarma, a great entomologist genius, truly reflects the spirit radiated from these famous lines by Robert Frost. Dr. Sen-Sarma was one of those rare scientists who by their energetic character not only practically educate the budding researchers in habits of industry, but by the example of diligence and perseverance which they set before them, largely influence the scientific activity in all directions and contribute in a great degree to form the national character. The national progress, it is aptly said, is after all the sum of individual industry, energy and uprightness and Dr. Sen-Sarma was a glorious example to the young and enthusiastic researchers. He was a total entomologist, having exceptionally diverse and deep knowledge of entomology, though forest entomology has been his greatest forte.

Prof. (Dr) Parimal Kumar Sen-Sarma was born on 1st April 1929 in the East Bengal (now Bangladesh) then a part of India. He was barely two-and-a-half years old when mother passed away, to be thus brought up by his aunt (father's elder sister) whose husband was a medical practitioner. Frequent transfers of his uncle had an adverse effect on Parimal's early education which started in a village primary school. His uncle, after retirement, settled down in Khulna (in Bangladesh) where Parimal completed his schooling. He passed his Matriculation examination in 1945 in 1st division from B. K. Union Institution, Khulna, with letters in Science and Sanskrit. Thereafter; he was admitted in Daulatpur Hindu Academy (renamed Brojolal Academy) in Khulna, from where Parimal, notwithstanding his deep involvement in Students' Union activities, passed I.Sc. Examination with distinction in 1947. Partition of the country brought the family to the West Bengal. In spite of the disturbance brought about by the partition, he pursued his studies with more vigour and zeal at the University College of Science and Technology and City College, and not only passed B.Sc. Zoology (Hons.) examination in 1st Class but also topped the list of the successful candidates, in 1949. For this unique distinction, he was awarded Jubilee Post-Graduate Scholarship for obtaining 7th position in the Order of Merit in the combined list of the successful candidates of B.Sc. (Hons.), B.A. (Hons) and B.Com. Examinations. He passed M.Sc. examination in Zoology and Comparative Anatomy in 1951 in 1 st Class from the Calcutta University and also stood first among the successful candidates for which he was awarded University gold medal. He was also the recipient of the Brahma Mohan Mallik Gold Medal for securing the highest aggregate marks in the Faculty of Science. Seemingly he was the first biologist to achieve this.

Prof. Sen-Sarma was unassuming and retiring in nature but quite indulgent with his pupils, students and co-workers for whose betterment he would often go out-of-the-way. He was a man of firm convictions. His endearing behaviour had won him friends not only among the scientists but also among people belonging to other professions. He possessed a devastating sense of humour and would not mind cracking jokes at his own expense. Dr. Sen-Sarma was a unique forest entomologist whose ceaseless energy and industry, along with deep sense professional integrity and service to the nation, have made Division of Forest Entomology the cradle of learning of this great discipline – in the line of finest tradition of great Indian forest entomologists like Drs. C.B.C. Beeson, M. Lefroy and M.L. Roonwal, and go to his rest when it is done;

चाह नहीं मैं सुरबाला के, गहनों में गूँथा जाऊँ,
चाह नहीं प्रेमी–माला में, बिंध प्यारी को ललचाऊँ,
चाह नहीं, सम्राटों के शव, पर, हे हरि, डाला जाऊँ,
चाह नहीं, देवों के शिर पर, चढ़ूँ भाग्य पर इठलाऊँ,
मुझे तोड़ लेना वनमाली, उस पथ पर देना तुम फेंक,
मातृभूमि पर शीश चढ़ाने, जिस पथ जाएँ वीर अनेक।

Prof. Sen-Sarma began his research career soon after passing B.Sc. (Hons.) examination and submitted a thesis on the "Anatomy and affinity of the elephant louse *(Haematomyzus elephantis* in lieu of Entomology special paper. His first paper on the existence of spermetheca in sucking louse was published in the internationally acknowledged journal *Nature,* London, in 1951. His M.Sc. Thesis on anatomy and affinity of elephant louse" under the guidance of celebrated entomologist late Prof. Durga Das Mukerji, was published in a prestigious journal *Parasitology* of the Cambridge University in 1955 which brought him instant fame. However, he was not destined to continue his research pursuit on louse and had to shift to another fascinating field of research, the termites, after joining Forest Research Institute and Colleges, Dehra Dun. Almost by a divine call, Prof. Sen-Sarma joined Forest Entomology Branch of the Forest Research Institute & Colleges, Dehra Dun in a humble capacity under the great mentorship of Dr. M.L. Roonwal in 1951 and rose to the post of Director of Biological Research of the Institute. Dr. Roonwal, trained in the Indian Army's characteristic discipline, was a hard taskmaster. Therefore, when the young and naïve P.K. Sen-sarma approached him for his permission to be enrolled for doctoral research, Dr. Roonwal initially put him to some litmus tests of scientific patience, resilience and, above all, the habit of perseverance, in which Sen-Sarma not only exhibited affirmatively an extraordinary character but also impressed his future mentor for a humble employment under his tutelage. During this period he occupied various positions of responsibility as follows: Research Officer (1953-63), Senior Research Officer (1963-76), Forest Entomologist (1969-73, 1976-82), Head, Forest Research Laboratory, Bangalore (1973-76), and Director, Biological Research (1980-1987). He was the Principal Investigator of a PL 480 Project entitled, '*Studies on wood-destroying termites in relation to natural termite resistance in timbers*'. He taught Forest Entomology to the probationers of the Indian Forest College, Dehra Dun for a long time during his service career at the Forest Research Institute and Colleges, Dehra Dun.

Prof. Sen-Sarma earned international fame primarily due to his original research on white ants or termites (Isoptera) and other forest entomological subjects. In 1962, he was awarded a FAO Fellowship for advance research on termites at the Institute of Applied Zoology, University of Wuerzburg (West Germany) under the celebrated termitologist Prof. Karl Gosswald. During this period he had the fortune of visiting and working for short durations in the laboratories of several internationally known specialists on termites, such as Prof. P. P. Grasse (Paris), Prof. M. Luescher (Berne), Prof. Carlo Jucii (Pavia), Prof. G. Becker (Berlin), Prof. H. Schmid (Hamburg) and Dr. W. V. Harris (London). He also developed abiding friendship with some of the pupils of these Professors such as Dr. E. Ernst (Basel), Prof A. Springhetti (Ferrera), Prof. Ch. Noirot (Dijon), Prof. H. Schmidt (Hannovar), Prof. W. Kloft and Prof. Kneitz (Bonn) etc. which continued unabated in the most fruitful manner for several decades. He collaborated in research with renowned Zoologists/ Entomologists such as Prof D. Mukerjee, Dr. M. L. Roonwal, Dr R. N. Mathur, Dr. Baini Prasad, Dr. P. N. Chatterjee, Prof W. Kloft (Bonn) and Prof. Elizabeth A. McMahan (North Carolina).

Recognition in scientific arena came quite early in Dr. Sen-Sarma's life, though not without often a string of hardships. He was an elected Fellow of following societies: National Academy of Sciences, India; Indian Academy of Sciences; National Academy of Agriculture Sciences; Indian Academy of Wood Sciences; Indian Academy of Horticultural Sciences; Entomological Society of India; West Bengal Academy of Science and Technology and Mendelian Society of India. He was also life member of Indian Science Congress Association, Indian Institute of Public Administration, Ecological Society of India, Society of Indian Foresters, Society of Tree Scientists of India, Rotary Club, Kanke (Ranchi), etc. Prof. Sen-Sarma was an excellent orator, with a great sense of calling established quotations with rich meaning. He was thus elected as Chairperson to the section *Tropical Forest Pest Management* in the 18th Congress of the International Union of Forestry Research Organization (IUFRO) held in September 1986 at Ljubljana, Yugoslavia. He was also elected special speaker in the IXth Congress of International Union of Study of Social Insects (IUSSI), Colorado (USA) and a Panel Speaker at IXth Congress of IUSSI held in Munich in 1986. He presided over the section Zoology, Entomology and Fisheries at the Platinum Jubilee Session of the Indian Science Congress Association, held in Pune in 1983. A fable goes round about Dr. Sen-Sarma's infinite confidence on his knowledge of forest entomology and zoology. Around 1980 Dr. Sen-Sarma had filed his candidature for the post of Director, Zoological Survey of India and was called to appear for an interview before the expert committee in the Union Public Service Commission, Delhi. During his interview one of the experts had commented on his publications' quality with particular reference to his books and the various chapters written in many others! Dr. Sen-Sarma was so infuriated by his comment that he rebutted a barrage of queries to him asking how much he knew about science that he was compelled to rate his works inferior? He did not stop at this and gave before the expert committee a series of plausible statistical judgments in favour of his far greater superiority of publications, a large proportion of which appeared in journals like *Nature*, and also figured in world's highly rated books such as that of K. Krishna & F. M. Weesner (*Biology of Termites*, Vol. 1 (1969) & 2 (1970), Academic Press, Inc., New York). Understandably the expert committee facing this grand old and seasoned expert of forest entomology would have had a harrowing time, and certainly might have felt they were wrong in undermining this genius's unquestionably superior skills. It reminds us of what Burns famously said;

"Though losses and crosses
Be lessons right severe,
There's wit there, you'll get there,
You'll find no other where."

Prof. Sen-Sarma was member of several committees/institutions such as Scientific Panel in Entomology of the Indian Agricultural Research Institute, Peers Committee of Central Building Research Institute, Roorkee, Review Committee of Indian Lac Research Institute, Ranchi, Pesticide Evaluation Committee, Government of India, Screening Committee for Importation of Bio-control Agents of the Government of India, Planning, Subgroup, Forestry and Wildlife, Government of Bihar, Research Degree Committee in Zoology Garhwal University. Srinagar (Garhwal), Scientific Panel in Agro-Forestry of the Indian Council of Agricultural Research He was an adopted adviser to the Agricultural Scientists Recruitment Board, New Delhi, and referee for Research Projects submitted to Indian Council of Scientific and Industrial Research (CSIR), Indian Council of Agricultural Research (ICAR), Department of Science & Technology (DST). Prof. Sen-Sarma's editorial skills were exceptionally brilliant. He was Hony. Editor in-Chief of *Annals of Entomology* published from Dehra Dun and member of International Editorial Board of *Science and Technology of Termites* published from Hong Kong and Beijing.

Dr. Sen-Sarma was a great visionary. He followed science very meticulously. A fearless entomologist, he would never let the opportunity slip away from his grip if once he had been convinced for its genuine original nature. It was sheer due to his indefatigable habit of industry that he was able to describe as new to the science a series of new species and genera, besides one whole new family of termites, as annotated below:

I. New Order:

RYNCOPHTHIRAPTERA

II. New Family:

1. **Indotermitidae** Roonwal & Sen-Sarma, 1960. *Contrib. Syst. Orient. Termites. Indian Counc. Agric. Res. Entomol.*, New Delhi, Mongr. No.1, 40.

Type genus: **Indotermes** Roonwal & Sen-Sarma, 1960. *Contrib. Syst. Orient. Termites. Indian Counc. Agric. Res. Entomol.,* New Delhi, Mongr. No.1, 41.

III. New Genera

1. ***Afrograllatotermes*** Sen-Sarma, 1966. *J. Bombay Nat. Hist. Soc., Mumbai*, 63 (1): 180.

Type species: *Afrograllatotermes africanus* (Harris, 1954). *Afrograllatotermes* has been synonymized under ***Grallatotermes*** Holmgren by Chhotani, 1997. Fauna of India, Isoptera, Vol. II, p.750 following Krishna, 1970, p.149 (Krishna, In: *Biology of Termites*, (Eds Krishna & Weesner, 2, p. 149).

Type locality: East Africa: Ngameni NE Tanganyika.

2. ***Emersonitermes*** Mathur & Sen-Sarma, 1959. *Z. Ang. Ent.*, 45(1): 67. (Also see 45(3): 344 for errata).

Type species: *Emersonitermes thekadensis* Mathur & Sen-Sarma, 1959. *Z. Ang. Ent.*, 45(1): 68.

3. ***Fletcheritermes*** Sen-Sarma, 1965. *Proc. Zool. Soc.*, Calcutta 18: 143.

Type species: *Fletcheritermes salemensis* Sen-Sarma, 1965. Proc. Zool. Soc., Calcutta 18: 144. Synonymised under ***Nasutitermes.***

4. ***Indograllatotermes*** Sen-Sarma, 1966. *J. Bombay Nat. Hist. Soc.*, Bombay, 63 (1): 176.

Type species: *Indograllatotermes grallatoriformis* (K & N Holmgren, 1917). Synonymised under ***Grallatotermes***.

5. ***Indotermes*** Roonwal & Sen-Sarma, 1960. *Contrib. Syst. Orient. Termites. Indian Counc. Agric. Res. Entomol.*, New Delhi, Mongr. No.1, 41.

Type species: *Indotermes maymensis.* Placed in new family **Indotermitidae** Roonwal & Sen-Sarma, 1960. *Contrib. Syst. Orient. Termites. Indian Counc. Agric. Res. Entomol.*, New Delhi, Mongr. No.1, 40.

6. ***Philippinitermes*** Sen-Sarma, 1966. *J. Bombay Nat. Hist. Soc.*, Bombay, 63 (1): 172.

Type species: *Philippinitermes admirabilis* (Light, 1930). Synonymised under ***Grallatotermes***.

7. ***Thailanditermes*** Sen-Sarma, 1966. *Indian J. Entomol.*, 28: 202.

Type species: *Thailanditermes thailandicus* Sen-Sarma, 1966. *Indian J. Entomol.*, 28: 203. Synonymised under ***Lacessitermes***.

IV. New Species

(Abbreviation: COLL. = Type specimens deposited, ZSIK= Zoological Survey of India, Kolkatta, NFIC= National Forest Insects Collection, Forest Research Institute, Dehra Dun, AMNH- American Museum of National History, New York).

1. ***Anacanthotermes rugifrons*** Mathur & Sen-Sarma, 1958. *Entomologist*, 91: 233.

Type specimens: Holotype - soldier; Morphotype-worker (both in spirit in a single vial), ex. small mound ca 5" in height, 18.iii. 1958, leg. R.N. Mathur. COLL. NIFC.

Paratypes- 8 soldiers; Paramorphotypes- 22 workers (in spirit in a vial), ex. small mound ca 5" in height, 18.iii. 1958, leg. R.N. Mathur. COLL. NIFC-7 soldiers, 20 workers and in collection of Prof. Alfred E. Emerson, Chicago (U.S.A.)-1 soldier, 2 workers.

Type locality: Pallayamcottah (latitude 8°44'N., longitude 77°44'E), Tinnevelly District (Madras State).

Distribution: India: Pallayamcottah (Tinnevelly district); Nagarcoil (Tamil Nadu).

2. ***Angulitermes acutus*** Mathur & Sen-Sarma, 1960. *Ann. Mag. Nat. Hist.*, (13), 31: 401-406.

Type specimens: Holotype - soldier, Morphotype-worker (both in spirit in single vial), ex. root of *Phoenix* sp., 17.i.1958, leg. R.N. Mathur. COLL. NIFC.

Paratypes- 2 soldiers and Paramorphotype-11 workers (in spirit, in a vial), ex. root of *Phoenix* sp., 17.i. 1958, leg. R.N. Mathur. COLL. NFIC-1 soldier and 10 workers and in collection of Prof. Alfred E. Emerson, Chicago (U.S.A.)-1 soldier and 1 worker.

Type locality: Pallayamcottah (latitude 8° 44' N., longitude 77 ° 44'E), Tinnevelly District (Madras State).

Distribution: India: Tinnevelly (Tamil Nadu).

3. ***Angulitermes mishrai*** Sen-Sarma & Thakur, 1979. *Indian Forest Records* (n.s.) 13(1):18-20.

Type specimens: Holotype - soldier; Morphotype - 1 worker (both in spirit in a vial M58/23.1.1971), 23.i.1971, leg. S.C. Mishra. COLL. NFIC.

Paratypes-6 soldiers; Paramorphotypes-a few workers, 23.i. 1971, leg. S.C. Mishra. COLL. NFIC.

Type locality: India: Zoelchara, Ambasa (Tripura).

Distribution: India: Ambasa and Tiliamura (Tripura).

4. ***Capritermes dunensis*** Roonwal & Sen-Sarma, 1960. *Contri. Syst. Orient. Termites. No. 5. Indian Counc. Agric. Res. Entomol.*, New Delhi, Mongr. No. 1, 28-33.

Type specimens: Holotype – 1 soldier (in a vial), Morphotype-1 worker (in a vial), 18.ii.1955, ex. under a log in sal (*Shorea robusta*) area, leg. Om Bahadur.

Paratypes-3 soldiers, paramorphotypes-6 workers, 18.ii.1955, ex. under a log in sal (*Shorea robusta*) area, leg. Om Bahadur. COLL. NFIC - 1 soldier, 2 workers in a single vial; ZSIK-2 soldiers, 4 workers in single vial.

Now synonymized with ***Precapritermes dunensis*** (Roonwal & Sen-Sarma, 1960).

Type locality: India: New Forest, ca 610 m., Dehra Dun (U.P.).

Distribution: India: Dehra Dun (Uttaranchal).

5. ***Capritermes orientalis*** Mathur & Sen-Sarma, 1961. *Bull. Entomol.*, 2: 1-4.

Type specimens: Holotype – soldier (in spirit in a vial), ex. mound, 24-27.ii. 1950, leg. Silviculturist, COLL. NFIC.

Paratypes-11 soldiers, ex. mound, 24-27.ii. 1950, leg. Silviculturist. COLL. NFIC-7 soldiers and in collection of Prof. Alfred E. Emerson, Chicago (U.S.A)- 4 soldiers.

Now synonymized with ***Precapritermes tetraphilus*** (Silvestri).

Type locality: Myanmar- Maymyo (altitude 22°5'N longitude 96°30'E).

Distribution: Myanmar- Maymyo; China-Tho-li (Yunnan Province).

6. ***Discuspiditermes tripurensis*** Sen-Sarma & Thakur, 1979. *Indian Forest Records* (n.s.) 13: 21.

Type specimens: Holotype - soldier; Morphotype-worker (in spirit in a vial), ex. unknown, 21.i.1971, leg. R.C. Yadav. COLL. NFIC.

Paratypes-7 soldiers; Paramorphotype - several workers, ex. unknown, 21.i.1971, leg. R.C. Yadav. COLL. NFIC-6 soldiers and several workers and ZSIK-1 soldier and 1 worker.

Now synonymized with ***Discuspiditermes laetus*** (Silvestri).

Type locality: India- Tiliamura (Tripura).

Distribution: India- Tiliamura (Tripura).

7. ***Doonitermes buxensis*** Thakur & Sen-Sarma, 1973. *Bull. Entomol.*, 14: 85-90.

Type specimens: Holotype –1 soldier

Now synonymized with genus ***Indotermes rongrensis*** Roonwal & Chhotani, 1962. Not valid species.

Type locality: India -West Bengal.

Distribution: India: Buxer, West Bengal.

8. ***Emersonitermes thekadensis*** Mathur & Sen-Sarma, 1959 *Z. Ang. Ent.*, 45(1): 68. (Also see vol . 45(3): 344 for errata).

Type specimens: Holotype-soldier; Morphotype-worker (both in spirit in a vial), ex. decayed wood in soil, 19.x. 1956, leg. Gunther Becker, COLL. NFIC.

Paratypes-2 soldiers; Paramorphotypes-5 workers, ex. decayed wood in soil, 19.x. 1956, leg. Gunther Becker, COLL. NFIC -1 soldier, 4 workers; Hamburg Museum, Hamburg (Germany)-1 soldier, 1 worker.

Type locality: India: Thekaday (Travancore) (Kerala).

Distribution: India: Thekaday (Travancore) (Kerala).

9. ***Eremotermes dehraduni*** Roonwal & Sen-Sarma, 1960. *Contrib. Syst. Orient. Termites. Indian Counc. Agric. Res. Entomol.*, New Delhi, Mongr. No. 1, 53-63.

Type specimens: Holotype-soldier; Morphotype -1 worker (both in a single vial, NFIC No. 19937/1), ex roots of *Ocimum kilimandscharicum* plantation, 14.viii. 1953, leg. Ajab Kishor.

Paratypes-9 soldiers; paramorphotypes-13 workers, ex roots of *Ocimum kilimandscharicum* plantation, 14.viii. 1953, leg. Ajab Kishor. COLL. NFIC- 6 soldiers; 6 workers (3 major, 3 minor) in a single vial, also slide no. 161 having body parts of a soldier paratype and slide no. 170 having body parts of a major worker paramorphotype; in Collection of Prof. Alfred E. Emerson, Chicago (U.S.A.)-1 soldier, 3 workers in spirit, in a vial; ZSIK- 1 soldier, 2 workers in spirit, in a vial and in Pusa collection, New Delhi- 1 soldier, 2 workers in spirit, in a vial.

Type locality: India: Dehra Dun (Bainkhala Block, Malhan Forest Range, about 12 miles north-west of Dehra Dun), Kalsi (ca.750 m.).

Distribution: India; Dehra Dun, Uttaranchal.

10. ***Eremotermes madrasicus*** Roonwal & Sen-Sarma, 1960. *Contrib. Syst. Orient. Termites. Indian Counc. Agric. Res. Entomol.*, New Delhi, Mongr. No. 1, 75-79.

Type specimens: Holotype- soldier; Morphotype- 2 workers (1 major, 1 minor) in spirit in a single vial (NFIC No. M.C. 19939/1), ex soil, leg. M.L. Roonwal, 9.ii.1955.

Paratypes-9 soldiers; paramorphotypes-8 workers, ex soil, 9.ii.1955, leg. M.L. Roonwal. COLL. NFIC -6 soldiers, 6 workers, also 2 slides (nos 164, 166) having body parts; in Collection of Prof. Alfred E. Emerson, Chicago (U.S.A.)-1 soldier & 1 major and 2 minor workers; ZSIK- 1 soldier, 3 minor workers and Pusa collection, New Delhi- 1 soldier, 2 workers.

Type locality: India: Madras (Villivakkam, 10 miles north of Madras city), Tamil Nadu.

Distribution: India – Chennai, Tamil Nadu.

11. ***Fletcheritermes salemensis*** Sen-Sarma, 1965. Proc. Zool. Soc., Calcutta, 18: 143.

Type specimens: Holotype - soldier; Morphotype-1 worker, ex. unknown, 21.xii. 1913, leg. T.B. Fletcher. COLL. NFIC.

Paratypes- 12 soldiers; Paramorphotypes- 13 workers, ex. unknown, 21.xii. 1913, leg. T.B. Fletcher. COLL. NFIC-6 soldiers, 6 workers; PUSA - 4 soldiers, 5 workers and ZSIK-2 soldiers, 2 workers.

Now this species is placed with ***Nasutitermes salemensis*** (Sen-Sarma, 1965) (vide Krishna, 1970).

Type locality: Vercaud, Salem district, Madras.

Distribution: Vercaud, Salem district (Tamil Nadu).

12. ***Glyptotermes nigrifrons*** Mathur & Sen-Sarma, 1960. *Entomologist*, 93: 79.

Type specimens: Holotype - soldier; Morphotype, 1 nymph of alate in spirit, in a vial, ex. rotton wood, 6.i.1958, leg. R.N. Mathur, COLL. NFIC.

Paratype-8 soldiers, Paramorphotype-11 nymphs of alates, ex. rotton wood, 6.i.1958, leg. R.N. Mathur. COLL. NRC-7 soldiers, 10 nymphs of alates, and in Collection of Prof. Alfred E. Emerson, Chicago (U.S.A.)-1 soldier, 1 nymph of alates.

Now synonymized with ***Glyptotermes coorgensis*** Holmg. & Holmg. (vide Chhotani, 1997. Fauna series).

Type locality: India: Top slip near Pollachi (latitude 10°39' N; longitude 77°3' E), Coimbatore S. Division, Madras State.

Distribution: India: Coimbatore (Tamil Nadu).

13. ***Hospitalitermes ataramensis*** Prashad & Sen-Sarma, 1960. Indian Council of Agricultural Research Monograph No. 10.29: 5-10.

Type specimens: Holotype - soldier (major) & Morphotypes -1 worker (major) 1 worker (minor), (all in spirit in a vial), ex. Wood of *Xylia dolabiformis*, on 11.xii. 1938, leg. P.F. Garthwhite (No. 914),. COLL. NFIC (MC 20090/1).

Paratypes-16 soldiers; Paramorphotypes-3 workers (major), 2 worker (minor), ex. Wood of *Xylia dolabiformis*, on 11.xii. 1938, leg. P.F. Garthwhite.COLL. NFIC-10 soldiers, 3 workers (major) and 2 workers (minor); ZSIK-4 soldiers, and in Collection of Prof. Alfred E. Emerson, Chicago (U.S.A.)-2 soldiers.

Type locality: Myanmar: Pakabo Reserve, Ataram.

Distribution: Myanmar: Ataram.

14. ***Hospitalitermes blairi*** Roonwal & Sen-Sarma, 1956. *Indian J. Agric. Sci.*, 26(1): 29-36.

Type specimens: Holotype – 1 soldier; Morphotype, 1 worker (major), 1 worker (minor), ex. a marching column of ants on the jungle path, 15.xii. 1934, leg. H.S. Rao. COLL. ZSIK (vial No. 95).

Paratypes-3 soldiers, Paramorphotypes-6 workers, ex. a marching column of ants on the jungle path, 5.xii. 1934, leg. H.S. Rao. COLL. NFIC-1 soldier, 1 worker (major) and 1 worker (minor); ZSIK - 1 soldier, 1 worker (major) and 1 worker (minor) and in collection of Prof. Alfred E. Emerson, Chicago (U.S.A.)-1 soldier, 1 worker (major) and 1 worker (minor).

Type locality: India: S. Coroyn' cave, Port Blair (Andaman Islands).

Distribution: India: Port Blair (Andaman Islands).

15. ***Hospitalitermes brevirostratus*** Prashad & Sen-Sarma, 1960. Indian Council of Agricultural Research Monograph No.10.29:4: 14-17.

Type specimens: Holotype - 1 soldier (major); Morphotypes -1 worker (all in spirit in a vial), ex *Glochidion* sp., 20.xii. 1939, leg. R.H.O. (No. 1408), COLL. NFIC (MC 20093/1).

Paratypes-5 soldiers, ex *Glochidion* sp., 20.xii. 1939, leg. R.H.O. (No. 1408). COLL. NFIC (MC 20093/2).

Type locality: Myanmar: Maymo.

Distribution: Myanmar: Maymo.

16. ***Hypotermes nongpriangi*** Roonwal & Sen-Sarma, 1956. *Indian J. Agric. Sci.*, 26: 21.

Type specimens: Holotype – soldier; Morphotype-worker, ex. under bark of dead tree, 20.xi. 1923, leg. S.L Hora, COLL. ZSIK.

Paramorphotypes-4 workers, ex. under bark of dead tree, 20.xi. 1923, leg. S.L Hora. COLL. ZSIK-1 worker, COLL. NFIC-2 workers vial No. 19867, 1 worker dissected and mounted on slide No. 138.

Type locality: India: Bank of Nong-priang River, 1200' altitude, Khasi hills, North Frontier Agency, (Assam).

Distribution: India -? Assam.

17. ***Indotermes maymensis*** Roonwal & Sen-Sarma, 1960. *Contrib. Syst. Orient. Termites. Indian Counc. Agric. Res. Entomol.*, New Delhi, Mongr. No.1, 42.

Type specimens: Holotype – 1 soldier in sprit in a vial (NFIC), ex unknown, 23.ii.1950, leg. Silviculturist, Myanmar in Botanical Gardens at Maymyo, lat 22°5' N., long 96° 30'E (Central Maynmar).

Paratypes-8 soldiers, ex unknown, 23.ii.1950, leg. Silviculturist, Myanmar in Botanical Gardens at Maymyo, lat 22°5' N., long 96° 30'E (Central Maynmar). COLL. NFIC-7 soldiers in two vials; ZSIK- 3 soldiers in a vial and in collection of Prof. Alfred E. Emerson, Chicago (U.S.A.)-1 soldier.

Type locality: Maymyo, Myanmar (former Burma).

Distribution: Myanmar- Maymyo.

18. ***Microcerotermes labioangulatus*** Sen-Sarma & Thakur, 1979. *Indian Forest Records* (n.s.) 13: 16.

Type specimens: Holotype – 1 soldier; Morphotype – 1 worker (in spirit, in a vial 35/23.1.71), ex. unknown, 23.i.1971, leg. R.C.Yadav. COLL. NFIC.

Paratype-11 soldiers; Paramorphotypes-a few workers, 23.i.1971, leg. R.C.Yadav. COLL. NFIC-10 soldiers and a few workers; ZSIK-1 soldier and 1 worker.

Type locality: India-Zoelchara, Ambasa (Tripura).

Distribution: India- Zoelchara, Ambasa (Tripura).

19. ***Microcerotermes manjikuli*** Sen-Sarma, 1966. *Indian J. Entomol.*, 28: 197.

Type specimens: Holotype - soldier (in spirit in a vial), ex. sugarcane, date of collection unknown, leg. A Manjikul, COLL. NFIC.

Paratypes-6 workers (in spirit in a vial), ex. sugarcane, date of collection not mentioned, leg. A Manjikul, COLL. NFIC.

Type locality: Thailand: Pathaloong.

Distribution: Thailand-Pathaloong

20. *Macrotermes serrulatus hopini* Roonwal & Sen-Sarma, 1956. *Indian J. Agric. Sci.*, 26: 15-21.

Type specimens: Holotype - soldier; Morphotype-worker (in spirit, in a vial 10a), ex. found under rotten tree, 13. x. 1926, leg. B. Chopra, COLL. ZSIK.

Paratypes-3 soldiers; Paramorphotypes-8 workers, ex. found under rotten tree, 13.x. 1926, leg. B. Chopra, COLL. NFIC-1 soldier and 4 workers vial No. 19868; ZSIK-1 soldier and 3 workers and in Collection of Prof. Alfred E. Emerson, Chicago (U.S.A.)-1 soldier, 1 worker.

Now this species is renamed as ***Macrotermes hopini*** Roonwal & Sen-Sarma.

Type locality: Hopin, Myitkyina district, Upper Burma.

Distribution: Myanmar-Hopin (Myitkyina district).

21. *Nasutitermes beckeri* Prashad & Sen-Sarma, 1959. *Indian Counc. Agric. Res.*, No. 10.23: 8.

Type specimens: Holotye -1 soldier (in spirit in a vial); Morphotype-1 imago, 1 worker (in spirit in a vial), ex. cartoon nest in a tree, 18.ix. 1956, leg. Gunther Becker, COLL. NFIC (No. MC 20028/1 & 20028/2 respectively).

Paratypes-21 soldiers; Paramorphotypes-9 imagoes and 20 workers, ex. cartoon nest in a tree, 18.ix. 1956, leg. Gunther Becker. COLL. NFIC-9 soldiers, 4 imagoes, 5 wokers; Hamburg Museum, Hamburg (Germany)-5 soldiers, 2 imagoes, 5 workers and in Collection of Prof. Alfred E. Emerson, Chicago (U.S.A.)-2 soldiers, 1 imago, 5 workers.

Now this species is synonymized with ***Nasutitermes indicola*** (Holmg. & Holmg.). (vide, Chhotani, 1997).

Type locality: India - Neriamangalam (Western Ghats).

Distribution: India - Neriamangalam, Thekaday, Thekaday game sanctuary (Western Ghats).

22. *Nasutitermes guptai* Sen-Sarma & Thakur, 1980. *Indian Forest Records* (n.s.), 14(1): 25.

Type specimens: Holotype-soldier; Morphotype-worker (in spirit in a vial), ex. galleries on the trunk of *Shorea robusta*, 30.v. 1971. leg. B.K. Gupta. COLL. NFIC.

Paratypes-11 soldiers; Paramorphotypes-11 workers (in spirit in a vial), ex. galleries on the trunk of *Shorea robusta*, 30.v. 1971. leg. B.K. Gupta. COLL. NFIC-10 soldiers,10 workers and ZSIK-1 soldier, 1 worker.

Now synonymized with ***Nasutitermes dunensis*** Chatterjee & Thakur (vide, Chhotani, 1997).

Type locality: India-Dholakot, Askot (Kumaon Hills).

Distribution: India-Dholakot (Kumaon Hills, Uttaranchal).

23. *Nasutitermes jalpaigurensis* Prashad & Sen-Sarma, 1959. *Indian Counc. Agric. Res.* Monogr, No. 10.23: 39.

Type specimens: Holotype-soldier; Morphotype - worker (in spirit in a vial), ex. not known, on 26.x. 1933, leg. N. C. Chatterjee., COLL. NFIC (MC. 20027/1 & 20027/2 respectively).

Paratypes- 6 soldiers; paramorphotypes-10 workers (all in spirit), ex. not known, on 26.x. 1933, leg. N. C. Chatterjee. COLL. NFIC-4 soldiers, 6 workers; ZSIK-1 soldier, 2 workers, and in Collection of Prof. Alfred E. Emerson, Chicago (U.S.A.)-1 soldier, 2 workers.

Type locality: India-Jalpaiguri, Teesta Forest Division (West Bengal).

Distribution: India-Jalpaiguri (West Bengal), Tiliamura (Tripura).

24. *Nasutitermes suknensis* Prashad & Sen-Sarma, 1959. *Indian Counc. Agric. Res.*, No. 10.23: 64.

Type specimens: Holotype-soldier (in spirit), ex. rotton wood on ground, date not known, leg. S.K. Chatterjee. COLL. NFIC (M.C. 20059/1).

Paratype-soldier, ex. rotton wood on ground, date not known, leg. S.K. Chatterjee. COLL. NFIC (M.C. 20059/2).

Type locality: India-Sukna (West Bengal).

Distribution: India-Sukna (West Bengal).

25. *Nasutitermes thanensis* Prashad & Sen-Sarma, 1959. *Indian Counc. Agric. Res.*, No. 10.23: 61.

Type specimens: Holotype -1 soldier & Morphotype -1 worker (in spirit, in a vial), ex. not known, on 21.IV.1914, leg. N. C. Chatterjee, COLL. NFIC (No. MC 20026/1 & 20026/2 respectively).

Paratypes-3 soldiers; paramorphotypes-10 workers, ex. not known, on 21.iv. 1914, leg. N. C. Chatterjee. COLL. NFIC-2 soldiers, 8 workers; ZSIK-1 soldier, 2 workers.

Type locality: India- Thano (Dehra Dun, Uttaranchal).

Distribution: India-Thano (Dehra Dun, Uttaranchal).

26. *Neotermes buxensis* Roonwal & Sen-Sarma, 1960. *Contrib. Syst. Orient. Termites. Indian Counc. Agric. Res. Entomol.*, New Delhi, Mongr. No.1: 161-166.

Type specimens: Holotype-soldier in spirit in a vial, ex. *Macaranga denticulate*, 6.i.1915, leg. Unknown, COLL NIFC; Morphotype- 1 imago and a psuedoworker in spirit in a single vial, ex rotten wood, 25.vi.1945, leg. N. C. Chatterjee.COLL NIFC.

Paratypes-2 soldiers, ex rotten wood, 25.vi.1945, leg. N.C. Chatterjee. COLL. NFIC- 1 soldier and ZSIK- 1 soldier.

Type locality: India-Buxa Forest Division (North Bengal).

Distribution: India-North Bengal; Bagdogra.

27. ***Neotermes kemneri*** Roonwal & Sen-Sarma, 1960. *Contrib. Syst. Orient. Termites. Indian Counc. Agric. Res. Entomol.*, New Delhi, Mongr. No.1: 188-190.

Type specimens: Holotype-soldier; Morphotype-psuedoworker in spirit, ex *Semecarpurs* sp. (Anacardiaceae), 13.ix.1928, leg. F.P. Jepson. COLL. (Zoological Institute, Lund University, Lund (Sweden).

Paratypes- more than 2 soldiers; paramorphotypes-more than 4 psuedoworkers, ex *Semecarpurs* sp. (Anacardiaceae), 13.ix.1928, leg. F.P. Jepson. COLL. NFIC- 1 soldier, 3 psuedoworkers, also paratype body parts mounted on slide No. 136; ZSIK- 1 soldier, 1 psuedoworker and some paratype soldiers and psuedoworkers in Zoological Institute, lund University, Lund (Sweden).

Type locality: Sri Lanka-Gannoruwa (Peradeniya).

Distribution: Sri Lanka-Gannoruwa (Peradeniya).

28. ***Neotermes mangiferae*** Roonwal & Sen-Sarma, 1960. *Contrib. Syst. Orient. Termites. Indian Counc. Agric. Res. Entomol.*, New Delhi, Mongr. No.1: 208-210.

Type specimens: Holotype- 1 soldier (male); morphotype – 1 psuedoworker, both in spirit in a single vial, ex. felled log of *Mangifera indica* lying in a timber depot, 28.xii. 1953, leg. M.L. Roonwal, COLL. NFIC.

Paratypes-9 soldiers; paramorphotypes-13 psuedoworkers, ex. felled log of *Mangifera indica* lying in a timber depot, 28.xii. 1953, leg. M.L. Roonwal. COLL. NFIC- 7 soldiers, 7 psuedoworkers, also one slide (no. 131) having body parts of paratype; ZSIK- 1 soldier, 3 psuedoworkers, and in Collection of Prof. Alfred E. Emerson, Chicago (U.S.A.) -1 soldier, 3 psuedoworkers.

Type locality: India-Eastern India.

Distribution: India-Eastern India.

29. ***Neotermes megaoculatus*** Roonwal & Sen-Sarma, 1960. *Contrib. Syst. Orient. Termites. Indian Counc. Agric. Res. Entomol.*, New Delhi, Mongr. No.1: 214.

29a. ***Neotermes megaoculatus megaoculatus*** Roonwal & Sen-Sarma, 1960. *Contrib. Syst. Orient. Termites. Indian Counc. Agric. Res. Entomol.*, New Delhi, Mongr. No.1: 214-217.

Type specimens: Holotype-1 soldier in spirit in a vial; morphotype-psuedoworker in spirit in a vial, ex rotten mango log, 8.xii.1939, leg. JCM. Gardner. COLL. NFIC.

Paratypes-8 soldiers; paramorphotypes-6 psuedoworkers, ex rotten mango log, 8.xii.1939, leg. JCM. Gardner. COLL. NFIC- 6 soldier; 4 psuedoworkers; ZSIK- 1 soldier, 1 psuedoworker, and in Collection of Prof. Alfred E. Emerson, Chicago (U.S.A.) -1 soldier, 1 psuedoworkers.

Type locality: India: New Forest (2000 ft.), Dehra Dun (Uttaranchal).

Distribution: India: New Forest, Dehra Dun (Uttaranchal).

29b. ***Neotermes megaoculatus lakhimpuri*** Roonwal & Sen-Sarma, 1960. *Contrib. Syst. Orient. Termites. Indian Counc. Agric. Res. Entomol.*, New Delhi, Mongr. No 1: 217-220.

Type specimens: Holotype-soldier with partly dissected right mandible and right antenna with basal half only (COLL NFIC MC 19949/1); Morphotype-1 psuedoworker, ex wood of unknown plant, 11.i.1936, leg. CFC. Beeson. COLL NIFC.

Paratypes-5 soldiers; paramorphotypes-4 psuedoworkers, ex wood of unknown plant, 11.i.1936, leg. CFC. Beeson. COLL. NFIC- 3 soldiers; 2 psuedoworkers (M.C. 19949/3); ZSIK-1 soldier,1 psuedoworker, and in Collection of Prof. Alfred E. Emerson, Chicago (U.S.A.)-1 soldier, 1 psuedoworker.

Type locality: India-Makun, Kakhimpur (Assam).

Distribution: India-Makun, Kakhimpur (Assam).

30. ***Neotermes microculatus*** Roonwal & Sen-Sarma, 1960. *Contrib. Syst. Orient. Termites. Indian Counc. Agric. Res. Entomol.*, New Delhi, Mongr. No. 1: 220-223.

Type specimens: Holotype-soldier with broken antennae and legs (NFIC MC 19951/1); Morphotype-psuedoworker, ex. dead *Salix alba*, 24.i. 1933, leg. GD Bhasin. COLL. NIFC.

Paratypes-1 soldier; paramorphotypes-3 psuedoworkers, ex. dead *Salix alba*, 24.i. 1933, leg. GD Bhasin.COLL. NFIC- 1 soldier paratype with broken body parts and left antenna mounted on slide (No. 198) and 3 psuedoworkers.

Type locality: India-Dehra Dun, U.P.

Distribution: India-Dehra Dun (Uttaranchal).

31. ***Neotermes paratensis*** Sen-Sarma & Thakur, 1979. *Indian Forest Records (n.s.)* 13: 6.

Type specimens: Holotype-soldiers; Morphotype -1 dealated imago and 1 pseudoworker (in spirit, in a vial), ex. standing tree of *Morus* sp., 16.i. 1971, leg. S.C.Mishra. COLL. NFIC.

Paratype-1 soldier; Paramorphotypes-1 dealated imago and several pseudoworkers, ex. standing tree of *Morus* sp., 16.i. 1971, leg. S.C.Mishra. COLL. NFIC.

Type locality: India-Forest Guest House, Paratia (Tripura).

Distribution: India-Paratia (Tripura).

32. ***Odontotermes paralatigula*** Chatterjee & Sen-Sarma, 1962. J. Bombay Nat. Hist., 59: 822.

Type specimens: Holotype-soldier (in spirit in a vial), ex. mound, date (not mentioned), leg, Range officer. COLL. NFIC.

Paratypes-35 soldiers, (in spirit in a vial), ex. mound, date (not mentioned), leg, Range officer. COLL NFIC-20 soldiers; ZSIK-5 soldiers; PUSA-5 soldiers, and in Collection of Prof. Alfred E. Emerson, Chicago (U.S.A.)-5 soldiers.

Type locality: Burma-Hlegu, Insein Forest Division.

Distribution: Myanmar-Hlegu (Insein Forest Division).

33. *Parrhinotermes khasii* Roonwal & Sen-Sarma, 1956. *Indian J. Agric. Sci.*, 26: 6.

Type specimens: Holotype-soldier; Morphotype- worker (in spirit in a vial No. 20a (i)), ex. under bark of a dead tree, 20.xi. 1923, leg. S.L. Hora, COLL. ZSIK.

Paratypes-3 soldiers; Paramorphotype-5 workers (in spirit in a vial), ex. under bark of a dead tree, 20.xi. 1923, leg. S.L. Hora. COLL NFIC-1 soldier, 3 workers in spirit in a vial No. 18651; ZSIK-1 soldier,1 worker, and in Collection of Prof. Alfred E. Emerson, Chicago (U.S.A.)-1 soldier,1 worker.

Type locality: Bank of Nong-Priang stream, 1200 'altitude, Khasi Hills, North Frontier Agency (Assam).

Distribution: India:? Assam.

34. *Speculitermes sighalensis* Roonwal & Sen-Sarma, 1960. *Contrib. Syst. Orient. Termites. Indian Counc. Agric. Res. Entomol.*, New Delhi, Mongr. No. 1: 15, 23-26.

Type specimens: Holotype – 1 worker in spirit in a single vial (NFIC), ex unknown, leg. COLL NFIC.

Paratypes-8 workers2 workers in a vial; ZSIK- 2 paratype workers; in Colombo Museum, Colombo, Sri Lanka- 2 paratype workers in a vial and in Collection of Prof. Alfred E. Emerson, Chicago (U.S.A.) -2 workers paratype.

Type locality: Sri Lanka-Vavuniya (North Province).

Distribution: Sri Lanka-Vavuniya, Peradeniya, Kurunegala and Maha Illuppallama. India: Shevaroy Hills and Chitteri (Salem district, Tamil Nadu).

35. *Speculitermes triangularis* Roonwal & Sen-Sarma, 1960. *Contrib. Syst. Orient. Termites. Indian Counc. Agric. Res. Entomol.*, New Delhi, Mongr. No.1: 15: 26-28.

Type specimens: Holotype – 1 worker in spirit in a single vial (NFIC), ex unknown, 28.ix.1929, leg. JCM Gardner, COLL NIFC.

Paratypes-10 workers, ex unknown, 28.ix.1929, leg. JCM Gardner. COLL. NFIC- 5 workers; ZSIK- 4 workers, and in Collection of Prof. Alfred E. Emerson, Chicago (U.S.A.) -1 worker.

Type locality: India-Dehra Dun (Uttar Pradesh).

Distribution: India-Dehra Dun (Uttaranchal).

36. *Synhamitermes colombensis* Roonwal & Sen-Sarma, 1960. *Contrib. Syst. Orient. Termites. Indian Counc. Agric. Res. Entomol.*, New Delhi, Mongr. No.1: 1-7.

Type specimens: Holotype-soldier (head & thorax only); Morphotype -1 worker in separate two vials, ex unknown, 24.ix.1952, leg. Sandarsagara (COLL NFIC No. MC 19950/1).

Paratypes-2 soldiers; paramorphotypes-2 workers, ex unknown, 24.ix.1952, leg. Sandarsagara. COLL Colombo Museum, Colombo, Sri Lanka-1 soldier (head), 1 worker (head); ZSIK- 1 soldier (head), 1 worker (head).

Type locality: Sri Lanka-Colombo, west Province.

Distribution: Sri Lanka-Colombo

37. *Thailanditermes thailandicus* Sen-Sarma, 1966. *Indian J. Entomol.*, 28: 203.

Type specimens: Holotype – soldier; Morphotype-1 imago, 1 worker (in spirit both in a single vial), ex. decaying wood, date of collection not mentioned, leg. A Manjikul, COLL. NFIC.

Paratypes-8 soldiers; Paramorphotypes-4 imagoes, 8 workers (in spirit both in a single vial), ex. decaying wood, date of collection not mentioned, leg. A Manjikul, COLL. NFIC.

Now known as ***Lacesitermes thailandicus*** (Sen-Sarma, 1966) (vide Krishna, 1970).

Type locality: Thailand-Chanthaburi.

Distribution: Thailand-Chanthaburi.

38. *Trinervitermes nigrirostris* Mathur & Sen-Sarma, 1959. *Z. Ang. Ent.*, 45(1): 71.

Type specimens: Holotype-1 soldier (major); Morphotypes-1 soldier (minor), 1 imago,1 worker (both in spirit in a vial), ex. mound, 30.x. 1956, leg. Gunther Becker, COLL. NFIC.

Paratypes-9 soldiers (major); Paramorphotypes-6 soldier (minor), 3 imagoes, 5 workers, ex. mound, 30.x. 1956, leg. Gunther Becker. COLL. NFIC-6 soldiers (major), 5 soldier (minor), 2 imagoes, 3 workers; Hamburg Museum, Hamburg (Germany)-2 soldiers (major), 1 soldier (minor), 1 imago,1 worker, and in Collection of Prof. Alfred E. Emerson, Chicago (U.S.A.)-1 soldier (major), 1 worker.

Type locality: India-Mandapam (Madras).

Distribution: India-Mandapam and Tirikalikundram Gate (Chennai, Tamil Nadu).

Prof. Dr. P.K. Sen-Sarma is immortalized in the annals of science, especially Isopterology, wherein several species have been named after him, in profuse reverence and honour for his enormous contribution to the science of termites, as annotated below:

Glyptotermes sensarmai Maiti, 1974.

Glyptotermes sensarmai Maiti, 1974. Proc. Zool. Soc., 27: 117-122. Holotype – Soldier (ZSI No. 574/H11) in avail in spirit, Ghagribara, Atiamochar Beat, Cooch behar Forest range, Koch Bihar district, West Bengal, 11.11, 1966. coll. P.K. Maiti, ex. live sisso tree (*Dalbergia sissoo*). COLL. ZSIK (2 soldiers, 6 Pseudoworkers), NIFC (1 soldier, 2 Pseudoworkers) and AMNH (1 soldier, 2 Pseudoworkers).

Trinervitermes sensarmai Bose, 1984

Trinervitermes sensarmai Bose, 1984. Termite fauna of southern India. *Records of the Zoological Survey of India. Occasional paper*, 49: 248. Holotype - Soldier, W. Western slope of Hill colony, North Vijayapuri, Andhra Pradesh, 18.viii. 1962, coll. I.N. Maligai. COLL. ZSIK.

Thrips sensarmai Vijay Veer & Srivastava, 1985.

Thrips sensarmai Vijay Veer & Srivastava, A.K. 1985. Two new species of Thripidae (Insecta: Thysanoptera) from India. *Entomon*, 10(1): 78. Holotype-female, West Bengal, Darjeeling, ex. Flowers of *Citrus* sp., 4.iii.1980, coll. Vijay Veer. COLL. In collection of Dr.Vijay Veer.

Neococcidencyrtus sensarmai Singh, 2007

Neococcidencyrtus sensarmai Singh, 2007. Two New species of Genus *Neococcidencyrtus* Compere (Hymenoptera: Chalcidoidea: Encyrtidae) parasitising diaspidid scales (Hemiptera: Diaspididae) from Assam, India. In this volume. *Holotype*: ♀, dissected and mounted on a slide under six covers; INDIA; Assam: Jorhat, Teok; 7.ix.1994; coll. Sudhir Singh; ex. Diaspidid scales under leaf sheaths of *Bambusa balcooa*; deposited in National Forest Insect Collection- Forest Research Institute, Dehra Dun (NIFC-FRI) with accession no. 21463. *Paratypes*: 3♀♀, 2♂♂; one male and female dissected and mounted on slide; 2♀♀ and 1 ♂ on cards; data same as for holotype.

Dr. Sen-Sarma had made a great service to the science of forest entomology during his tenure of Forest Entomologist, in particular. It is generally difficult to enumerate his endless contributions, but a sincere analytical approach will possibly highlight his research achievements as summarized below:

1. Louse: His research on the sucking lice led to the discovery of existence of the spermetheca in sucking lice. Studies on the detailed anatomy and affinity of elephant louse also resulted in the discovery of two new organs, i.e., Farrish Plate and Weber's Organ, and the creation of a **new order Rhyncophthiraptera**, to accommodate the highly specialized elephant louse. These have recently received due recognition from the specialists of the world.

2. Termites: Dr. Sen-Sarma had a life long association with termite research and worked on taxonomy, biology, biochemistry, ecology, phylogeny, physiology, polymorphism, polyethism and control.

(i) *Taxonomy*: He discovered and described 38 new species, 7 new genera and a new family, Indotermitidae, of termites. He published five monographs on revision of some termite genera and faunistic surveys of some areas in the Indian region which were hitherto *terra incognita* from the termite viewpoint. His elucidation of phylogeny of the world genera of the subfamily Nasutitermtinae (1968) still remains the standard reference work on the subject. His work on the oriental termites was highly appreciated by the termite researchers worldover.

(ii) *Biology & Ecology*: Dr. Sen-Sarma worked out the biology and ecology of several species of termites. His contributions on the ecological homeostasis with particular reference to temperature and humidity responses of various species of termites, termite-fungi interactions, population ecology including worker soldier proportions etc. have received global acclaim.

(iii) *Polymorphism & Polyethism*: He was the first to undertake the study on polymorphism and temporal polyethism in India which include mechanism of social coordination and cohesion among various polymorphic forms that constitute a colony, including division of labour among workers based on age and sex. He also elucidated the developmental pathways among the neuter castes of higher termites, e.g. *Microcerotermes, Odontotermes* and *Nasutitermes.*

(iv) *Biochemical & Physiological Studies*: Dr. Sen-Sarma and his collaborators made significant contributions in the field of biochemical and physiological researches in hard wood feeding (xylophagous) insects. Their research on the digestive physiology of several phylogenetically related species of borer and higher termites has yielded interesting and useful information. They proved conclusively that both larvae of some coleopterous species as well as worker caste in termites are capable of producing the enzyme cellulase and the salivary glands are one of its major sources. The termites are dependent on the hindgut microbiota primarily for the digestion of hemicellulose group. Worker caste of termites has also acquired the capability of digesting a portion of lignin, chitin, fat, etc. The acquired digestive enzymes in fungus growing termites have resulted in the loss of chitinase which is present in sound wood feeding lower termites during the course of evolution.

(v) *Termites Laboratory Culturing and Laboratory Testing of Material*: Prof. Sen-Sarma had been the first in India to develop methods of culturing four species of Indian termites, *Neotermes bosei, Heterotermes indica, Coptotermes heimi* and *Microcerotermes beesoni.* He and his collaborators further improved upon the laboratory testing methods for various treated and untreated materials including synthetic fibers, plastics, cables etc. to test the degree of termites' resistance of these materials. He further devised a non-destructive method (NDM) of testing of the natural termite resistance of wood of trees with superior genetic qualities. This assisted in screening elite trees in a forest stand for genetic improvement.

(vi) *Termites Control*: In addition to using chemicals (soil sterilization) for the control of termites, he and his collaborators developed non-toxic control measures which include use of inert dust, e.g. fluoridated silica aerogel, unbalancing of ratio of food gatherers and non-food gatherers, etc. This has opened up a new vista of research in termites' control.

3. Forest Entomological Problems: He had rendered able leadership in giving directions in solving several forest and timber entomological problems, thus saving the scarce natural resources. His emphasis for forest insect management had always been silvicultural-cum-biologcial methods, integrating therein the IPM concept. He was the first to discover the vector *(Nophotettix virecens)* of dreaded spike disease of sandal *(Santalum album)* which eluded scientists for more than 80 years. He developed an ecological approach for managing spike disease of sandal trees through elimination of vector-host plants by ecological manipulations.

Prof. Sen-Sarma was a well-traveled person. He had visited many countries in America and Europe, such as Berlin, Bonn, Munich, Eberswalde (GDR), Wuerzburg (FGR), Paris, Pavia, Rome, Zurich, Berne, London, Boulder, North Carolina, Washington and New York. A prolific researcher he had published about 194 original scientific papers in various national and international journals as well as contributed Chapters in some multi-authored books published from Holland and USA, and also edited books. Several of his research results are being referred to in books and in reviews. After attaning superannuation in 1987 from the Forest Research Institute, Dehra Dun he had joined Birsa Agricultural University, Ranchi (now capital of Jharkhand) as Dean, Faculty of Forestry and Principal, College of Forestry till 1992. He then joined in 1992 the Centre for Afforestration and Ecodevelopment, North-Eastern Hill University, Shillong (Meghalaya) as an Emeritus Scientist. During about nearly forty years of highly chequered career he had guided several research scholars for the Ph.D. degree some of whom, including both Dr. Vijay Veer and Dr. B.K. Tyagi, are currently holding positions of great responsibility.

Prof. Sen-Sarma was married to a highly cultured and soft spoken young lady named Aparna Sen Gupta from Kolkata in the year 1958. Both Prof. Sen-Sarma and his wife were generous hosts and no visiting friend and associate could ever think to leave their warm hospitality including having a meal with them. They have a son, Shantanu Sen-Sharma, an IT professional, and two daughters, Sharmistha Bhattacharya and Sudeshna Sen-Sharma, all married and well settled. Prof Sen-Sarma passed away on 27 July, 2002 in Noida near New Delhi after a brief illness.

Scientific Publications of Dr. P.K. Sen-Sarma

Technical:

1951

Mukerji, D. and Sen-Sarma, P.K. 1951. Spermatheca in sucking lice. *Nature*, London, **168**: 612.

1953

Sen-Sarma, P.K. 1953. Easy technique of dissecting the mouth parts of sucking-lice (Insecta: Anoplura). *Current Science*, Bangalore, **22**: 47.

1954

Sen-Sarma, P.K. 1954. On sectioning lice. *Zool. Anz.*, **153** (7/8): 194-195.

1955

Mukerji, D. and Sen-Sarma, P.K. 1955. Anatomy and affinity of the elephant louse, *Haematomyzus elephantis* Piaget (Insecta: Rhyncophthiraptera), *Parasitology*, Cambridge, **45**: 5-30.

Roonwal, M.L. and Sen-Sarma, P.K. 1955. Biology and ecology of Oriental termites (Isoptera). No. 3. Some observations on *Neotermes gardneri* (Snyder) (Family Kalotermitidae). *Journal of Bombay Natural History Society*, Mumbai, **53**: 234-239.

1956

Sen-Sarma, P.K. 1956. A systematic catalogue of the main identified entomological collection at the Forest Research Institute, Dehra Dun. Pt. 30-Order Lepidoptera

(Contd…). Suborder Heteroneura (contd…), Superfamily Bombaycoidea. *Indian Forest Leaflet (Entomology)*, Dehli, No. **121** (4): 285-294 (Printed in1963).

Roonwal, M. L. and Sen-Sarma, P.K. 1956. A systematic catalogue of the main identified entomological collection at the Forest Research Institute, Dehra Dun. Pt. 36-Order Lepidoptera (Contd…).Suborder Heteroneura (contd…), Superfamily Noctuoidea (in part), families Notodontidae, Syntomiidae and Arctiidae. *Indian Forest Leaflet (Entomology)*, Dehli, No. **121** (4): 463 – 480. (Printed in 1963).

Roonwal, M.L and Sen-Sarma, P.K., 1956. Systematics of Oriental termites (Isoptera). No. 3. Zoological Survey of India collections from India and Burma with four new termites of the genera *Parrhinotermes*, *Macrotermes*, *Hypotermes* and *Hospitalitermes*. *Indian Journal of Agriculture Sciences*, New Delhi, **26**: 1-37.

1957

Sen-Sarma, P.K. and Mathur, R.N. 1957. Further records of occurrence of *Cryptotermes dudleyi* (Banks) in India (Insecta: Isoptera: Kalotermitidae). *Current Science*, Bangalore, **26**: 399.

1958

Mathur, R.N. and Sen-Sarma, P.K. 1958a. A new species of *Anacanthotermes* from South India (Insecta: Isoptera: Hodotermitidae). *Entomologist*, London, **91**(1146): 233-241.

Mathur, R.N. and Sen-Sarma, P.K. 1958b. Hitherto unknown imago caste of the species of *Globitermes*, *Microcerotermes* and *Odontotermes* from Indo-Malayan Region (Isoptera: Termitidae). *Indian Forest Bulletin (Entomology)*, Delhi, No. **219**: 1-9, 3 pls.

1959

Mathur, R.N. and Sen-Sarma, P.K. 1959. Notes on the habits and biology of Dehra Dun termites, Pt – 1. *Journal of Timber Dryers' and Preservers' Association of India*, Dehra Dun, **5** (3): 3-9, 2 pls.

Mathur, R.N. and Sen-Sarma, P.K. 1959. Two new termites *Emersonitermes thekadensis* gen. et sp. nov. and *Trinervitermes nigirirostris* sp. nov. from India (Insecta: Isoptera). *Zeit Ang. Ent.*, Hamburg, **45**(1): 66-78, 5 text figs.

Prashad, Baini. and Sen-Sarma, P.K. 1959. Revision of the termite genus *Nasutitermes* Banks (Isoptera: Termitidae: Nasutitermitinae) from the Indian Region. *Indian Council of Agriculture Research*, New Delhi, Monograph No. **10.23**: 1-66.

1960

Mathur, R.N. and Sen-Sarma, P.K. 1960. Notes on the habits and biology of Dehra Dun termites, Pt-2. *Journal of Timber Dryers' and Preservers' Association of India*, Dehra Dun, **6**(2): 23-27, 1pl.

Mathur, R.N. and Sen-Sarma, P.K. 1960. A new species of *Angulitermes* from South India (Insecta: Isoptera: Termitidae: Termitinae). *Ann. Mag. Nat. Hist.*, London, **13**: 401-406.

Mathur, R.N. and Sen-Sarma, P.K. 1960 *Glyptotermes nigifrons* sp. nov. from South India (Insecta: Isoptera: Kalotermitidae). *Entomologist*, London, **93**: 79-85.

Prashad, Baini and Sen-Sarma, P.K. 1960. Revision of the genus *Hospitalitermes* Holmgren (Isoptera, Termitidae, Nasutitermitinae) from the Indian region. *Indian Council of Agriculture Research*, New Delhi, Monograph No. **10.29**: 1-32.

Roonwal, M.L and Sen-Sarma, P.K. 1960. Systematics of Oriental termites, No 4. A new species of *Synhamitermes* Holmg. from Ceylon and the imago of *Speculitermes quadriceps* (Wasm.) (Termitidae: Amitermitinae): 1-14. In: Contributions to the systematics of Oriental termites (Eds. Roonwal, M.L and Sen-Sarma, P.K). *Indian Counc. Agric. Res.*, New Delhi, Entomological Monograph No. **1**, xiv+407 pp., 65 pls.

Roonwal, M.L. and Sen-Sarma, P.K. 1960. Systematics of Oriental termites. No. 5. New and little known species of genera *Speculitermes*, *Capritermes* and *Odontotermes*: 15-39. In: Contributions to the systematics of Oriental termites (eds. Roonwal, M.L and Sen-Sarma, P.K). *Indian Counc. Agric. Res.*, New Delhi, Entomological Monograph No. **1**, xiv+407 pp., 65 pls.

Roonwal, M.L and Sen-Sarma, P.K. 1960. On a new termite, *Indotermes maymensis* Roonwal and Sen-Sarma gen. et sp. nov., from Burma (Family Indontotermes, nov.): 40-45. In: In: Contributions to the systematics of Oriental termites (eds. Roonwal, M.L and Sen-Sarma, P.K). *Indian Counc. Agric. Res.*, New Delhi, Entomological Monograph No. **1**, xiv+407 pp., 65 pls.

Roonwal, M.L and Sen-Sarma, P.K., 1960. Revision of the termite genus *Eremotermes* Silverstri (Isoptera: Termitidae: Amitermitinae): 46-63. In: In: Contributions to the systematics of Oriental termites (eds. Roonwal, M.L and Sen-Sarma, P.K). *Indian Counc. Agric. Res.*, New Delhi, Entomological Monograph No. **1**, xiv+407 pp., 65 pls.

Roonwal, M.L and Sen-Sarma, P.K., 1960. Systematics monograph of Oriental Kalotermitidae (Isoptera). Pt. 1. Genera *Kalotermes* and *Neotermes*: 94-407. In: In: Contributions to the systematics of Oriental termites (eds. Roonwal, M.L and Sen-Sarma, P.K). *Indian Counc. Agric. Res.*, New Delhi, Entomological Monograph No. **1**, xiv+407 pp., 65 pls.

1961

Mathur, R.N. & Sen-Sarma, P.K. 1961. *Capritermes orientalis*, a new species from Burma. *Bulletin of Entomology*, Madras, **2**: 1- 4.

Sen-Sarma, P.K. and Mathur, R.N. 1961. *Trinervitermes biformis* (Wasmann), a mound building termite in South India. *Indian Forester*, Dehra Dun, **87**(4): 252, 1 pl.

1962

Chatterjee, P.N.and Sen-Sarma, P.K. 1962. Seasonal incidence of wood destroying subterranean termites. *Indian Forester*, Dehra Dun, **88**(2): 139-141.

Chatterjee, P.N. and Sen-Sarma, P.K. 1962. *Odontotermes paralatigula*, a new species of termite from Burma (Isoptera: Termitidae: Macrotermitinae). *Journal of Bombay Natural History Society*, Bombay, **59**(3): 822-826.

Mathur, R.N. and Sen-Sarma, P.K. 1962. Notes on the habits and biology of Dehra Dun termites. Pt. 3. *Journal of Timber Dryers' and Preservers' Association of India*, Dehra Dun, **8**(1); 1-18, 2pls.

Mathur, R.N. and Sen-Sarma, P.K. 1962. Imago caste of *Odontotermes assmuthi* (Holmgren) (Isoptera: Termitidae). *Bulletin of Entomology*, Madras, **3**: 7-12.

Sen-Sarma, P.K. 1962. Some observations on swarming in nature and colony foundation under laboratory conditions in *Odontotermes assmuthi* (Holmgren) at Dehra Dun. *Beitr. Zur Entomol.*, **12** (3-4): 292-297.

1963

Chatterjee, P.N. and Sen-Sarma, P.K. 1963. Antitermite characteristics of insecticides I. Results of field experiments using commercial samples of shell liquid of *Anacardium occidentale*, Aldrin and Dieldrin. *Indian Forester*, Dehra Dun, **89**(4): 280-285.

Sen-Sarma, P.K. 1963. Studies on the natural resistance of timbers to termites. I. Observations on the longevity of the test termites, *Heterotermes indicola* (Wasm.) in the saw - dust from forty common Indian timbers. *Indian Forest Bulletin (New Series) (Entomology)*, **220**: 1-3. (FRI 14.220).

Sen-Sarma, P.K. 1963. Studies on the natural resistance of timbers to termites. II. Preliminary result of accelerated laboratory tests of 37 common Indian timbers against *Heterotermes indicola* (Wasm.). *Indian Forest Bulletin (New Series) (Entomology)*, **220**: 4-10. (FRI-14.220).

Sen-Sarma, P.K. 1963. Methods of testing of termite resistance of materials in European laboratories. *Journal of the society of Indian Forester*, Dehra Dun, **3**(2): 57-65.

Sen-Sarma, P.K. 1963. Studies on the natural resistance of timbers to termite attack. III. Results of accelerated laboratory tests of 9 species of Indian woods against the Mediterranean dry wood termite (Yellow necked dry wood termite) *Kalotermes flavicollis* Fabr. *Holzfor. Holzverw.*, Vienna, **15**(3): 51-56.

1964

Sen-Sarma, P.K. 1964. The effect of temperature and relative humidity on the longevity of pseudoworkers of *Kalotermes flavicollis* (Fabr.) (Isoptera) under starvation conditions. *Proc. National Inst. Sci. of India*, Calcutta **30 (B)**(5/6): 300 – 314.

1965

Sen-Sarma, P.K. 1965. A new termite genus and species from South India (Insecta: Isoptera: Termitidae). *Proceedings of Zoological Society*, Calcutta, **18**: 143 -148.

Sen-Sarma, P.K. and Chatterjee, P.N. 1965. Colony foundation through substitute reproductives in *Heterotermes indicola* (Wasmann) under laboratory conditions (Insecta: Isoptera). *Journal of Timbers Developer Association India*, Dehra Dun, **11**(3): 9-11, 1pl.

Sen-Sarma, P.K. and Chatterjee, P.N .1965. Studies on the natural resistance of timbers to termite attack. IV. Qualitative and quantitative estimations of resistance of sixteen species of Indian woods against *Neotermes bosei* Snyder (Isoptera: Kalotermitidae) based on laboratory tests. *Indian Forester*, Dehra Dun, **91**(11): 805-813.

Sen-Sarma, P.K. and Kloft, W. 1965. Trophallaxis in pseudoworkers of *Kalotermes flavicollis* (Fabricius) (Insecta: Isoptera: Kalotermitidae) using radioactive I^{131} . *Proceedings of Zoological Society*, Calcutta, **18**(7): 41-46.

1966

Prashad, B. and Sen-Sarma, P.K. 1966. Review of the genus *Trinervitermes* Holmgren from the Indian Region (Isoptera: Termitidae: Nasutitermitinae). *Indian Forest Bulletin (New Series) (Entomology)*, Delhi, **247**: 1-42, 14 plates.

Prashad, B. and Sen-Sarma, P.K. 1966. Studies on the genera *Ceylonitermes*, *Grallatotermes, Emersonitermes* and *Ceylonitermellus* from the Indian region (Isoptera: Termitidae: Nasutitermitinae). *Indian Forest Bulletin (New Series) (Entomology)*, Delhi, **248**: 1-17. 6 plates. (FRI-14248).

Sen-Sarma, P.K. 1966. Threes new genera of *Grallatotermes* complex (Isoptera: Termitidae: Nasutitermitinae). *J. Bombay Nat. Hist. Soc.*, Mumbai, **63**(1): 167-184, 8 text figs.

Sen-Sarma, P.K 1966. Two new termites, *Microcerotermes manjikuli* sp. nov. and *Thailanditermes thailandicus* gen. et sp. nov. from Thailand (Isoptera: Termitidae). *Indian Journal of Entomology*, New Delhi, **28**: 197-209.

Sen-Sarma, P.K and Chatterjee, P.N. 1966. Humdity behaviour of termites. I. Effect of relative humidity on the longevity of workers of *Microcerotermes beesoni* Snyder (Insecta: Isoptera: Termitidae) under starvation. *Indian Forest Bulletin (New Series) (Entomology)*, Delhi, **255**: 1-6.

Sen-Sarma, P.K and Chatterjee, P.N. 1966. The effect of preconditioning on the humidity reactions of workers of *Microcerotermes beesoni* Snyder (Isoptera: Termitidae). *Insectes Sociaux*, Paris, **13**(4): 267-276.

Sen-Sarma, P.K and Chatterjee, P.N. 1966. Tolerance of pseudoworkers of *Neotermes bosei* Snyder (Insecta: Isoptera: Kalotermitidae) to prolonged starvation. *Science and Culture*, Calcutta, **32**: 472.

1967

Prashad, B., Thapa, R.S. and Sen-Sarma, P.K. 1967. Revision of the Indian species of the genus *Microcerotermes* Silvestri (Isoptera: Termitidae: Amitermitinae). *Indian Forest Bulletin (New Series) (Entomology)*, Delhi, **246**: 1-56, 35 text figs. (FRI 14.246).

1968

Chatterjee, P.N and Sen-Sarma, P.K, 1968. Important current problems of forest entomology in India. *Indian Forester*, Dehra Dun, **94**(1): 112-117.

Sen-Sarma, P.K, 1968. Phylogenetic relationship of the termite genera of the subfamily Nasutitermitinae (Isoptera: Termitidae). *Oriental Insects*, Delhi, **2**: 1-34.

Sen-Sarma, P.K and Chatterjee, P.N. 1968. Studies on the natural resistance of timbers to termite attack. V. Laboratory evaluation of the resistance of three species of Indian wood to *Microcerotermes beesoni* Snyder (Termitidae: Amitermitinae). *Indian Forester*, Dehra Dun, **94**: 694-704.

Sen-Sarma, P.K. and Gupta, B.K. 1968. Responses of termites (Insecta: Isoptera) to gravity. *Current Science*, Bangalore, **37**(17): 501-502.

1969

Sen-Sarma, P.K, 1969. Effect of relative humidity on the longevity of starving workers and soldiers of *Heterotermes indicola* (Wasm.) (Insecta: Isoptera): 263-266 In: Proceedings of the Congress of the International Union for the study of social insects, Bern, 15-20 Sept., 1969.

Sen-Sarma, P.K and Mishra, S.C. 1969. Seasonal variations of nest population in *Microcerotermes beesoni* Snyder. Proc. Nat. Inst. Sci. India, New Delhi **35, B** (5): 361- 367.

1970

Mathur, R. N., Chatterjee, P. N. and Sen-Sarma, P.K. 1970. Biology, ecology and control of *Alianthus* defoliator, *Atteva fabriciella* Swed. (Lepidoptera: Yponomeutidae) in Madhya Pradesh. *Indian Forester*, Dehra Dun, **96**(7): 538-552.

Sen-Sarma, P.K and Chatterjee, P. N. 1970. Recent trends in laboratory evaluation of termite resistance of wood and wood products. *Journal of Indian Academy Wood Science*, Bangalore, **1** (2): 119-128.

Sen-Sarma, P.K., Mishra, S.C. and Gupta, B.K. 1970. Studies on the natural resistance of timbers to termite attack. VI. Laboratory evaluation of resistance of 'Mundani', 'Anjan' and 'Banati' to *Microcerotermes beesoni* Snyder. *Indian Forester*, Dehra Dun, **96**(2): 75-84.

1972

Sannasi, A., Sen-Sarma, P.K., George, C.J. and Basalingappa, S. 1972. Juvenile hormone activity from various sources of termite castes and their fungus garden. *Insectes Sociaux*, Paris, **19**(2): 81-85.

Sen-Sarma, P.K 1972. Laboartory testing of natural resistance of Indian woods: 8-14. In: *Termite Problems in India* (ed. M.L. Roonwal), CSIR, New Delhi.

Sen-Sarma, P.K, 1972. Insect factors in the management of man-made forests. *Proc. & Tech. papers of Symp. on Man-made Forests in India*, Dehra Dun, part 4: 16-19. Society of Indian Foresters, New Forest, Dehra Dun.

Sen-Sarma, P.K and Mishra, S.C., 1972. Colony foundation in *Microcerotermes beesoni* Snyder by budding (Insecta: Isoptera: Termitidae). *Current Science*, Bangalore, **41**(7): 257-258.

Sen-Sarma, P.K and Mishra, S.C. 1971. Seasonal fluctuations of colony composition and population in *Neotermes bosei* Snyder (Insecta: Isoptera: Kalotermitidae). *J. Indian Acad. Wood Sci.*, Bangalore, **3**(2): 43-48.

Sen-Sarma, P.K and Thapa, R. S. 1972. Deterioration of timbers by insect borers. Proc. Seminar quality, Construction and Cost aspects of Building and timber structures, Dehra Dun (National Building Organization), New Delhi: 1-10 (Mimeographed)

1973

Sen-Sarma, P.K., Gupta, B.K. and Kapur, G. 1973. Water-loss in workers of *Micrpcerotermes beesoni* Snyder (Insecta: Isoptera: Termitidae) under constant temperature and humidity. *J. Indian Acad. Wood. Sci.* Bangalore, **4**(2): 92-95.

Sen-Sarma, P.K, Thakur, M.L., Mishra, S.C. and Gupta, B.K., 1973. Studies on natural resistance of wood to termites. *Van Vigyan*, **11**(3/4): 23-26.

Thakur, M. L. and Sen-Sarma, P.K, 1973. A new species of termite genus *Doonitermes* Chatterjee and Thakur (Isoptera: Termitidae: Amitermitinae) from West Bengal, India. *Bulletin of Entomology*, New Delhi, **14**: 85-90.

1974

Sen-Sarma, P.K, 1974. Ecology and biogeography of the termites of India: 421-472. In: *Ecology and biogeography in India* (Ed. M.S. Mani), Monographiae Biologicae, vol. **23**, 773 pp+xv. Published by Dr W. Junk, The Hague.

Sen-Sarma, P.K., Arun Kumar and Ahuja, S.S, 1974. Biometrical studies on the sexual dimorphism in *Hoplocerambyx spinicornis* Newman (Coleoptera: Cerambycidae). *Journal of Indian Academy of Wood Science*, Bangalore, **5**(1): 41-53.

Arun Kumar and Sen-Sarma, P.K.1974. Studies on the genitalia of sixteen Indian species of the tribe Cyphicerinii (Coleoptera: Curculionidae: Eremninae) with special reference to their taxonomic importance. *Journal of Indian Academy of Wood Science*, Bangalore, **5**(2): 86-105.

Sen-Sarma, P.K and Thakur, M. L., 1974. Biology, distribution and economic significance of some termites attacking buildings. *Indian Plywood Industry Research Institute Journal*, Bangalore, **4** (3): 115-125 (Paper read in National Building Organization symposium on quality construction and cost aspects of building, timbers and timber structures held on at 1-3 June, 1972, Dehra Dun).

1975

Sen-Sarma, P.K, 1975. Termites-our foes and friends. *Everyman's Science*, Calcutta, **10**(2): 67-72.

Sen-Sarma, P.K, Thakur, M.L., Mishra, S.C. and Gupta, B.K., 1975. Studies on wood destroying termites in relation to natural termite resistance of timber. Final techn. Report. PL- 480 Project A-7-58 (1968-73), Dehra Dun, viii + 1-187, 10 text figs, 26 pls.

1976

Sen-Sarma, P.K and Mishra, S.C., 1976. Bioassay of silica aerogel against wood inhabiting termite, *Neotermes bosei* Snyder. *Indian Journal of Entomology*, New Delhi **38**(1): 11-17.

1977

Sen-Sarma, P.K and Mishra, S.C., 1977. The influence of temperature and relative humidity on the survival of workers of *Microcerotermes beesoni* (Snyder). *J. Indian Acad. Wood Sci.*, Bangalore, **8**(1): 21-26.

Sen-Sarma, P.K 1977. Evaluation of termite resistance of wood. Proc. National Seminar Optimum Utilization of Wood Resources, New Delhi, 111-112.

Sen-Sarma, P.K 1977. Saving wood destruction from insects. Proc. National Seminar Optimum Utilization of Wood Resources, New Delhi, 113-115.

1978

Gupta, B. K. and Sen-Sarma, P.K 1978. Antitermite properties of some anthraquinone derivatives. *Holzforschung und Holzverwertung*, Vienna, **30**(3): 57-58.

Sen-Sarma, P.K 1978. Insect pests and their control in rural housing. Indian J. Entomol., New Delhi, **39** (3): 284-287.

Sen-Sarma, P.K 1978. Spike disease of sandal - an yellows type of disease. *Indian Journal of Forestry*, Dehra Dun, **1**(2): 100-104.

Sen-Sarma, P.K, Ananthanarayana, A.K., Parthasarthi, K., Deekshatulu, B.L., Perinbam, J.B.P, Naga Bhusana, S.R. and Sharma, R.D. 1978. Application of remote sensing technique in the enumeration of sandal trees. *Indian Forester*, Dehra Dun, **104** (2): 123-132.

Sen-Sarma, P.K. and Gupta, B.K. 1978. Natural resistance of six Indian timbers to termites under laboratory conditions. *Holzforsch und Holzverw.*, Vienna, **30**(4): 88-91.

Sivaramakrishnan, V.R. and Sen-Sarma, P.K.1978. Experimental transmission of spike disease of sandal, *Santalum album* by the leafhopper *Nephotettix virescens* (Distant) (Homoptera: Cicadellidae). *Indian Forester*, Dehra Dun, **104**(3): 202-205.

Thakur, M.L. and Sen-Sarma, P.K 1978. Macrotermitinae (Isoptera: Termitidae) from Madagascar, with remarks on the origin of the sub-family in Malagasy Region. *J. Entomol. Res.*, New Delhi, **2**(2): 172-180.

Sen-Sarma, P.K. and Thakur, M.L. 1978. Population ecology of Indian termites. *Bull. Entomol.*, New Delhi **19**: 89-99.

Thakur, M.L. and Sen-Sarma, P.K. 1978. A flourishing colony of *Coptermes heimi* (Wasm.) (Insecta: Isoptera) in a naval boat. *J. Bombay Nat. Hist. Soc.*, Mumbai, **76** (1): 188-189.

1979

Sen-Sarma P.K and Thakur, M.L. 1979. Termites of Tripura (Insecta: Isoptera) *Indian Forest Records (New Series) (Entomology)* **13**(1): 1-67, 11 pls

Agarwal, V. B. and Sen-Sarma, P.K 1979. Circadian and seasonal fluctuations of temperature and relative humidity inside the mound of *Odontotermes microdentatus* (Isoptera: Termitidae). *Acta Entomologia Bohemoslovaca*, Prague, **76**: 379-386.

Gurumurti, K, Raju, P.S. and Sen-Sarma, P.K 1979. Evidence for the existence of toxin like substances in spiked sandal. *Indian Journal Forestry*, Dehra Dun, **2**(1): 43-45.

Mishra, S. C. and Sen-Sarma, P.K 1979. Deterioration of wood by insects, III. Chemical composition of faecal matters, nest material and fungus comb of some Indian termites. *Material und Organismen*, Berlin, **14**: 1-14.

Mishra, S. C. and Sen-Sarma, P.K 1979. Studies on deterioration of wood by insects. V. Influence of temperature and relative humidity on wood consumption and digestibility in *Neotermes bosei* Snyder (Isoptera: Kalotermitidae). *Material und Organismen*, Berlin, **14**: 278-286.

Sen-Sarma, P.K and Thakur, M.L. 1979. Relative termite resistance of heartwood of teak trees from known seed sources. *Holzforsch. und Holzverw.*, Vienna, **31**(1): 14-16.

Shankaranarayana, K.H., Shivaramakrishnan, V.R., Ayyar, K.S. and Sen-Sarma, P.K. 1979. Isolation of a compound from the bark of sandal, *Santalum album* L. and its activity against some lepidopterous and coleopterous insects. *J. Entomol. Res.*, New Delhi, **3**(1): 116-118.

Sen-Sarma, P.K and Gupta, B.K. 1979. Antitermite characteristics of wood of *Lophopetalum wightianum. Holzforsch. und Holzverw.*, Vienna, **31**(5): 100-101.

Sen-Sarma, P.K and Gupta, B.K. 1979. Food preference in laboratory colonies of *Microcerotermes beesoni* Snyder. *Indian Journal of Entomol*ogy, New Delhi, **41**(3): 183-185.

Thakur, M.L. and Sen-Sarma, P.K. 1979. Revision of termite genus *Heterotermes* Froggatt(Isoptera: Rhinotermitidae: Heterotermitinae) from the Indian region. *Indian Forest Records, (New Series) (Entomology)*, New Delhi, **13**(2): 1-27, 9 pls. (FRI 23.xiii.2).

Mishra, S.C. and Sen-Sarma, P.K. 1979. Seasonal fluctuations of colony composition, nest population and foraging in *Nasutitermes dunensis* Chatterjee and Thakur (Isoptera : Termitidae). *Indian Journal Entomology*, New Delhi, **41** (4): 360-365.

1980

Mishra, S.C. and Sen-Sarma, P.K., 1980. Studies on deterioration of wood by insects. VI. Degradation and digestion of lignin in the digestion tract of termites. *Material und Organismen*, Berlin, **15**(2): 119-124.

Mishra, S.C. and Sen-Sarma, P.K., 1980. Humic acids in faecel mattar, nest material and fungus comb of termites. *Bulletin of Entomology*, New Delhi, **21**: 122-125.

Mishra, S.C. and Sen-Sarma, P.K., 1980. Studies on deterioration of wood by insects. VII-Consumption and assimilation of wood and utilization of major wood components by the larvae of *Buprestis geometrica* C. and G. (Coleoptera: Buprestidae). *J. Indian Acad. Wood Sci.*, Bangalore, **11** (1): 16-20.

Sen-Sarma, P.K. and Thakur, M.L., 1980. On a collection of termites of Kumaon Hills, Uttar Pradesh, India (Insecta: Isoptera). *Indian Forest Records (New Series) (Entomology)*, New Delhi, **14** (1): 1-30.

Thakur, M.L. and Sen-Sarma, P.K., 1980. Current status of termites as pests of forest nurseries and plantations in India. *J. Indian Acad. Wood Sci.*, Bangalore, **11**(2): 7-15.

Tyagi, B.K. and Sen-Sarma, P.K.1980. Inter-colonial aggression in workers of the wood carton nest building termite *Microcerotermes beesoni* Snyder (Insecta: Isoptera; Termitidae). *Abstr. Pap. Symposium on the Ecology of Insect Communities,* Dehradun, p. 35.

Sen-Sarma, P.K.1980. Some current problems of forest insect pests and the research programmes. Proc. 2nd For. Conf., Dehra Dun, 1980, 1-12.

1981

Mishra, S.C. and Sen-Sarma, P.K., 1981. Chitinase activity in the digestive, tract of termites (Isoptera). *Material und Organismen*, Berlin, **16** (2): 157-160.

Mishra, S.C. and Sen-Sarma, P.K., 1981. Neutral lipid in different castes of termites. *Indian Journal Entomology*, New Delhi, **43**(1): 80-82.

Mishra, S.C. and Sen-Sarma, P.K., 1981. Hydrogen ion-concentration in the digestive tract of three species of Indian termites (Insecta: Isoptera). *Entomon*, Trivandrum, **6** (2): 131-134.

Sen-Sarma, P.K., 1981. Recent advances in termite research in India and future strategy: 236- 253. In "Progress in Soil Biology and Ecology in India", (Ed. G.K. Veeresh) University of Agricultural Sciences, Bangalore.

Sen-Sarma, P.K., 1981. Insect pests of forest plants. Proc. All India Workshop on pest management, recommendations and future guidelines of research, Udaipur, 1980,

Sen-Sarma, P.K., 1981. Sandal spike, a vector transmitted tree disease. Proc. Indian Nat. Sci. Acad. New Delhi, (B), **46**: 817-821.

Sen-Sarma, P.K. and Gupta, B.K., 1981. Insect pests of poplars: 121-122. In: *Proc. Silviculture, Management and Utilization of Poplars* (October 16-18, 1979), Srinagar (J & K). (Edited R.V. Singh, Govt. of India Press).

Sen-Sarma, P.K. and Mishra, S.C., 1981. Fluctuations of brood population in termite colony (Insecta: Isoptera). *Indian J. Soil Biol. Ecol.*, Bangalore, **1**: 47-55.

Sen-Sarma, P.K. and Singh, S., 1981. Biodegradation of pulp wood and its control. *Paper pulp and For. Prod. J.,* Dehra Dun, **2**: 15-19.

Sen-Sarma, P.K. and Thapa, R.S., 1981. Recent advances in Forest Entomology in India: 21-36. In: *Recent Advances in Entomology in India*, (Ed. T. N. Ananthakrishnan). Loyala College, Madras, 161 pp.

Thakur, M.L. and Sen-Sarma, P.K., 1981. Flight and colony foundation in termites with special reference to Indian Region. *Van-Vigyan*, Dehra Dun, **19**(1): 39-43. (Fourth all India congress Zool., Gaya, 18-20 Oct., 1978).

Mishra, S.C. and Sen-Sarma, P.K, 1981. Movement of food material in the digestive tract of *Neotermes bosei* Snyder and *Odontotermes distans* Holmgren (Insecta: Isoptera). *J. Entomol. Res.*, New Delhi, **5**(1): 66-68.

1982

Sen-Sarma, P.K. & Mishra, S.C. 1982. Natural termite resistance of some Indian woods. *Proc. Natl. Sem. Build. Mat.*, New Delhi, III, (**18**): 1-4.

Sen-Sarma, P.K, 1982. Economically important Oriental termites of the Oriental regions and their management: 142-146. In: *The Biology of Social Insects* (Eds. M.D. Breed, C. D.Michener & H. E. Evans), Westview Press, Boulder (Colorado).

Sen-Sarma, P.K, 1982. Wood destroying insects: 161-175. In: Handbook of Indian Wood and Wood Panels (ed. V. Ranganathan). Indian Acad. Wood. Sci., Dehra Dun…?

Sen-Sarma, P.K., 1982. Insect vectors of sandal spike disease. *European Journal Forest Pathology*, Hamburg, **12**: 297-299.

Sen-Sarma, P.K 1982. Sandalwood : Its cultivation and utilization: 395-405. In: Cultivation and Utilization of Aromatic Plants. (Ed. Atal, C.K. and Kapur, B.M), xxiii + 815 pp. Regional Research Laboratory, Jammu-Tawi.

Gupta, B.K., Vijay Veer & Sen-Sarma, P.K., 1982. Laboratory evaluation of some insecticides against third instar larvae of *Selpa celtis* Moore (Noctuidae: Lepidoptera). *Indian Forester*, Dehra Dun, **108**(7): 481- 486.

Sen-Sarma, P.K., Gupta, B.K. & Vijay Veer, 1982. Relative toxicity of some insecticides against third instar larvae of *Dirades adjutaria* Walker (Epiplemidae: Lepidoptera). *Indian Journal of Forestry*, Dehra Dun, **5** (1): 42- 46.

Sen-Sarma, P.K., Singh, S. and Vijay Veer, 1982. Strengthening plant quarantine in India and South East Asia. FAO Publication.

Sen-Sarma, P.K., Singh, Y., Mishra, S.C. and Goswami, N.L., 1982. Studies on the anti-termite characteristics of some organic fibrous material. *Proc. Natn. Sem. Building Material*, New Delhi, III, (1): 1-3.

Sen-Sarma, P.K. and Mishra, S.C., 1982. Population of soldiers in termite colonies (Insecta: Isoptera): 23-30. In: *Proc. Symp Ecology Animal Population.* Zoological Survey of India, Pt. 4, pp 23-30.

Tyagi, B.K., Pandey, P.C. Rehill, P.S. and Sen-Sarma, P.K, 1982. Termite-fungi interactions. I. Bioassay of decayed wood to *Coptotermes heimi* (Wasmann), *Neotermes bosei* Snyder and *Microcerotermes beesoni* Snyder. *Assiut. J. Agricultural Sciences, Assiut (Egypt)*, 13 (3): 139-147. (Paper presented at the Second *International seminar on the Management of tropics in Arid Ecosystems, 6-11 March, 1982,* Assiut (UAR), p. 41).

Tyagi, B.K. and Sen-Sarma, P.K. 1982. Ecological studies on termites with reference to their role in the process of wood decomposition in the Dehra Dun forests

(Northern India): 23-24. *Symposium on Insect Ecology and Resource Management*, Muzaffarnagar (Ed. S.C. Goel).

1983

Vijay Veer and Sen-Sarma, P.K., 1983. Descriptions of the third instar larvae of some species of the genus *Anomala* (Coleoptera: Scarabaeidae: Rutelinae) from India. *J. Soil Biol. Ecol.*, Bangalore, **3** (2): 124-135.

Tyagi, B.K., P.C. Pandey, P.S. Rehill & P.K. Sen-Sarma, 1983. Termite-fungi interactions. III. Laboratory testing of *Diplodia pinea* attacked wood-blocks to *Microcerotermes beesoni* Snyder (Isoptera: Termitidae). *Annals of Entomology*, Dehra Dun, **1**: 88-89

Tyagi, B.K. and Sen-Sarma, P.K, 1983. Inter-colonial aggression in carton nest building termite *Microcerotermes beesoni* Snyder (Termitidae): 15-22. In: Insect Interrelations in forest and agro-ecosystems (Eds P. K. Sen-Sarma, S. K. Kulshrestha and S. K. Sangal), Jugal Kishore & Co., Dehra Dun.

Sen-Sarma, P.K., Ahmed, S.I. and Ahuja, S.S. 1983. Influence of temperature and relative humidity on the survival period of starved adults and larvae of *Calopepla leayana* Latr. (Chrysomelidae: Coleoptera): 59-77. In: Insect Interrelations in Forest and Agroecosystems (Eds. P.K. Sen-Sarma, S.K. Kulshrestha and S. K. Sangal). Jugal Kishore & Co., Dehra Dun.

Sen-Sarma, P.K., 1983. Some aspects of ecology of termites and resource management: 10-16. In: Proceedings of the symposium on Insect ecology and resource management, (ed. S.C. Goel), Santam Dharam College, Muzaffarnagar (U.P., India).

Ahmed, S.I. and Sen-Sarma, P.K. 1983. On seasonal variations in population *Calopepla leayana* Latr. (Coleoptera: Chrysomelidae): 83-88. In: Proceedings of the symposium on Insect Ecology and Resource Management, (ed. S. C. Goel), Mazaffarnagar). Santam Dharam College, Muzaffarnagar (U.P., India).

McMahan, E.A., Sen-Sarma, P.K. and Sushilkumar. 1983. Biometric, polyethism and sex- ratio studies of *Nasutitermes dunensis* Chatterjee and Thakur (Isoptera: Termitidae). *Ann. Entomol.*, Dehra Dun, **1** (1): 15-25.

Ahmed, S.I. and Sen-Sarma, P.K. 1983. Investigation on a newly recorded nuclear polyhedrosis virus of *Pygaera fulgurita* Wlk. (Notodontidae: Lepidoptera I. Nature of disease and insect-virus interactions. *Journal of Entomological Research*, New Delhi, **7**(1): 150-153.

Sen-Sarma, P.K, 1983. Host as an environment in the ecological succession of insect borers in freshly felled tree trunks: 109-115. In: *Proc. Symp. Host. Environ. Zool. Surv. India.* Zoological Survey of India, Kolkata.

Sen-Sarma, P.K., 1983. Economically important termites and their management in the Oriental region: 69-102. In: Economic Importance and Control of Social insects. (Ed. S.B. Vinson), Praeger Press, New York.

Sen-Sarma, P.K., 1983 Forest Entomology in India. *Indian Review of Life Sciences*, Jodhpur 3: 89-103.

Sen-Sarma, P.K., 1983. Some aspects of forest pest management. Presidential address All India Symposium on Pest Management Section Forest Pest Udaipur (Rajasthan), December 1978, pp. 25-30.

Sen-Sarma, P.K., 1983. Mycoplasma Spiroplasma diseases of forest trees in India. International Union Forestry Research Organization (IUFRO) Conference, Tokyo.

Sen-Sarma, P.K., 1983. Important factors in the rationalisation of methodology in laboratory evaluation of termite resistance of wood and wood products. Workshop on Testing of Wood and Wood Products.-Rationalisation of Methodologies and Procedures in Wood Testing, Dehra Dun, pp. 1-7.

Sen-Sarma, P.K. and Rehill, P.S., 1983. Mycoplasma and allied diseases and constraints for conservation of forest productivity in India. Proceedings of the Symposium held in Jaipur, Rajasthan, 1982.

Sen-Sarma, P.K. and Singh, S., 1983. Perspective manpower requirements in Forest Biology during the next two decades. Presented at Seminar on "Forest Education, Research and Extension.

Sen-Sarma, P.K. and Thakur, M.L., 1983. Insect pests of *Eucalyptus* and their control. *Indian Forester*, Dehra Dun, **109** (12): 864-881.

1984

Sen-Sarma, P.K. and Verma, S.C. 1984. Review of the subfamily Nasutitermitinae from India. *Oriental Insects*, New Delhi, **17**: 79-108.

Sen-Sarma, P.K. and Ahmed, S.I. 1984. A polyhedral virus infecting poplar defoliator, *Pygaera fulcurita* [sic] Walk. (Noctuidae: Lepidoptera). *Indian Journal of Entomology*, New Delhi, **46** (1): 112-114.

Sen-Sarma, P.K. and Ahmed, S.I. 1984. Investigations on the nuclear polyhedrosis virus, *Pygaera fulgurita* Wlk. (Lepidoptera: Notodontidae). Nature of disease and insect-virus interactions. *J. Entomol. Res.*, New Delhi, **7**(1): 150-153.

McMahan, E.A., Sushilkumar and Sen-Sarma, P.K. 1984. Male/female (size) polyethism in workers of *Odontotermes distans* Holmgren and Holmg. & Holmg. (Isoptera : Termitidae: Macrotermitinae). *Ann. Entomol. Soc. Am.*, **77**(4): 429-434.

Sen-Sarma, P.K, 1984. Mycoplasma and allied diseases of forest trees in India and vectors-hosts-pathogens interactions. *Proc. Indian Acad. Sci, (Anim Sci.)*, Bangalore, **93**(4): 323-333.

Sen-Sarma, P.K. and Ahmed, S.I., 1984. Studies on the digestive physiology of *Calopepla leayana* Latr. (Chrysomelidae: Coleoptera). I. Digestibility and chemical composition of food and faecal matters. *Annals of Entomology*, Dehra Dun, **2**(1): 11-15.

Sen-Sarma. P.K. and Thakur, M.L, 1984. War against termites in buildings. *Newsletter Indian Acad. Wood Sci.*, Dehra Dun, **9** (4): 1-2.

Tyagi, B.K., Pandey, P.C., Rehill, P.S. and Sen-Sarma, P.K. 1984. Termite-fungi interactions. 2. Laboratory testing of the decayed wood blocks to the wood carton nest building termite, *Microcerotermes beesoni* Snyder (Insecta: Isoptera). *Material und Organismen,* Berlin, **19**(1): 69-75

1985

Mishra. S.C., Sen-Sarma, P.K. and Rajpal Singh. 1985. Chemical changes in wood during the digestive process in the larvae of *Hoplocerambyx spinicornis. Material und Organismen*, Berlin, **20**(1): 53-64.

Mishra, S.C. and Sen-Sarma, P.K 1985. Nutritional significance of fungus comb and *Termitomyces albuminosus* (Berk.) Heim in *Odontotermes obesus* (Rambur). (Isoptera: Termitidae). *Material und Organismen*, Berlin, **20** (3): 205-213.

Mishra, S.C. and Sen-Sarma, P.K.1985. Carbohydrases in xylophagous coleopterous larvae (Cerambycidae and Scarabaeidae) and their evolutionary significance. *Material und Organismen*, Berlin, **20**(3): 221-230.

Mishra, S.C. and Sen-Sarma, P.K.1985. Degradation and digestion of wood in the gut of faunated and partially defaunated pseudoworkers of *Neotermes bosei* Snyder (Isoptera: Kalotermitidae). *Material und Organismen*, Berlin, **20**(4): 293-300.

Mishra, S.C. and Sen-Sarma, P.K. 1985. The origin and distribution of carbohydrases (cellulase and non-cellulases) in higher termites *Odontotermes distans* (Holmg.) and *Speculitermes cyclops* (Wasm.). *Indian Journal of Entomology*, New Delhi, **47**(1): 83-91.

Ahmed, S.I. and Sen-Sarma, P.K. 1985. Insect vectors of sandal spike diseases and the possible use of light trap for their control: 103-107. In: Proceedings of the National Seminar on Use of Traps in Vector Research & Control, Kalyani, West Bengal.

Sen-Sarma, P.K. 1985. Soil insect pests of forest and their management. Souv and Abst. National Symp. Soil Pests & Soil Organisms, (1984), pp. 32-36.

Sen-Sarma, P.K. 1985. Ecological aspects of vectors of yellows diseases with special reference of Sandal spike disease. Indo-US Workshop on Mycoplasma diseases of crops.-Basic and applied aspects, Dehra Dun, p. 19.

Sen-Sarma, P.K. and Mishra, S.C. 1985. Origin and distribution of proteinases and esterases in lower and higher termites (Insecta: Isoptera). *Proc. Natl. Acad. Sci.* India., New Delhi, (B) **55**: 18-21.

Sen-Sarma. P.K. and Thakur, M.L. 1985. Pest management in Indian Forestry. *Indian Forester*, Dehra Dun, 111(11): 956-964.

Sen-Sarma, P.K. and Vijay Veer.1985. Vertebrate pests of forest trees in India: 224-230. In: *Non-insect pest and predators*. (Eds. R. K. Bhanotar, Shashi Verma, and I Farooqui). All India Scientific Writers' Society of India, New Delhi, 279 pp.

Sen-Sarma, P.K. and Thakur, M.L.1985. Termite problems in plantation crops and their management in the Oriental Region. *Journal Indian Acad. Wood Sci.*, Bangalore **16** (2): 99-108.

Sen-Sarma, P.K. 1985. Novel approach to termite control. *PTI Science Service*, February , 1-15, 1 p.

1986

Sushilkumar and Sen-Sarma, P.K. 1986. Swarming, sex-ratio and sexual dimorphism, weight and moisture content in *Microcerotermes beesoni* Snyder (Isoptera). *J. Entomol. Res*, New Delhi, **10**(2): 180-183.

Sen-Sarma, P.K., 1986. Forest entomological problems in India and their management in the year 2000 A.D. *Proc. Indian Natn. Sci. Acad.*, New Delhi (B) **52** (1): 134-147.

Sen-Sarma, P.K. and Mishra, S.C. 1986. Population dynamics of food-gatherer in a termite colony and its significance in environmentally acceptable control strategy: 228-232. In: Proceedings of symposium Pest Residues and Environmental Pollution,.....

Sen-Sarma, P.K. and Mishra, S.C. 1986. Biological control of forest weeds in India. Retrospect and Prospects. *Indian Forester*, Dehra Dun, **112**: 1088-1093.

Sen-Sarma, P.K. & Thakur, M L. 1986. Impact of deforestation and monoculture plantations in Western Ghats: 238-243. In: Proc. Eco-development of Western Ghats, 17-18 Oct., 1984, PEECHI (Kerala).

Sen-Sarma, P.K. & Thakur, M.L., 1986. Insect pests of Dipterocarpaceae and their management in India. *Indian Journal Forestry*, **9** (1): 29-50.

Sen-Sarma, P.K., Thakur, M.L. and Seghal, H.S., 1986. Protection of forest seeds against pests and fungi during storage: 137-146. In: Proceedings of the national seminar on Forest Tree Seed (28-30 Jan., 1986, Hyderabad).

Sen-Sarma P.K. 1986. Some important problems of forest insect pests and research programmes: 879-883. In: Proceedings of the Second Forestry Conference on 16-19 June, 1980, FRI, Dehra Dun (Ed. N.K. Mathur).

Thakur, M.L and Sen-Sarma P.K. 1986. Current status of termites as pests of forest nurseries and planations in India: 883-887. In: Proceedings of the Second Forestry Conference on 16-19 June, 1980, FRI, Dehra Dun (Ed. N.K. Mathur).

Sen-Sarma P K and Thakur, M L. 1986. Termite problems in fodder and fuel wood plantation and their management in India. *Proc. Symp. Afforestation of Wastelands (28-30 May, 1986).* Dehra Dun., 15 pp

Mishra, S.C. and Sen-Sarma, P.K, 1986. Host specificity test and a note on life history of *Leptobyrsa decora* Drake (Hemiptera: Tingidae) on teak. *Bulletin of Entomology*, New Delhi, **27**: 81-86.

Mishra, S.C. and Sen-Sarma, P.K., 1986. Certain digestive enzymes (carbohydrases, proteinases, chitinases, lignase) and the transit of food material in the digestive tract of larvae of *Stromatium barbatum* Fab. (Coleoptera: Cerambycidae), a dry wood borer. *Indian Journal of Entomology*, New Delhi, **48**: 339-345.

1987

Sushilkumar and Sen-Sarma, P.K. 1987. Some observations on the swarming behaviour, sex-ratio, sexual dimorphism and moisture content study in *Odontotermes distans* (Isoptera: Macrotermitinae). *Annals of Entomology*, Dehra Dun, **5**(2): 55-58.

Mishra, S.C. and Sen-Sarma. P.K.1987. Origin and distribution of digestive enzymes in soldiers of *Neotermes bosei* Snyder with a note on gut morphology (Isoptera: Kalotermitidae). *Material und Organismen*, Berlin, **22**(2): 127-138.

Mishra, S.C. and Sen-Sarma. P.K., 1987. pH trends in the gut of xylophagous insects and their adaptive significance. *Material und Organismen*, Berlin **22**(4): 311-319.

Sen-Sarma, P.K., 1987. Insect pest problems in social forestry plantations and their management. *Indian Journal Forestry*, Dehra Dun, **10** (4): 239-244.

1988

Mishra S.C. and Sen-Sarma, P.K., 1988. Digestion of food in scarabaeid larvae *Holotrichia insularis* Brenske and *Anomala marginipennis* Ar. (Insecta: Coleoptera: Scarabaeidae). *Proc. Natn. Acad. Sci. India*, Allahabad (B) **58**: 23-31.

Sen-Sarma, P.K., 1988. Homeostasis in Termite Society. Presidential Address in Seventy fifth Indian Science Congress (Platinum Jubilee Session). Section, Zoology, Entomology and Fisheries Pune, 1988. pp. 1-38.

Sen-Sarma, P.K. 1988. Homeostasis in Termite Society. *Current Science*, Bangalore, **56**(7): 323.

Sen-Sarma, P.K., 1988. Protection of cultural heritage against insect borers and termites. Proc. Symp. Cultural Heritage, Khajuraho (In press)

Sen-Sarma, P.K. and Thakur, M.L., 1988. Termite problems in fodder and fuel wood plantations in India and their management: 71-77. In: Wasteland development for fuel wood and fodder production (A compilation), Controller of Publications, Delhi, 340 p. (Also presented in National Seminar on fuel wood and fodder from wasteland, 30-31 May, 1986, Dehra Dun).

Sen-Sarma P.K and Thakur, M.L. 1988. Natural termite resistance as a test for rational utilization of timbers: 1-7. In: Proceedings of Second Forest Products Conference, 26-28 May, 1986, FRI, Dehra Dun.

Sen-Sarma, P.K. and Thakur, M.L. 1988. Insect factor in the management of forest resources. *My Forest*, **24**(2): 99-113.

1989

Mishra, S.C. and Sen-Sarma, P.K., 1989. Quantitative relationship between food consumption and excretion of dry matter by larvae of *Anomala polita* Blanch and

Xylotrupes gideon Linn. (Coleoptera: Scarabaeidae). *Annals of Entomology*, Dehra Dun, **7**: 48-52

1990

Ahmed, S.I. and Sen-Sarma, P.K. 1990. Bionomics of *Calopepla leayana* Latr. (Coleoptera: Chrysomelidae), a serious defoliator of *Gmelina arborea* Roxb. plantations in India. *Indian Forester*, **116**: 71-82.

1991

Sushilkumar and Sen-Sarma, P.K.1991. Polymorphism of sterile castes in *Odontotermes obesus* (Rambur) (Isoptera: Termitidae: Macrotermitinae). *Annals of Entomology*, Dehra Dun, **10**(2): 77-83.

1992

Vijay Veer, Gupta, B.K. and Sen-Sarma, P.K. 1992. Ovicidal action of three synthetic pyrethroids against poplar defoliator, *Pygaerea fulgurita* Walker (Lepidoptera: Notodontidae). *J. Entomol. Res.*, New Delhi, **16**(1): 54-56.

Sushilkumar and Sen-Sarma, P.K., 1992. Polyethism in workers of the termite *Odontotermes obesus* (Rambur) (Isoptera: Termitidae: Macrotermitinae). *Annals of Entomology*, Dehra Dun, **10**(1): 57-62.

Sen-Sarma, P.K. 1992. Forest insect pests, their management and environmental considerations. *Proc. Zool. Soc.*, Calcutta, 45 (Suppl. B): 81-102. (Paper presented in the symposium on zoological research in relation to man and environment, Madras, 1-4, 1992, The Zoological society, Calcutta)

1994

Sushilkumar, Sen-Sarma, P.K. and Jha, L.K., 1994. Insect pest management for increased forest production in India. *Annals of Forestry*, Dehra Dun, **2**(1): 191-205.

Sen-Sarma, P.K, Thakur, M.L. and Sehgal, H.S. 1994. Protection of forest seeds against insect pests and fungi during storage: 1-10. In: Forest Entomology. Ashish Publishing Hosue, New Delhi.

Thakur, M.L and Sen-Sarma, P.K. 1994. Termites as pests of forest nurseries and plantations: 11-21. In: Forest Entomology. Ashish Publishing Hosue, New Delhi

Mathur, R.N., Chatterjee, P.N. and Sen-Sarma, P.K. 1994. Biology, ecology and control of Alianthus defoliator, Atteva fabriciella Swed.: 99-120. In: Forest Entomology. Ashish Publishing Hosue, New Delhi

Sen-Sarma, P.K and Thakur, M.L. 1994. Pests of Dipterocarpaceae and their management: 165-186. In: Forest Entomology. Ashish Publishing Hosue, New Delhi

Sen-Sarma, P.K. 1994. Pests problems in rural housing and their management: 313-320. In: Forest Entomology. Ashish Publishing Hosue, New Delhi

Sen-Sarma, P.K. 1994. Pest problems in social forestry plantation and their management: 321-330. In: Forest Entomology. Ashish Publishing Hosue, New Delhi

Sen-Sarma, P.K. 1994. Mycoplasma and allied disease of forest trees and vector host-pathogens interactions: 331-347. In: Forest Entomology. Ashish Publishing Hosue, New Delhi

Sen-Sarma, P.K. 1994. Sandal spike disease: An integrated approach to control: 349-359. In: Forest Entomology. Ashish Publishing Hosue, New Delhi

Sen-Sarma, P.K. 1994. Economic importance of termites in forestry and agro-ecosystem: 361-365. In: Forest Entomology. Ashish Publishing Hosue, New Delhi.

1996

Sen-Sarma, P.K. 1996. Impact of termites in the forest ecosystem: 11. In: Impact of diseases and insect pests in tropical forests Proceedings of the IUFRO symposium, 23-26 Nov., 1993, Peechi, India (eds. K.S.S Nair, J.K. Sharma & R.V. Varma), 521 pp.

1997

Sen-Sarma, P.K. 1997. Principles, practices and importance of extension education in forestry: 1-35. In: A manual of forestry extension education. Ashish Publishing Coporation, New Delhi.

Jha, L.K., Roy, Punam and Sen-Sarma, P.K. 1997. Learning in groups: 75-90. In: A manual of forestry extension education. Ashish Publishing Coporation, New Delhi.

Sen-Sarma, P.K. and Jha, L.K. 1997. Propagation techniques for forest trees: 279-336. In: A manual of forestry extension education. Ashish Publishing Coporation, New Delhi.

Sen-Sarma, P.K. 1997. Forest insects pests and their management: 341-374. In: A manual of forestry extension education. Ashish Publishing Coporation, New Delhi.

1999

Tyagi, B.K. and Sen-Sarma, P.K. 1999. Morphology of the spermatozoa of *Microcerotermes beesoni* Snyder with reference to termite phylogeny (Isoptera: Termitidae). *Uttar Pradesh Journal of Zoology*, **17**: 245-246.

2000

Sen-Sarma, P.K. 2000. Some aspects of biology and ecology of termites relevant to their management. *Cobweb*, Vadodara, **1**(2): 19-22.

General

Sen-Sarma, P. K., 1963. Book Review – "The Insect factor in wood decay" by Norma E. Hickin, 336 pp. + 263 illustrations. Hutchinson of London. *Journal of the Society of Indian Foresters*, **3**(1): 1-2.

Sen-Sarma, P. K., 1972. Book Review –"Pesticide Application Equipment" by O. S. Bindra and Harcharan Singh, Oxford & IBH Publishing Co, New Delhi, 350 pp., *Indian Forester*, 98 (9): 575.

Sen-Sarma, P. K., 1974. Book Review –"Termite Problems in India" edited by M.L. roonwal, Indian Council of Scientific and Industrial Research, 1972, 81pp, *Indian Forester*, 100 (3), p. 227.

Sen-Sarma, P. K.and V.B.S.1978. Obituary-Dr. R. N. Mathur (1903-1977). *Indian Forester*, 104 (3): II.

Sen-Sarma, P. K., 1979. Book Review –"Termite life and termite control in tropical south Asia", by M. L. Roonwal, Scientific Publishers, Jodhpur, xi+ 177pp +8 Plates,. *Indian Forester*, 105 (12): 917.

Sen-Sarma, P. K. 1985. Biographical Memoir of late Dr. Cyril Frederick Cherrington Beeson, D.Sc. (1889-1975). Foundation Fellow along with a list of his publications.- Biographical Memoirs of Fellows of the Indian National Science Academy, New Delhi (INSA), 10: 34-40.

Sen-Sarma, P. K., 1986. Biographical Memoir of late Dr. Nibaran Chandra Chatterjee, D Sc. (1890-1950) Elected fellow 1942 along with a list of publications. Biographical Memoirs of Fellows of the Indian National Science Academy, New Delhi (INSA) 11: 106-111.

Sen-Sarma, P. K., 1992. Biographical Memoir of late Sir (Dr) Charles Gerald Trevor Kt. CIE (1882-1959) Foundation Fellow along with a list of publications. Biographical Memoirs of Fellows of the Indian National Science Academy, New Delhi (INSA), 15: 67-74.

Sen-Sarma, P. K. 1993. Biographical Memoir of late Prof. Harry G. Champion, D.Sc. (1891-1979) Foundation Fellow along with a list of publications. Biographical Memoirs of Fellows of the Indian National Science Academy, New Delhi (INSA), 17: 86-101.

Sen-Sarma, P. K. 1994. Biographical Memoir of late Prof. (Dr) Mithan Lal Roonwal D.Sc. (1908-1990) Elected Fellow, 1945 along with a list of publications. Biographical Memoirs of Fellows of the Indian National Science Academy, New Delhi (INSA), 19: 29-62.

Sen-Sarma, P. K. 1995. Biographical Memoir of late Dr Vishwa Gopal Jhingaran D.Sc. (1919-1991) Elected Fellow, 19984 along with a list of publications. Biographical Memoirs of Fellows of the Indian National Science Academy, New Delhi (INSA), 20: 167-193.

Books written/edited

Roonwal, M.L and Sen-Sarma, P.K. 1960. Contributions to the systematics of Oriental termites. *Indian Counc. Agric. Res.*, New Delhi, Entomological Monograph No. 1, xiv+407 pp., 65 pls.

Sen-Sarma, P.K., Kulshrestha, S.K. and Sangal, S.K. (edited).1983. Insect Interrelations in forest and agro-ecosystems. Jugal Kishore & Co, Dehra Dun, 262 pp.

Trivedi, R.N., Sen-Sarma, P.K. and Singh, M.P. (edited).1991. Environmental assesment [sic] and management: social forestry in tribal regions. Proceedings of the third conference of Mendelian Society of India, Patna. Today and Tomorrow's Printer and Publishers, New Delhi, x+453 pp.

Jha, L.K and Sen-Sarma, P.K 1993. Agroforestry in Indian Prospectives. Ashish Publishing House, New Delhi, India, 343 pp.

Jha, L.K. and Sen-Sarma, P.K. (edited). 1994. Forest Entomology. Ashish Publishing Hosue, New Delhi, 387 pp.

Jha, L.K. and Sen-Sarma, P. K. 1994. Forestry for the people. Ashish Publishing House, New Delhi, x+410 pp.

Tuyen, Vu Van and Sen–Sarma, P.K. 1994. Termites damaging dikes & dams and their management. Surya International Publications, Dehra Dun, vi+ 81 pp.

Jha, L.K. and Sen-Sarma, P.K. (edited). 1997. A manual of forestry extension education. Ashish Publishing Coporation, New Delhi, pp. xii + 386pp.

2. Dr. B.K. Tyagi

Dr. B.K. Tygai receiving the ICMR's Dr. M.O.T. Iyengar Memorial Award (2007) for 'outstanding work in the field of biomedical research', from Shri Ghulam Nabi Azad, Union Minister for Health, along with Dr. V.M. Katoch, Secretary, DHR and Director General, ICMR, and other dignitaries on dais.

कर्मण्येवाधिकारस्ते मा फलेषु कदाचन,
मा कर्मफलहेतुर्भूर्मा ते सङ्गोऽस्त्वकर्मणि।

– Bhagavad Gita, Chapter 2, Verse: 47

The life of Dr. B.K. Tyagi, an utterly indefatigable medical arthropodologist-scientist and originator of the conferences of Medical Arthropodology in the country, about whom Prof. T.N. Ananthakrishnan, PhD, DSc, FNA, FASc, FNAAS, Jawaharlal Nehru Fellow and the INSA Emeritus Scientist, and arguably the world's one of the greatest entomologists ever, had once commended as "... an outstanding Medical Entomologist", truly reflects the spirit radiated from these famous verses of the great Indian scripture 'Shrimadabhagavadgeeta'. For more than a decade he served as the Director in-Charge of the Centre for Research in Medical entomology (ICMR), Madurai and it is no surprise, therefore, that under his leadership the Centre networked with a large number of scientific laboratories and academic institutions the world over, all along developing new research collaborations in nanotechnology, stem cell research, molecular and biotechnology, besides conventional disciplines – all however with a focused objective of increasing knowledge on arthropod vectors, for their effective control along with the diseases they transmit! Dr. Tyagi's irresistible combative spirit without even giving an iota of thinking to any rewards, whatsoever, reminds us of William Wordsworth's famous ode in "Character of the happy warrior."

"Who comprehends his trust, and to the same
Keeps faithful with a singleness of aim;
And therefore does not stoop, nor lie in wait
For wealth, or honour, or for worldly state;
Whom they must follow, on whose head must fall,
Like showers of manna, if they come at all."

Dr. Brij Kishore Tyagi was born in Agra on 23rd July, 1951, to a highly illustrious parents. His father late Shri Sheonath Singh Tyagi was a gazetted Police Officer in the Uttar Pradesh's Provincial Arms Constabulary and widely commended for his strict sense of discipline, prudence, frugality and good management in both his professional and personal life. Mother, Smt. Basanti Devi was a home maker and highly devotional and systematic in all domestic chores; never ever forgetting to light a *deeva* (the earthen cup with oil to keep the cotton wick lighted), a symbol of remembering the Almighty God to bring wisdom to all, and assembling the whole family to do *Aaarti*, the prayer during the dusk – the time when day meets the night, and thanks offered once again to the omnipresent God for a highly educative, and safe and sound day! Parents formed the first school of learning for him: "*Pahel haque hey so maa-baap ka, jo badumka haque hey wo ustaad ka, gar maa-baap ki khidmatt bajaalaogey, to khadim se maqudoom ban jaogey.*" Obviously, in a family where everything, however trifle, will go by strict observance to the rule of good living, Dr. Tyagi's early education took place at home only; rising early, readying for prayer, essentially going outdoors for play, and in the night, just before all would retire to sleep, to hear the various spellbinding, long and seemingly never ending, educational and moral boosting tales of great legendry warriors like Aalah-Udal, Nal-Damyanti etc. as well as episodes from the great Hindu epics such as Ramayan and Mahabharat, narrated by father. All lights in home would be switched off at 10 pm sharp. Thus, the rule was: *Early to bed and early to rise, keep a man happy and healthy.* Even in those times when electricity, despite fewer generating resources and far too inexpensive, as a rule only those couple of rooms would be lit where children would undertake to their lessons.

"If every one would see
To his own reformation,
How very easily
You might reform a nation."

Dr. Tyagi took his early education in Christian Intermediate College, Bareilly (1962-64), Parker Intermediate College, Moradabad (1964-67), Government Intermediate College, Meerut (1967-69) and later graduated from D.N. Degree College, Meerut (1969-72) and postgraduated from J.V. College, Baraut in Meerut district (1972-74). It was during his MSc that he developed interest towards research and completed a highly commendable project on freshwater fishes, thereby also securing highest marks in the practicals as well as first division in MSc. This sparked a turning point in his life and he yearned to undertake higher education by registering for the PhD degree. He, by recommendation of Prof. P.P. Singh, who taught him during the MSc, took Prof. S.K. Sangal, a noted odonatologist, of Dayanand Anglo Vedic (Post Graduate) College, Dehradun as his guide for PhD in 1974. Inspired by Dr. Sangal's work on dragonflies, particularly, *Crocothemis servilia* Drury, Dr. Tyagi came in contact with some of the greatest entomologists and cytogeneticists, such as Prof. P.K. Sen-Sarma, an internationally acclaimed termite expert and the Director, Biology Research, Forest Research Institute & Colleges, Dehradun, Mr. Pratap Singh, Forest Entomologist, *ibid.*, Prof. T.N. Ananthakrishnan, an unparalleled thysanopterist and the Director, Zoological Survey of India, Prof. G.P. Sharma, an internationally reputed cytogeneticist and Head, Zoology Department, Punjab University, Chandigarh, Prof. J.S. Yadav, a celebrated geneticist on Coleoptera, in particular, and Principal, Choturam College, Jind in Haryana, and of course Dr. S.K. Kulshrestha, a colleague of Dr. Sangal in the same department, Dr. Santokh Singh, noted high altitude entomology expert and Head, School of Entomology, St. John's College, Agra, Dr. H.N. Baijal, a famous collembologist and Head of Zoology Department, Agra College, Agra, Prof. J.M. van Brink, Editor of GENETICA and GENEN EN FENEN, and a senior professor at the Utrecht University's Institute of Genetics, and many more...! But, one person who made the most significant impression on his persona is Prof. Bastian Kiuta, Department of Animal Cytotaxonomy & Cytogenetics, Institute of Genetics, Utrecht University, The Netherlands and also the Executive Editor, ODONATOLOGICA as well as the Head, Societas Internationalis Odonatologica (SIO) and the Director, Khumbu Himal Research Expedition of the Netherlands Centre for Alpine Biological Research (N.C.A.B.R.), Utrecht in collaboration with Nepal Research Centre, Kathmandu. It was the early teachings of Prof. Kiauta who had not only guided Dr. Tyagi towards accomplishing successfully his doctoral dissertation, but also taught him the lessons in performing science in the best tradition! Prof. Kiauta, a father-like figure to Dr. Tyagi, played the role of a mentor and this strong scientific bond continued ever till date. Incidentally, Prof. Kiauta editorially abstracted his thesis in the international journal ODONATOLOGICA and expertly commented on Dr. Tyagi's PhD thesis (Dec., 1978; (*cf.* OA No. 2683) titled "*Studies on the chromosomes of Odonata of Dun Valley (Dehradun, India)*" as:

> "*The comprehensive volume is based on original observations on 45 spp. (the main cytotaxonomic data on 13 of which were subsequently published elsewhere (cf. OA No. 2665), leaving as new Anisopleura lestoides [n=13, m] only on a thorough literature study. It is organized into thirteen chapters, incl. a comprehensive bibliography and a synopsis of chromosome numbers recorded in the order up to 1977, and presents several novelties for the regional and/or Indian fauna. The book was not issued commercially, but a substantial abstract is available from the author.*"– **(<u>Abstractor's note</u>: The author should be congratulated on the production of this work, which has been carried out under often hard personal circumstances and under a great financial stress).**

At the same time Prof. G.P. Sharma, his first PhD thesis examiner, had commented ".... a phenomenal research work" while another evaluator, Dr. P.K. Sen-Sarma wrote ".... such a massive research work must be published as a priority". Dr. Tyagi's doctoral research brought out many novelties about odonate cytotaxonomy, notably (i) the discovery of neo-neoX/neo-neoY sex determining mechanism, a new mechanism altogether in the dragonfly biology, through species like *Onychogomphus saundersi duaricus* Freaser from the Doon Valley in Dehradun, northern India, and (ii) conservation knowledge generated on endangered dragonflies whose existence was threatened due to habitat destruction in view of the booming unplanned urbanization and industrialization (e.g., brick kilns, lime kilns, quarry and sand mining etc.) in the valley, consequently helping listing of the endemic species, *Burmagomphus siwalikensis*, in the Red Data Book of endangered fauna. This association with dragonfly research and many a world famous odonatologists from allover the world went on for a long winded journey which obviously culminated in Dr. Tyagi's dedication book, "ODONATA: *Biology of* dragonflies", in the honour of 70th birthday celebration of his teacher, guide and mentor, Prof. Dr. B. Kiauta. The infinite sense of respect towards teachers was ingrained in him very early in life:

गुरु गोविन्द दोऊ खड़े काको लागूं पायं।
बलिहारी गुरु आपने जिन गोविन्द दियो बताय।।

There are many anecdotes about Dr. Tyagi when he was engaged in accomplishing his dissertation work in the DAV College, Dehradun.

Command on English language was one major subject that occupied his attention all along his formative research days. Dr. Tyagi soon realized the importance of English in science and discourses. One particular instance stirred him good enough to persevere to learn the language a hard way. After Dr. Sangal had taken him as a research scholar, the first thing Dr. Sangal did, as it was customary in those days, was to dispatch him for two weeks to consult the world famous entomology library and museum at the Forest Research Institute and colleges, Dehradun. There Dr. Tyagi was to meet Dr. Pratap Singh, Forest Entomologist for permission to consult library and museum. Dr. Pratap Singh inquired about many things than Tyagi could have hoped in this first encounter; and, the result was, during the entire 15 min of interaction he could hardly utter a few words, that too quite incoherently. It made him ashamed of his ignorance, on one hand, and the awareness of his dwindling inner strength to meet the literary challenges in future, on the other! With the dragonflies as an object of his doctoral dissertation, Dr. Tyagi soon found in these elegant and evolutionarily highly important creatures his best accomplices with whom he would chat in English a variety of subjects, illustrating information from the timeless odonte literary classics like Tillyard's *The Biology of Dragonflies* (1927), Fraser's *The Fauna of British India* volumes I-III (1933, 1934 and 1936) and *A Reclassification of the order Odonata*, Needham and Westfall's *Manual of the Dragonflies of North America* (1955), Corbet's *A Biology of Dragonflies* (1962), Longfield's *The Dragonflies of the British Isles,* and Kiauta's *Cytotaxonomy of Dragonflies* (1975).

Dragonflies literally taught Dr Tyagi a whole lot of diction and articulation. Another story goes like this: since, during the later part of his dissertational research Dr. Tyagi was not supported by a fellowship, not to mention the unfortunate incident in which he had lost all his photographic and microscopic evidences for his entire three years' research work during his brief Zoological Survey of India fellowship at Calcutta (now Kolkata) in late 1977, on his precocious return from Calcutta, he took up teaching in schools like Guru Nanak Academy and Cambrian Hall at Dehradun to help his research needs of preparing the thesis. He would then work in the school, teaching students during the day, and, thanks to courteous permission by Dr. M.B. Lal, HOD, carry out his research in his zoology department's laboratory locked from outside, from the scratch all over once again, in the night! This went on for nearly one year before he was able to write his thesis anew and submit to the university, just a few weeks before expiry of period, in December 1978, with the financial help from his guide guide Dr. S.K. Sangal. During the period of his doctoral research on odonate cytotaxonomy, Dr. Tyagi, by virtue of his association with Prof. Ananthakrishnan, Dr. Vijay Veer, Dr. P.K. Sen-Sarma, Dr. A.K. Malhotra, Dr. H.N. Baijal etc., had also acquired eco-biological knowledge about thrips, termites, ants, fungi, mayflies and springtails etc. On one occasion, Dr. O.P. Mittal, a leading authority on the karyosystematics and phylogeny of spiders (Acarina), quoting Prof. J.S. Yadav, a renowned coleopterologist, had written to Dr. Tygai, "You have a long way to go in insect sciences."

After becoming exposed to dragonfly cytotaxonomy, ecology, biology and phylogeny under the tutelage of Prof. Bastian Kiauta, Dr. Tyagi received his postdoctoral training from the legendary termite systematics and evolutionary biologist Dr. P.K. Sen-Sarma at the Forest Research Institute & Colleges, Dehradun, and began working on ecology and control of termites using certain fungi. Here he came in for a brief contact of Prof. R.N. Roonwal, one of the greatest zoologists India ever has produced! One incident involving Dr. Tyagi, Dr. Sen-Sarma and Prof. Roonwal exemplifies significance of acute and subtle observation in science. In 1980 Prof. Roonwal had once visited Dehradun to offer a lecture in the Zoology Department of DAV College and Dr. Tyagi, who had an inherent penchant to converse with great scientists, made a full recording of his speech on termite taxonomy and biodiversity. In course of his lecture he had highlighted as to how he tumbled upon a new termite species in the backyard of his house in Jodhpur, reminding the ingenuity and subtlety of his mind:

Wise haveth their eyes, in the head;
Fools waketh through forest, and see no firewood!

After the lecture, Prof. Roonwal, accompanied by Dr. Sen-Sarma, had visited the FRI & College's Forest Entomology Division, his *Alma mater*. Dr. Tygai had meanwhile also come to the institute for consulting entomology library. When Prof. Roonwal and Dr. Sen-Sarma took time to make rounds of the Division and entered library, Dr. Sen-Sarma got Dr. Tyagi introduced to Prof. Roonwal who asked him about his research title, earlier work on odonate cytotaxonomy and cytophylogeny as well as current interest in Isoptera. Sooner, Prof. Roonwal and Dr. Sen-Sarma got engaged into a lengthy discourse pertaining to the evolutionary aspects of the order Isoptera which had ever intrigued biologists as explained by Krishna Kumar in Biology of Termites (1969, 1970). Always a keen learner and patient listener, Dr. Tyagi soon set on finding out the roots of the origin of various termite families on the basis of chromosomal characteristics. He subsequently published many papers on different aspects of termite biology and ecology, but the one paper which had a tint of conversation between Dr. Roonwal and Dr. Sen-Sarma and constituted the thesis of correlation between morphology and quantum dynamics of spermatozoa with that of the evolutionary position of different families, was published following an arduous peregrination of nearly 20 years in 2001!

In 1980, Dr. Tyagi married Ajita, a dedicated teacher and plant scientist, who remained the greatest source of encouragement and support throughout his tumultuous, meandering, and challenging, almost maddening, scientific pursuits. Both even authored a few scientific publications.

In 1981 Dr. Tyagi joined as Senior Research Officer on a WHO-TDR funded research project, "A laboratory and field study of malathion resistance in *Anopheles culicifacies*", under Malaria Research Centre (ICMR) in collaboration with the NMEP, Delhi. After obtaining necessary training and exposing himself to a variety of laboratory and field related malariology in the field of vector ecology, biology, insecticide resistance, cytogenetics, mosquito colonization and malaria epidemiology for three months, both in Delhi and Haldwani, UP, he was transferred to the project site at Ukai in Songarh tehsil in Surat district, south Gujarat State, as its in-Charge. During course of this project (1981-84) he mastered in ecology of vector species, *A. culicifacies* complex, as well as other malaria vector species, in the heavily forested and mountainous terrain of southern Gujarat, simulated hut experiments with malathion-resistant strains of *A. culicifacies*, larval breeding of various anopheline mosquitoes, and susceptibility status of major malaria vectors against insecticides in vogue then. He was principally responsible in sampling the newly described 'Species B' of *An. culicifacies* sibling species complex, from the Ukai and surrounding areas.

It was during this period that, in recognition of his sustained dragonfly research achievements, the International Society of Odonatology (Societas Internationalis Odonatologica, S.I.O.) in its General Body Meeting held during the 5th International Symposium of Odonatology, held in Chur, Switzerland in 1981, has decided to open a National Office in India to promote odonatological research in the subcontinent and appointed Dr. Tyagi as its first National Representative in-Charge in India with effect from 1981 (later on, with the scope of the work of the office ever expanding over to other Asian countries as well, promoted to the rank of Regional Representative in-Charge, Regional Office in Southern Asia, SIO-ROSA, in 1988). During his tenure as the Head of the SIO-ROSA he organized four national symposia in Madurai (1984), Dehra Dun (1986), Erode (1988) and Allahabad (1990) and one international conference in Madurai (1988). He brought out several significant publications on the activities of the SIO-ROSA (Tyagi, B.K. and Ajita Tyagi, 1986. Directory of odonatologists from the Indian subcontinent. *Occ. Publs S.I.O. Natl. Office India*, No.1, 20 pp.; Tyagi, B.K., A. Tyagi and S.K. Sangal, 1986. An up-to-date catalogue and bibliography of dragonfly fauna of the Dehradun valley (India)

Occ. Publ. SIO Natl. Off. India 2:1-14; Tyagi, B.K., 1987. An introduction to the activities of the S.I.O. National Office in India. IX International Symposium of Odonatology, Madurai 18-21, 1988. Pondicherry, pp. 37; Tyagi, B.K. (Editor), 1988. Odonatological bibliography of Frederick Charles Fraser {originally manuscripted by Basil Elwood Montgomery. *Occ. Publs. SIO Natl. Off. India* No.3: 1-24). He also published the first ever odonatological journal in India, *Indian Odonatology* (published annually since 1988) and was the originator of the SIO National Office in India/ROSA newsletter, *Fraseria* (published semi-annually since 1982).

In 1984, Dr. Tyagi got the scientific cadre in the Indian Council of Medical Research at its institute of Vector Control Research Centre, Pondicherry. Here researched on a variety of subjects like evaluating different types and insecticides, both conventional and candidate compounds, as larvicides, adulticides, JHs/IGRs/IDIs, repellents/attractants etc. (important being the evaluation of a new amide group larvicide, VCRC/INS/A-23, and determination of mosquitocidal properties of ARTILIN paint, fenozycarb, arosurf, bendiocarb and alphamethrin etc.). Dr. Tyagi was also in-Charge of various important outstation field projects such as (i) the Koraput project on control of malaria and ecology of vectors (1985-1987) whereupon he co-ordinated research and administrative activities of the Koraput Field Station at Jeypore (Orissa) with VCRC Directorate. His field work in Jeypore, Orissa, a tribal belt, was a very ambitious programme since the region had a history of hyper endemic malaria for decades, with *P. falciparum* contributing as high as 90% in certain areas. During course of malariological investigations in the Koraput district, he also worked on the tribal population in the seclusive Bonda hills and others in Malkangiri district, (ii) Bangalore Mosquito Control Project, where he was chiefly involved in planning, organization and implementation (by personally joining with the staff at the field station at Bangalore), and finally authoring country's one of the first and significant reports on the MASTER PLAN OF MOSQUITO CONTROL IN BANGALORE METROPOLITAN CITY (1987), which proved a landmark for other metropolises as well.

In 1988 Dr. Tyagi was transferred to Desert Medicine Research Centre in Jodhpur and got promoted to the post of Assistant Director in July 1990 and to the post of Deputy Director in July 2000. Most of these years, i.e., between 1991 and 1999, he also looked after the work of DMRC as Local in-Charge. In The Thar Desert he conducted many novel research activities, while also winning for the Centre its first WHO (2001: Identification and classification of larval breeding habitats for determining adult population estimates, US$ 55,300.-; 2001: Distribution of Thar malaria mosquitoes belonging to the *Anopheles culicifacies* and *Anopheles stephensi* complexes) and IWMI (Control of vectors of malaria in relation to domestic/canal water supply strategies in Jaisalmer district, the Thar Desert (india, US$ 9,000.-) funded extramural research projects on malaria, in addition to ICMR Task Force (extra-mural) project (2001. Evaluation of disease burden due to leptospirosis, Rs. 1.60 Lakh). He thoroughly investigated ecology, biology and blood feeding behaviour of the principal malaria vectors, *A. stephensi* and *A. culicifacies* in the Thar Desert particularly under the impact of the extensive canalization by the Indira Gandhi Canal (640 km, with over 8000 km network of distributaries), and travelled all over the Thar Desert to understand the impact of anthropocentric behaviour over malaria exacerbation or conflagration. The culmination of his thirteen years long research work was the production of several internationally acclaimed research papers in high Impact Factor journals, such as *The Lancet*, topped with the unique monograph, "Malaria in the Thar Desert", published in 2002. Prof. P. Amerasinghe, an internationally acclaimed mosquito ecologist and malariologist of the International water Management Institute, Colombo, Sri Lanka, has once commended Dr. Tyagi as, ".... *Tyagi is best known internationally among malaria and medical entomology circles for his long years of research into mosquitoes, malaria and irrigated agriculture in the Thar Desert of Rajasthan, India.*" (*cf.* Dr. Felix P. Amerasinghe's *Foreword* dated 7th April, 2004 in Dr. B.K. Tyagi's monograph, "*The Invincible Deadly Mosquitoes: India's health and economy enemy # 1*", Scientific Publishers (India), pp. 276). He was the first to have offered a classification of the Thar Desert's physiography on the basis of malaria vectors' distribution, linking this with that of the type of dominant malaria parasite in the Thar. Innovatively, he designed a *"Tanka"* lid to prevent the breeding of vector, *Anopheles stephensi*, the species predominant in the Thar's xeric environments and best adapted inherently in the local, ageold underground reservoir, the Tanka. He was also the first to have linked the traditional man-made well-like water reservoirs in the belly of ponds, called *"Beri"*, with the malaria vector breeding (Tyagi, B.K. and S.P. Yadav, 1996. Malariological and sociological significance of "Tanka" and "Beri" in the Thar Desert, Western Rajasthan (India). *Journal of Arid Environments* (U.K.) 33: 497-501). It was during this period of his research that two eminent scientists visited him and worked in his collaboration: (i) Prof. Peter Miller from Zoology Department, Oxford University, United Kingdom, who came to India on a fellowship facilitated jointly by the Indian National Science Academy, The Trustees of the Aneurin Bevan Foundation and the Indian Council for Cultural Relations to work on Odonata in Madurai Kamaraj University, Madurai and at Jodhpur under the aegis of SIO-National office in India, and (ii) Dr. Manu Thomas, Zoology Department, Madras Christian College, Tambaram, Chennai, who received a Visting Fellowship from the Indian National Science Academy, Delhi under Dr. Tyagi's guidance at the DMRC, Jodhpur. With Prof. Miller, he even published a research paper (Tyagi, B.K. & Peter Miller, 1991. A note on the Odonata collected in south western Rajasthan, India. *Notulae odonatologicae* 3: 134-135).

The Thar Desert offered many challenges since some of the major vector-borne diseases were specific to this hostile environment, such as the guinea worm disease or dracunculiasis and cutaneous leishmaniasis, while some others were on the rise in incidence and intensity. He studied these diseases in great detail and reviewed them for the Thar (Tyagi, B.K., 2001. Dracunculiasis eradication in India, with special reference to the Thar Desert (Rajasthan). (In: *Helminthology in India.* (Ed.) M.L. Sood], International Books Distributors, Dehradun. pp. 543-555; Tyagi, B.K., 2003. Cutaneous leishmaniasis in the Thar Desert: a critical appraisal of the fast emerging "sand-disfiguring disease". In: *Advancement in insect biodiversity*. Ed. Prof. Rajiv Gupta, JNV Univ., Jodhpur, pp. 313-323). He served as the full-fledged Deputy Director & Officer in-Charge,

Desert Medicine Research Centre (ICMR), Jodhpur during 1999-2001. He received his first international prize, the First WHO-TDR Prize in providing a panoramic view to the achievements, as well as future course of action, of TDR, in 1995-96.

Having been recognized as a PhD Guide by many Universities, including Madurai Kamaraj University, Madurai University, Pondicherry University etc. he had by now examined several doctoral dissertations, notably of Dr. Somasundaram (a student of Prof. S. Chokalingam), Dr. R.J. Andrew (a student of Prof. D.B. Tembhare) and Dr. Manu Thomas (a student of Prof. Gladstone), and many a Masters Dissertations.

At CRME in Madurai, his real *karmakchetra*, where Dr. Tyagi joined in 2001, he became its Director in-Charge in May 2005, Dr. Tyagi has brought in phenomenal changes in its research capabilities and infrastructural development. A careful visionary and schemer of all his activities, he sees accomplishment of each and every assignment, however small and trifle looking, once he is committed. His leadership qualities are unmatched, being absolutely transparent and accessible in all his dealings. He keeps the benefits of the institute and its employees, especially scientists, in the forefront of all his functions. His dedication and intensity for high quality research are visible through a large number of research programmes funded by international bodies which he has initiated at the Centre during past nearly one decade. No wonder, therefore, that he soon became a scientist of many "firsts", in context with research in the science of medico-entomology, but transformed CRME into a formidable force of knowledge in medical entomology globally, as Prof. Dr. Graham B. White, an international authority in medical entomology in the Department of Entomology and Nematology, University of Florida, Gainesville, USA, while recommending Dr. Tyagi's nomination for the WHO's STAC membership wrote on 14th February, 2011 in his appreciation:

> "... For the past three decades, Dr. Tyagi has contributed greatly to research and leadership on vector-borne diseases across the spectra of biological science and organizational systems. During his career with the Indian Council of Medical Research (ICMR) his effectiveness and productivity transcend the norms of ICMR responsibilities: while zealously performing his formal research management duties, Dr. Tyagi emphasizes the pastoral side of institutional development, and somehow finds the time and energy to produce inspirational books on biological themes. His personal research began with insecticidal product evaluation and development, collaborating with many agencies and industries for innovative technology, progressing to implementation research and cost-benefit analysis of components of integrated management and prevention of vector-borne diseases. One cannot exaggerate the versatile skill that Dr. Tyagi brings from PI management of grants from TDR and other agencies, for projects spanning many States of India, various disciplines and groups of TDR client countries, on wide ranging questions regarding arboviruses, malaria, filariasis, genetic control, and VBDs generally. His well balanced enthusiastic stewardship of everything he takes on, the depth and breadth of his biological

knowledge, multi-disciplinary and visionary approach, administrative and international experience in fields currently acquired by STAC, make Dr. Tyagi highly suited and willing to serve TDR in his capacity."

This truly sums up the leaps forward Dr. Tyagi has made in both his research and professional arenas, taking all along the CRME's own unprecedented development and prosperity in research capability and infrastructure as his *numero uno* priority and responsibility. He always looked after the scientific development of all the employees of the Centre, be it scientific, technical and administrative. That is why, setting a rare example in the whole Council, he sent a large number of technical staff for overseas training. As if this was not sufficient, he even arranged training and/or conference participation abroad on several occasions for the scientists working in temporary and ad-hoc projects.

At CRME, Dr. Tyagi has brought in phenomenal changes in its research networking development. His diverse collaborative research projects with a great number of institutions overseas like Institute of Hygiene and Epidemiology, Hanoi (Vietnam), Brandeis University, Waltham (USA)U, Bamako University (Mali, AFRICA), Mahidol University at Salaya (Thailand), Institute of Medical Research, Kuala Lumpur (Malaysia), University of Oxford /Oxitec (UK), and still a large number of institutes within the country such as National Institute of Mental Health and Neurosciences, Bangalore; Osmania University, Hyderabad, Bharthair University, Coimbatore, Bharathidasan University and Anna University, Tiruchirapally; Vellore Institute of Technology, Vellore; Madurai Kamaraj University, Madurai; Mapko Schlenk, Sibkasi; International Clinical Epidemiology Network, New Delhi, just to name a few, in addition to several of ICMR institutes have been instrumental in executing several research programmes of international significance, some of which being significant are tabulated below:

NUMBER OF EXTRAMURAL PROJECTS GENERATED BY DR. B.K. TYAGI

1. R & D (Govt. Agencies)

SN.	Title of Project	Funding Agencies Govt. Agencies R&D Projects (ICMR /DST/DBT etc) and Amount	Level of Participation whether		
			PI	Co-PI	Others
1.	Asian training course on bio-safety for human health and the environment in relation to potential release of genetically modified vectors	WHO-TDR; US$ 150,500.- (Rs. 75,00,000.-)	Dr. B.K. Tyagi	-	-
2.	Genotypic characterization of Japanese encephalitis virus (JEV) in different endemic areas of	DST, New Delhi; Rs. 10,00,000.-	Dr. B.K. Tyagi	-	-

India and Vietnam: its implications on the vector-pathogen interactions.				
3. Community based dengue vector control under health system in one district (south India) and transfer of technology.	WHO-TDR; INR 18,00,000.-	Dr. B.K. Tyagi	-	-
4. Identification and classification of larval, breeding habitats for determining population estimates as a tool for future malaria epidemics in the Thar desert using remote sensing and geographic information system.	WHO, Geneva; Rs. 27,65,000.-	Dr. B.K. Tyagi	-	-
5. Distribution of Thar malaria mosquitoes belonging to the Anopheles culicifacies and Anopheles stephensi complexes.	WHO, Geneva; Rs. 25,00,000.-	Dr. B.K. Tyagi	-	-
6. In-Depth review on Japanese encephalitis (National Vector Borne Disease Control Programme, Delhi).	NVBDCP (Govt. of India), New Delhi; Rs. 49,89,325.-	Dr. B.K. Tyagi	-	-
7. In-Depth review on Japanese encephalitis (National Vector Borne Disease Control Programme, Delhi)	NVBDCP (Govt. of India), New Delhi; Rs. 40,07,300.-	Dr. B.K. Tyagi	-	-
8. Establishment of a Field Station in South Arcot district, Tamil Nadu for control trials of Japanese Encephalitis.	ICMR; Rs. 27 lakhs a year	B.K. Tyagi (as Director, CRME)	-	-
9. Control of vectors of malaria in relation to domestic/canal water supply strategies in Jaisalmer district, Thar desert (India).	International Water Management Institute; US$ 9,000/-	Dr. B.K. Tyagi	-	-
10. Assessment of disease burden due to leptospirosis (A ICMR Task-Force project awarded to RMRC Port Blair, with DMRC as a collaborator to carry out its part of research work in Rajasthan)	ICMR Task Force; Rs. 1,60,360.- (funding allotted to Dr. B.K. Tyagi, as the Co-PI, DMRC, Jodhpur).	-	Dr. B.K. Tyagi (Co-PI & Coordinator, DMRC)	-

11. Cloning and expression of PRM and NS1 gene of dengue virus and its application in the diagnosis of dengue. (Collaborative research project with Anna Univ., Trichy).	DST, New Delhi; 25,00,000.-	-	Dr. B.K. Tyagi	-
12. Eco-bio-social aspects of dengue in urban and peri-urban ecosystems in Tamil Nadu, India.	WHO-TDR; US$ 125,000.-	-	Dr. B.K. Tyagi	-
13. Population genetics of the predominant Japanese Encephalitis vector Culex tritaeniorhynchus	DST, New Delhi; Rs. 20,46,000.-	-	Dr. B.K. Tyagi	-
14. Population genetic analysis of dengue vectors, Aedes aegypti and *Aedes albopictus* using biochemical markers	DBT, New Delhi; Rs. 29,27,000.-	-	Dr. B.K. Tyagi	-
15. Preparation of Japanese encephalitis virus (JEV) antigen in cell culture	DST, New Delhi; Rs. 13,24,000.-	-	Dr. B.K. Tyagi	-
16. Prevalence of Wolbachia in filariasis vector *Culex quinquefasciatus* in the filaria endemic villages of Trikoilur	DST, New Delhi; Rs. 5,00,000.-	-	Dr. B.K. Tyagi	-
17. An epidemiological study of vector borne diseases and control strategies for tribal health development through Remote sensing and GIS for Sitheri hills, Dharmapuri district of Tamil Nadu	ICMR; Rs. 9,00,00.- (being the CRME's allocation)	-	-	Dr. B.K. Tyagi
18. Forecasting JE mosquito vectors abundance through geo-environmental risk determinants, using remote sensing & GIS. (VCRC-CRME joint project)	ICMR; 54,63,600.- (CRME Rs. 26,00,000.-)	-	Dr. B.K. Tyagi	-
19. Bangalore Mosquito Control Project	ICMR; Rs. 10 lakhs	-	-	Dr. B.K. Tyagi
20. National network for genotyping of human lymphatic filarial parasite, *Wuchereria bancrofti*, from different endemic areas. (VCRC-MMC Wardha-RMRC Dibrugarh- CRME, RMRC Bhubneshwar, Anna Univ., Chennai, NZICD, Lucknow joint project)	ICMR; Rs. 45.90.- lakhs (CRME Rs. 6.00 Lakhs)	-	-	Dr. B.K. Tyagi

21.	Multicentric evaluation of L3 stage specific RT-PCR assay for the detection of infective stage (L3) Wuchereria bancrofti in vector. (VCRC- RMRC Bhubneshwar, CRME, RMRC Port Blair joint project).	ICMR; Rs. 26,97,800.- (CRME Rs. 5,15,560.-)	-	-	Dr. B.K. Tyagi
22.	Burden and economic cost of dengue in India: A study at major teaching hospitals and nearby ambulatory installations. (Brandeis Univ, USA+ CRME+INCLEN+NIHFW coll. Research project).	Sanofi-Pasteur; Rs. 9,77,613.-	-	Dr. B.K. Tyagi	-
23.	Studies on the influence of various hydrophytes on structure and composition of mosquito vector population in Vaigai river basin of Tamilnadu, south India (MK Univ.+CRME joint research project).	ICMR; Rs. 19,79, 530.-	-	-	Dr. B.K. Tyagi
24.	Evaluation of NATULAR Formulations	CLARKE, Mumbai/USA; Rs. 28,00,000.-			Dr. B.K. Tyagi

Non-R&D (Sponsored/Commercial) projects generated

Sl. No.	Title of Project	Funding Agencies	Level of Participation whether		
			PI	Co-PI	Others
1.	Oviposition attracticide trials at Chennai, Tamil Nadu and Thiruvanantha-puram, Kerala.	DRDO, Gwalior; Rs. 3.25 lakhs	Dr. B.K. Tyagi	-	-
2.	Field efficacy trials of RK BIO VECTA Bti AS, an acquous solution formulation of *Bacillus thurengiensis* var. *Israelensis* against *Culex quinquefasciatus,* the vector of lymphatic filariasis.	M/S RK Biotech Product Pvt. Ltd., Chennai; Rs. 3,98,200.-	-	-	Dr. B.K. Tyagi
3.	Field bio-efficacy of DRDO Defender Net against mosquitoes. (*CRME, Madurai + DRL, Tezpur coll. Research project*).	DRL, Tezpur; Rs. 14.5 Lakhs	Dr. B.K. Tyagi	-	-

A recipient of all major research fellowships, viz., Junior Research Fellowship (Z.S.I., Calcutta; 1977), Senior Research Fellowship, Post Doctoral Fellowship and Research Associate ship, all of CSIR, Delhi at the FRI & Colleges, Dehradun; 1979-1981), Dr. Tyagi is a Fellow of Royal Entomological Society, Fellow of Indian Society of Malaria & Communicable Diseases, Fellow of Applied Zoologists Research Association, Cuttack Fellow of the Academy of

Zoology, Fellow of International Biotechnological Society, and Member of over a dozen national and international scientific societies, viz., Member of the International Odonatological Society (SIO), Holland, Member of the Indian Sci. Congr. Assn., Calcutta Fellow of the Indian Soc. Malaria & other Comm. Dis., Delhi, Member of the Zool. Soc. India, Gwalior, Member of the Odonata Specialist Group of the *SPECIES SURVIVAL COMM.* (IUCN), Gland, Switzerland, Member of the Z.O.O. Outreach Orgn., Coimbatore, Member, Indian Society of Parasitology, Lucknow, Founding Fellow of the Entomology Academy of India, Chennai, Member, Invertebrate Conservation & Information Network of South Asia (ICINSA).

Dr. Tyagi, who himself had pioneered in producing a dragonfly journal, *Indian Odonatology*, has been on the editorial/referee boards of about two dozen scientific periodicals published both nationally and internally, such as *Odonatological Abstracts* (Holland), *Notulate odonatologicae* (Holland), Fraseria, *Geobios New Reports*, Jodhpur, *Journal of Applied Zoological Research, Journal of Indian Society of Health Education and Environmental Research*, *Annals of Medical Entomology*, *Indian Biologia*, *Indian Journal of Environment and Eco-planning*, *Zoo Outreach Journal*, *Entomon*, *Indian Journal of Medical Research*, *Journal of Vector Borne Diseases*, *Journal of Environmental Biology*, *Journal of Parasitic Diseases*, *Advances in Medical and Dental Sciences*, *BioTechniques*, *Gene Research*, *Acta Tropica*, *BioMed Central Public Health*, *Tropical Biomedicine*, *Caspian Journal of Environmental Sciences*.

He has been a member of Executive Council of the International Odonatological Society (Societas Internationalis Odonatologica, SIO) 1981-95 (Holland), National Representative in-Charge, National Office of the Int. Odonatol. Soc. (SIO) in the Indian Subcontinent, 1981-88 Liaison Officer of Int. Odonatol. Soc. (SIO) in the Indian subcontinent 1983-93, Head, Regional office in Southern Asia of the International Odonatological Society (SIO: Central Office in Bilthoven, Holland), 1988-95; Member in-Charge, Odonata Specialist Group of the species Survival Commission 1985-95; Member, Executive Council of Nature conservators (NATCON), Muzaffarnagar, India 1988; Member, Executive Council of Applied Zoologists' Research Association (AZRA), Cuttack, India; Scientific Secy, 15th National Congress of Parasitology; Oct. 1-3, 2001, Jodhpur; President & Org. Secretary, 8th Int. Symp. of Vectors & Vector-borne Diseases (Oct. 13-15, 2006), Madurai; Member, Advisory Board, UMAA (Uttarakhand Mosquito Abatement Association). E-mail dt 3rd Feb., 2008; Member, Academic Council, Lady Doak College, Madurai 2008.

Dr. Tyagi has widely travelled across the globe (France, The Reunion Republic, Switzerland, USA, Czeck Republic, Canada, Australia, South Africa, Thailand, Malaysia, The Netherlands, Denmark, Mali Africa, Singapore etc.) in connection with either presenting invited lectures in international conferences, chairing scientific session in symposia, representing as an expert on behalf of Indian Council of Medical Research or as advisor to the WHO. He had several occasions to discuss or work with a large number of legendary entomologists. During 1995 he first met one of the greatest entomologists ever of the United States, Prof. Minter J. Westfall Jr. His range of knowledge on entomological disciplines was amazing, as he produced a highly referenced book, "*Dragonflies of North America*", with another North American great, Prof. James G. Needham.

Dr. Tyagi won his first international award/prize from the World Health Organization-Special Programme for Research & Training in Tropical Diseases (TDR) for a worldwide competition in 2005-06. Recently, he was awarded with the ICMR's most coveted Dr. M.O.T. Iyengar Memorial Award (2007) 'For Outstanding work in the field of Biomedical Research'. He received AZRA (Applied Zoologists Research Association) Award in 2002. He is the Founder President of both the Indian Dragonfly Society (2006) and the Jodhpur Natural History Society (2002).

Dr. Tyagi is a scientist with many 'firsts':

(i) He is the first entomologist to have filed a patent application under the ICMR in 1992 (NRDC, New Delhi; Patent Grant No. 191635 in the year 2005).

(ii) He was the first to have given a classification based on the distribution of malaria vectors, *Anopheles stephensi* and *An. culicifacies*, corresponding to that of prevalence of the malaria parasites, *Plasmodium vivax* and *P. falciparum* in the Thar Desert region, in 1995.

(iii) He has provided for the first time a correlation between age old water conserving underground storage wells called '*beri*' and '*tanka*' in the Thar Desert with that of the existence of *An. stephensi* the truly desert species, in 1996.

(iv) In the Thar Desert, he designed for the first time a 'Tanka Lid' to prevent breeding of *An. stephensi* in the tanka, its most common breeding habitat in the xeric ecosystem, in 2000.

(v) He first demonstrated that *Aedes albopictus* in the Kerala's Western Ghat region is able to transmit dengue in nature without the support of the principal vector of dengue in the country, *Aedes aegypti*, and emphasizing role of the former in Kerala as a primary vector, in 2002.

(vi) He ventured to describe a new anopheline species, *Anopheles pseudosundaicus* Tyagi et al., 2009, after a gap of nearly six decades of the last Anopheles species described in India.

(vii) He for the first time laid the foundation of the science of Medical Arthropodology in India through regularly held annual conferences since 2007.

(viii) He is possibly the only scientist in the country who brought out regularly biobibliography of prominent entomologists from across the country – an invaluable treasure, to be remembered and emulated by the future generations of entomologists. So far he has compiled 8 biobibliographies (Prof. T.N. Ananthakrishnan, Prof. C.F. Curtis, Dr. Rachel Reuben, Dr. M.O.T. Iyengar, Dr. Vijay Veer, Dr. R. Rajendran).

(ix) Dr. Tyagi has also additionally published original biobibliographies of several nationally and internationally zoologists/entomologists, both from India and overseas, such as Prof. Bastian Kiauta, Prof. T.N. Ananthakrishnan, Prof. P.K. Sen-Sarma, and Prof. I. Prakash.

(x) He was instrumental in executing the FIRST Transfer of Technology on "*Detection of antigen for Japanese encephalitis virus (JEV) from the wild-*

caught vector mosquitoes as a disease surveillance tool" by the CRME (ICMR), Madurai to the Directorate, Public Health & Preventive Medicine, Tamil Nadu, on July 25, 2010.

Dr. Tyagi has to his credit over 575 titles, many of which have been published in the international journals of high impact factor. He has authored/edited as many as 28 books, which include two monographs on 'Malaria in the Thar Desert' and 'The invincible deadly mosquitoes.' Other highly acclaimed books written by him include (i) Entomology: Ecology and biodiversity, (ii) Odonata: Biology of dragonflies, and (iii) Pests of forest importance and their management. He has even written a highly thought provoking book, '*Scientist*', which is widely read by yound and budding researchers and students alike.

Dr. Tyagi is functioning as a discipline expert/research project reviewer of various different departments/organizations, such as Department of Biotechnology, Ministry of Forests & Environment, Indian Council of Medical Resaerch, Department of Science & Technology (West Bengal, Kerala, Delhi, Tamil Nadu) and Indo-US Science & Technology Forum (IUSSTF), University Grants Commission etc. He is an expert member of the DBT-RCGM's Subcommittee for inspection of the infrastructural facility and for monitoring the R&D work of Asian outcrossed RIDL OX-513A Strain of the GMO *Aedes Aegypti* mosquito" at both the M/S Gangubhushan Bhikulal Investment & Trading Ltd (GBIT), Mumbai and the International Institute of Biotechnology & Toxicology (IIBAT), Chengalpet, Tamil Nadu. DBT has also appointed as their representative on the Institutional Biosafety Committees of (i) Department of Biotechnology, Lady Doak College, Madurai, and (ii) Mepco Schlenk, Sivakasi, Tamil Nadu. He is also a member of the Expert Subcommittee of the DBT-RCGM for formulating Biosafety Guidelines to conduct and monitor Confined Research Trials (CRTs) on the genetically engineered (GE) insects.

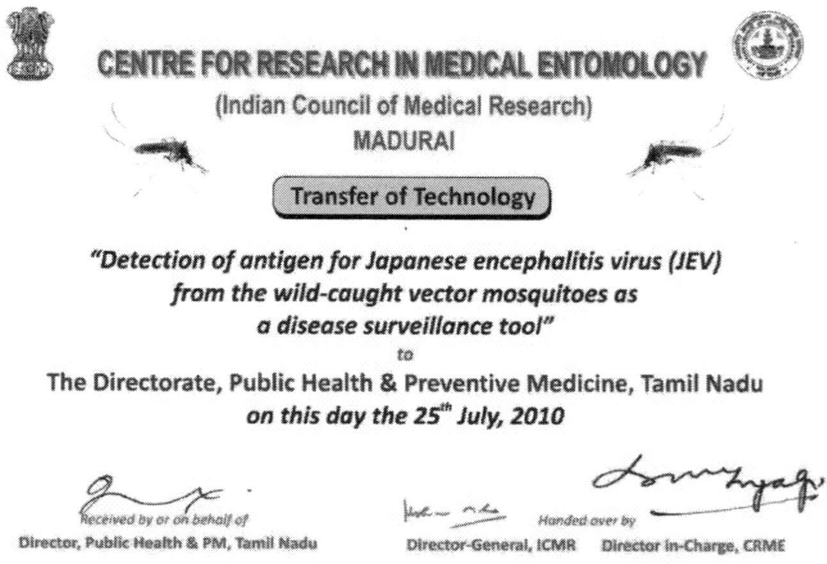

CENTRE FOR RESEARCH IN MEDICAL ENTOMOLOGY

(Indian Council of Medical Research)

MADURAI

Transfer of Technology

"Detection of antigen for Japanese encephalitis virus (JEV) from the wild-caught vector mosquitoes as a disease surveillance tool"

to

The Directorate, Public Health & Preventive Medicine, Tamil Nadu

on this day the 25th July, 2010

Received by or on behalf of — Director, Public Health & PM, Tamil Nadu

Director-General, ICMR

Handed over by — Director In-Charge, CRME

Having worked on a large number of insect orders and families, he is also one of very few entomologists who have an equal grip on more than one group of insects. Accordingly his publications numbering over 575 are based on a variety of subjects and insect orders such as mosquitoes, dragonflies, house flies, sandflies, cyclops, springtails, thrips, termites, fungi, malaria, filariasis, dengue, chikungunya, and Japanese encephalitis. In addition, he has authored/edited 28 books on as varied subjects as medical entomology, mosquitoes, malaria, biology of dragonflies, forest pests, desert ecology and biodiversity and entomology. He has written a biographic account of the world's all time great scientists, many of them being Indians, and his book, *Scientist*, became very popular among youth, in particular. In pursuing this diverse knowledge he has worked in almost all parts of the country. He has been always ready to move on to a new posting, and never considered one as matter of relegation; instead he considered such an opportunity to learn more and lead from the front! In this regard the saintly advise of Maharishi Patanchali in *Yogasuthra* comes to our mind:

"When you are inspired by some great purpose, some extraordinary project, all your thoughts break their bounds. Your mind transcends limitations, your consciousness expands in every direction, and you find yourself in a new, great, and wonderful world. Dormant forces, faculties, and talents come alive, and you discover yourself to be a greater person by far than you ever dreamt yourself to be."

Dr. Tyagi described his first insect species, *Calicnemia doonensis* Sangal & Tyagi, 1984 from the Dehradun Valley and his latest, *Anopheles pseudosundaicus* Tyagi et al. (2009) from the post-tsunami Kerala. His colleagues from the CRME have recently dedicated a new species in his honour, *Toxorhynchites tyagii* Krishnamoorthy et al., 2013.

Dr. Tyagi is a highly systematic planner and a careful visionary. A workaholic scientist, he arduously works every day for 14-16 hours, besides being an astute administrator magnificently looking after the varied responsibilities ranging from CRME's as well as scientists' research development and promotion at both national and international levels through extensive and intensive collaboration and networking with universities, medical colleges, government and non-government organizations to the Centre's infrastructure development, office management and staff welfare. He has won over several research projects from both national and international scientific bodies, and his WHO-TDR's Asian Biosafety Training Courses paertaining to Genetically Modified Vectors (2009-11) became known to be a landmark in science conduction and management at the highest level. Prof. T.N. Ananthakrishnan, the Founding Director of Entomology Research Institute, Chennai and Ex-Director, Zoological Survey of India, Kolkata, and unquestionably India's most outstanding and extraordinary entomologist in modern times, who was also the Chairman of the Centre's Scientific Advisory Committee in the CRME's formative days in mid-1980s, has chronicled Dr. Tyagi's growth in science since his PhD scholarship days and lauded his achievements (14th March, 2009) as:

"... I am privileged to be able to testify to the uncanny abilities of Dr. B.K. Tyagi, Director, CRME, as a medical entomologist of high calibre and whose dedication and perseverance has earned for CRME the status of internatinally acclaimed institution in the field of Medical Entomology. This is justified by the diversity of his numerous publications in journals of high standing. The solid foundation laid by the first Director, Rachel Reuben, has been further strengthened by his personal efforts, not to mention the interactions with his colleagues... . Friendly, alert, interested in diverse subjects concerned with Medical Entomology, Dr. Tyagi has been considered as an outstanding Medical Entomologist whose efforts to tide over many diseases like Dengue have been well known. The Annual Reports prepared by him reflect the diverse activities of CRME under his care and needless to say that in Dr. Tyagi we have a scientist with an uncanny ability to sustain the growth of an institution, which has adequately demonstrated the significance of the use of vector control technologies, some of which have had sufficient potentialities to end users. It is certainly a privilege to be associated with Dr. Tyagi whose sense of dedication is unparalleled ... ."

Dr. B.K. Tyagi, Conference President & Organizing Secretary, inaugurating the 4th Conference of Medical Arthropodology (Madurai, 20-21, Dec., 2010). Others seen on the dais include among others, Dr. P. Jambulingam and Dr. Reddyanaik.

Dr. Tyagi is a completely indefatigable and inexhaustible scientist with rare quality of leadership, as he professes "Leadership is in promoting team spirit and taking lead from the front, and not in domination." There is no Saturday or Sunday or holiday for him. There is no tomorrow for him since he invariably lays emphasis on completing all the work on the day, leaving for tomorrow no work of today! His quest for knowledge continues, reminding us of the most famous and inspiring lines by the Indian saint poet, Kabir –

काल करे सो आज कर आज करे सो अब
पल में परलय होएगी बहूरी करोगे कब

Dr. Tyagi retired in July 2013 as Scientist 'G' (Director) & Director in-Charge of the Centre for Research in Medical Entomology (ICMR) after more than 35 years of distinguidhed service. However, his service was extended for two more years till July, 2015. He is currently a Visiting Professor to the Bharathidasan University, Tiruchirapally (TN) and a Core Faculty to the SERB-DST School in Insect Biology at Hyderabad University, Hyderabad (Telangana). He is functioning from his home in Jodhpur (Rajasthan).

Scientific Publications by Dr. B.K. Tyagi

1977

1. Tyagi, B.K., 1977. A note on the karyotypes of *Burmagomphus pyramidalis Laidlaw and Onychogomphus saundersi duaricus* Fraser (Anisoptera: Gomphidae). *Odonatologica* (Holland) 6: (4): 277-282.

1978

2. Tyagi, B.K., 1978. The chromosome numbers and sex determining mechanisms newly recorded in thirteen Indian dragonflies (Odonata). *Chromosome Information Service* (Japan) 25: 5-7.
3. Tyagi, B.K., 1978. New dragonfly records from India and the Dehradun Valley (Northern India). *Current Science* 47 (18): 698.
4. Tyagi, B.K., 1978. Studies on the chromosomes of Odonata of Dun Valley (Dehra Dun, India). Ph.D. Thesis, Garhwal Univ., Srinagar, Garhwal. VIII+285 pp. (17 Tabs, 335 Figs.excl.).
5. Tyagi, B.K., 1978. A note on the karyotypes of *Burmagomphus pyramidalis* Laidlaw, *Onychogomphus saundersi duaricus* Fraser and *Onychogomphus schmidti* Fraser (Anisopter: Gomphidae). *Abstr. Ref. 5th Symp. Young Scientists' Society,* Dehradun, p.3.
6. Tyagi, B.K., 1978. The male germ cell chromosomes of *Macromia moorei moorei* Selys (Odonata: Cordullidae; Epophthalminae). *ibid.*, p.4.

1979

7. Tyagi, B.K. and Vijay Veer, 1979. A note on some entomogenous fungi from preserved dragonfly collections. *Journal of Bombay Natural History Society* 75: 946-947.
8. Tyagi, B.K. and Vijay Veer, 1979. More records of entomogenous fungi from preserved dragonfly collections. *ibid.* 77: 176-178.

1980

9. Tyagi, B.K. and P.K. Sen-Sarma, 1980. Inter-colonial aggression in workers of the wood carton nest building termite *Microcerotermes beesoni* Snyder (Insecta: Isopter; Termitidae). *Abstr. Pap. Symposium on the Ecology of Insect Communities,* Dehradun, p.35.

1981

10. Tyagi, B.K., 1981. Dragonflies feeding on houseflies. *Journal of Bombay Natural History Society* 77. 531.
11. Tyagi, B.K., 1981. Dragonfly crisis in some parts of India. *Selysia* (USA) 10: 12-13.
12. Tyagi, B.K., 1981. Adult dragonflies as human food in the Nagaland State, India. *Notulae odonatologiceae* (Holland) 1: 137-138.

13. Tyagi, B.K. and A. Tyagi, 1981. A note on the antennal asymmetry in some collembola. *Biologia* 27: 15-17.

1982

14. Tyagi, B.K. 1982. Cytotaxonomy of the Indian dragonflies. *Indian Review of Life Sciences* 2: 149-161.
15. Tyagi, B.K., 1982. Notes on some dragonflies from Dehradun valley (Uttar Pradesh, India). *Fraseria* 3: 9-10.
16. Tyagi, B.K. and S.K. Sangal, 1982. Spermatocyte chromosomes of *Anax immaculifrons* Rambur from India (Anisoptera: Aeshnidae). *Notulae odonatologiceae* 1: 154-155.
17. Tyagi, B.K. and P.K. Sen-Sarma, 1982. Ecological studies on termites with reference to their role in the process of wood decomposition in the Dehradun forests (Northern India). *Symposium on Insect Ecology and Resource Management,* Muzaffarnagar pp.23-24.
18. Tyagi, B.K., P.K. Sen-Sarma, P.S. Rehill, P.C. Pandey, 1982. Termite-fungi interactions. I. Laboratory testing of wood decayed blocks to *Coptotermes heimi* Easmann, *Neotermes bosei* Snyder and *Microcerotermes beesoni* Snyder. *Proceedings of the Second International Congress Termite Fauna of Arid Zone, Assiut* (UAR), p.41.

1983

19. Tyagi, B.K., 1983. Some interesting dragonflies from Dehradun Valley (Northern India) *Geobios New Reports* 2: 63-65.
20. Tyagi, B.K., 1983. Fifty years of karyological studies in Indian Odonata: Some highlights of achievements with emphasis on lacunae in studies. *Abstr. Pap. VII International Symposium of Odonatology,* Calgary, pp. 40-41.
21. Tyagi, B.K., 1983. Review of the m-chromosomes in Indian Odonata: Their significance for taxonomy geographical distribution at present. *ibid.*, pp. 39-40.
22. Tyagi, B.K., 1983. Cytogenetics, cytotaxonomy and cytophylogeny of dragonflies (Class Insecta: Order Odonata) *Abstr. Pap. Ist All India Symposium on Physiology of Insect Reproduction,* Nagpur, pp. 6-7.
23. Tyagi, B.K., 1983. A preliminary inventory of the endangered dragonflies of the north-western Himalayas (India). *Fraseria*, p. 9.
24. Tyagi, B.K., 1983. On a small collection of Odonata of Kanpur district (Uttar pradesh), India. *Geobios New Reports* 2: 70.
25. Tyagi, B.K. and P.K. Sen-Sarma, 1983. Inter-colonial aggression in the workers of termites (In : *Insect inter-relations in forest and agro-ecosystems.* Eds. P.K. Sen-Sarma, S.K. Kulshrestha and S.K. Sangal), Jugal Kishore & Sons, Dehradun, pp. 15-22.
26. Tyagi, B.K., P.C. Pandey, P.S. Rehill & P.K. Sen-Sarma, 1983. Termite-fungi interactions. III. Laboratory testing of *Diplodia pinea* attacked wood-blocks to *Microcerotermes beesoni* Snyder (Isoptera: Termitidae). *Annals of Entomology* 1: 88-89.

1984

27. Tyagi, B.K. 1984. List of odonata of the Chandigarh union Territory and the Chandigary Siwalik outcrops, India. *Notulae odonatologiceae* 2: 61-62.
28. Tyagi, B.K., 1984. Cytotaxonomy of the genus *Onychogomphus* Selys (Odonata: Anisoptera), with special reference to the evolution of the sex-determining mechanisms and the reduced chromosome number in the family Gomphidae. *Abstr. Pap Ist Indian Symposium of Odonatology*, Madurai, p.4.
29. Tyagi, B.K., 1984. Conservation of Odonata in India: Present situation, problems and some suggestions. *ibid.* pp. 14.

30. Sangal, S.K. and B.K. Tyagi, 1984. *Calicnemia miniata doonensis*, a new subspecies from the Dehradun Valley (Dehradun), India (Zygoptera: Platyenemididae). *ibid.*, p.2.

31. Tyagi B.K., 1984. Morphological variations in some anophelines from South Gujarat (India). *Abstr. Pap III Symposium of Oriental Entomology*, Trivandrum, p. 158.

32. Tyagi, B.K., 1984. Observations of the reflex immobilization behaviour of the mosquito *Anopheles subpictus* Grassi, 1899 in some villages of the south Gujarat, India. *Geobios new Reports* 3: 161-162.

33. Tyagi, B.K. 1984. A note on the new records of some anophelines from South Gujarat (India). *Geobios new Reports* 3: 149-151.

34. Tyagi, B.K., 1984. Seasonal prevalence of *Anopheles culicifacies* Giles in Ukai villages, Sough Gujarat (India). *Abstr. Pap. Proc. VI All India Congress of Zoology*, Gwalior, p.33.

35. Tyagi, B.K., 1984. Some considerations on the distribution of chromosome types in insect orders, with a note on phylogenetic significance of the chromosome numbers among insect orders. *ibid.*, p. 139.

36. Tyagi, B.K., 1984. A note on the variations in the pre-humeral and humeral dark spots on the wing costa of the mosquito, *Anopheles subpictus* Grassi, 1899 from South Gujarat (India). *Geobios New Reports* 4: 56-57.

37. Tyagi, B.K., P.C. Pandey, P.S. Rehill and P.K. Sen-Sarma, 1984. Termite fungi interactions. II. Laboratory testing of the decayed wood blocks to the wood carton nest building termite, *Microcerotermes beesoni* Snyder. *Material und Organismen* (Germany) 19: 69-75.

1985

38. Tyagi, B.K., 1985. Cytotaxonomy of the genus *Onychogomphus* Selys (Odonata: Anisoptera; Gomphidae), with special reference to the evolution of the sex-determiing mechanisms and the reduced chromosome number in the family Gomphidae. *Proc Ist Indian Symposium of Odonatology*, Madurai, pp. 217-228.

39. Tyagi, B.K., 1985. A review of the cytotaxonomy of the Indian Odonata, with reference to the Dehradun Valley fauna. *Abstr. Pap. VIII International Symposium of Odonatology,* Paris, p. 21.

40. Tyagi, B.K., 1985. Some investigations on the impact of chemical mosquito control agents on non-target organisms in aquatic ecosystems, with special reference to the dragonfly naiads. *ibid.*, p. 22.

41. Sangal, S.K. and **B.K. Tyagi**, 1985. *Calicnemia miniata doonensis*, a new subspecies from the Dhradun Valley (Dehradun, India), (Zygoptera: Platyenemididae). *ibid.*, pp. 193-198.

42. Tyagi, B.K., 1985. Dragonfly conservation in India: Present status and future prospects. *Odonata Specialist Group, IUCN Report* 6: 1-5.

43. Mariappan, T., M. Kalyanasundram, K.N. Panicker, N. Balakrishnan, **B.K. Tyagi** and P.K. Das, 1985. Evaluation of fenfluthrin (OMS 2013), a new synthetic pyrethroid, for insecticidal evaluation against mosquito vectors. *Indian Journal of Medical Research* 82: 1-8.

44. Mathavan, S., T.N. Ananthakrishnan, A. Kumar, J. Muthukrishnan, T.J. Pandian, D.B. Tembhare and **B.K. Tyagi**, 1985. *Proceedings of the First Indian Symposium of Odonatology*, Madurai pp. 226.

45. Tyagi, B.K., M. Kalyanasundaram, P.K. Das and N. Somachary, 1985. Evaluation of a new compound (VCRC/INS/A.23) with juvenile hormone activity against the mosquito vectors. *Indian Journal of Medical Research* 82: 9-13.

46. Tyagi, B.K. and P.K. Das, 1985. Use of pyrethroid insecticides in controlling vector mosquitoes and their impact on the environment. *Abstr. Pap. Symposium on Pesticide Residues and Environmental Pollution*, Muzaffarnagar, pp. 36-37.
47. Das, P.K. and **B.K. Tyagi**, 1985. Role of geo-climatic factors on the efficacy and adverse impact on the environment. *Abstr. Pap. International Conference on Pesticide Toxicity, Safety and Risk Assessment*, Lucknow, pp. 27-31.

1986

48. Tyagi, B.K. and Ajita Tyagi, 1986. Directory of odonatologists from the Indian subcontinent. *Occ. Publs S.I.O. Natl. Office India*, No.1, 20 pp.
49. Tyagi, B.K., 1986. Cytogenetics, karyosystematics and cytophylogeny of the Indian odonata. *Indian Review of Life Sciences* 6: 215-219.
50. Kalyanasundaram, M., Nisha George and **B.K. Tyagi**, 1986. Future prospects of plant materials for insect control. *Abstr. Pap. National Symposium on Insecticidal Plants and Control of Environmental Pollution* (Sec. A), Tiruchirapalli, p.2.
51. Tyagi, B.K. and P.K. Das, 1986. Use of pyrethroid insecticides in controlling vector mosquitoes and their impact on the environment. Proc. *Symposium on Pesticide Residues and Environmental Pollution*, Muzzafarnagar, pp. 137-149.
52. Tyagi, B.K., A. Tyagi and S.K. Sangal, 1986. An up-to-date catalogue and bibliography of dragonfly fauna of the Dehradun valley (India) *Occ. Publ. SIO Natl. Off. India* 2:1-14.
53. Das, P.K., **B.K. Tyagi** and M. Kalyasundaram, 1986. Vernacide, a new insecticide impregnated paint for controlling mosquito vector, *Culex quinquefasciatus* and cockroach, *Periplanata Americana. Indian Journal of Medical Research* 83: 268-270.
54. Das, P.K., **B.K. Tyagi**, N. Somachari and V. Vasuki, 1986. Efficacy of Arosurf, a molecular surface film in controlling *Culex quinquefasciatus* Say, *Anopheles stephensi* Liston and *Aedes aegypti* (L.). *Indian Journal of Medical Research* 83: 271-276.
55. Amalraj, D., M. Kalyanasundaram, T. Mariappan, K.D. Ramiah, N. Arunachalam, A.R. Rajavel, K.P. Paily, **B.K. Tyagi**, U.S. Rao, K. Narayana, T.B. Gaur, M.S. Mithyantha and P.K. Das, 1986. Field evaluation of FICAM (Bendiocarb), a carbamate adulticide, in two villages of Pondicherry. *Indian Journal of Medical Research* 84: 472-479.
56. Rajavel, A. R., V. Vasuki, K.P. Paily, K.D. Ramaiah, T. Mariappan, M. Kalyasundaram, **B.K. Tyagi** and P.K. Das, 1986. Evaluation of cyfluthrin (OMS 2012), a synthetic pyrethroid, for insecticidal activity against different mosquito species. *WHO/VBC/86.9*35.

1987

57. Rajavel, A.R., V. Vasuki, K.P. Paily, K.D. Ramaiah, T. Mariappan, M. Kalyasundaram, **B.K. Tyagi** and P.K. Das, 1987. Evaluation of a synthethic pyrethroid, cyfluthrin for insecticidal activity against different mosquito species. *Indian Journal of Medical Research* 85:168-175.
58. Tyagi, B.K., N. Somachari, V. Vasuki and P.K. Das, 1987. Evaluation of three formulations of a chitin synthesis inhibitor (fenoxycarb) against mosquito vectors. *Indian Journal of Medical Research* 85: 161-167.
59. Tyagi, B.K., 1987. An introduction to the activities of the S.I.O. National Office in India. IX International Symposium of Odonatology, Madurai 18-21, 1988. Pondicherry, pp. 37.
60. Amalraj, D., K.D. Ramaiah, M., A.R. Rajavel, T. Mariappan, V. Vasuki, K.P. Paily., B.K. Tyagi and P.K. Das, 1987. Evaluation of alphamethrin, a synthetic

pyrethroid, for insecticidal activity against mosquitoes. *Indian Journal of Medical Research* 86: 601-609.

61. Rajagopalan, P.K., P.K. Das, M. Kalyanasundaram, B.K. Tyagi, N. Arunachalam, N. Somachary. C.B.S. Reddy and C.M.R. Reddy, 1987. *Bangalore Mosquito Control Master Plan*. Vector control Research Centre, Pondicherry, pp. 309.

62. Tyagi, B.K., 1987. Review of the cytotaxonomy of Isoptera (Insecta), with a description of the male germ cell chromosomes of *Microcerotermes beesoni* Snyder from the Dehradun Valley, India. *Indian Review of Life Sciences* 7: 263-272.

1988

63. Tyagi, B.K. (Editor-author), 1988. Odonatological bibliography of Frederick Charles Fraser {originally manuscripted by Basil Elwood Montgomery (*f*); re-organised, updated and edited by Dr. B.K. Tyagi with the help and guidance of Prof. B. Kiauta, Exec. Editor, ODONATOLOGOCA}. *Occ. Publs. SIO Natl. Off. India* No.3: 1-24.

64. Amalraj, D., V. Vasuki, C. Sadanandane, M. Kalyanasundaram, **B.K. Tyagi** and P.K. Das, 1988. Evaluation of two new juvenile hormone compounds against mosquito vectors. *Indian Journal of Medical Research* 87: 19-23.

65. Amalraj, D., V. Vasuki, M. Kalyanasundaram and B.K. Tyagi, 1988. Laboratory and field evaluation of three insect growth regulators against mosquito vectors. *Indian Journal of Medical Research* 87: 24-31.

66. Tyagi, B.K., 1988. Management of vector mosquitoes in an urban set-up. *Abstr. Pap. Second Symposium on Vectors and Vector-Borne Diseases*, Trivandrum, p. 39.

1989

67. Jambulingam, P.K., K. Gunasekharan, S.S. Sahoo, P.K. Hota, B.K. Tyagi and M. Kalyanasundaram, 1989. Effect of permethrin-impregnated bednets in reducing population of malaria vector, Anopheles culicifacies in a tribal village of Orissa State (India). *Indian Journal of Medical Research* 89: 48-51.

68. Tyagi, B.K., 1989. Control of mosquito larval population by monomolecular surface films and their impact on non-target organisms including dragonfly naiads. Abstr. Pap. XI International Symposium of Odonatology, Tennessee, USA, p. 51.

1990

69. Tyagi, B.K., 1990. Annotated check-list of the anophelines from district Surat (Gujarat). *Journal of Applied Zoological Researches* 1: 73-76.

1991

70. Tyagi, B.K., 1991. Prevention from mosquito bites by using insecticide impregnated fabric and bednets. *Abstr. Pap. IV National Symposium on Growth, Development and Control Technology of Insect Pests*, Muzaffarnagar, p. 52.

71. Tyagi, B.K., 1991. Indoor resting habits of mosquitoes in Ganganagar (Rajasthan), assessed through abdominal condition and age composition. *ibid.*, p. 49.

72. Tyagi, B.K., 1991. A report on the indirect approach for conserving the dragonfly habitats in and around a metropolis in India. *Rep. Odon. Specialist Group Int. Un. Conserv. Nat.* 8: I-4.

73. Tyagi, B.K., 1991. Role of non-target organisms in determining the safe dose – limits of chemical agents for vector mosquito larval control. *Abstr. Pap. International Symposium of Odonatology,* Trevi, Italy, p. 31.

74. Tyagi, B.K., 1991. Control of vector mosquitoes by dragonflies – a myth or fact? ibid., p.31.

75. Tyagi, B.K. and K.V.S. Verma, 1991. Anopheline mosquitoes of Sri Ganganagar district (Rajasthan) transmitting malaria parasite. *Journal of Applied Zoological Researches* 2: 85-91.

76. Tyagi, B.K., S.K. Bansal and K.V.S. Verma, 1991. Intra-specific morphological variations in two ecotypes of *Anopheles Stephensi* Liston in north-western Rajasthan (Dipter: Culicidae). *Annals of Entomology* 9: 85-89.

77. Verma, K.V.S. and B.K. Tyagi, 1991. A brief report on mite infestation of *Anopheles Stephensi* Liston in the desert villages of Rajasthan. *Journal of Indian Society of Health and Environmental Research* 1: 73-75.

78. Tyagi, B.K. & Peter Miller, 1991. A note on the odonata collected in south western Rajasthan, India. *Notulae odonatologicae* 3: 134-135.

1992

79. Tyagi, B.K., 1992. Control of malaria vectors of India. *Indian Review of Life Sciences* 12: 211-238.

80. Tyagi, B.K., 1992. Prevention from mosquito bites by using insecticide impregnated fabric and bednets. *Bioecology & Control of Insect Pests. UPZS, Muzaffarnagar,* pp. 269-275.

81. Tyagi, B.K., 1992. Feasibility of using permethrin-treated bednets in reducing man-vector contact and malaria transmission, with special reference to the main rural vector in India, *Anopheles culicifacies. Abstr. Pap. First Conference on Impact of Applied Zoological Researches on Food Production and Environment*, Cuttack, pp. 53-54.

82. Tyagi, B.K., 1992. Malaria control in India, with particular reference to the desert region in Rajasthan State: Prospective alternate vector control strategies. *ibid.*, p. 52-53.

83. Tyagi, B.K., 1992. Prospective role of biological agents in controlling vector mosquitoes transmitting human infections. *Abstr. Pap. IV South Asian Symposium of Odonatology*, Allahabad, p.14.

1993

84. Tyagi, B.K., 1993. An indigenously developed mechanical device for sampling the resting adults of mosquitoes. *Journal of Applied Zoological Researches* 4: 180-184.

1994

85. Tyagi, B.K. 1994. Alternative control strategies for mosquito vectors of human diseases in India, pp.359-372 *(In: Perspectives in Entomological Research)* Scientific Publishers, Jodhpur/India, pp.406.

86. Brandson, R.R., R.K. Garg & B.K. Tyagi, 1994. Medical and Health Facilities, pp. 178-179 (In: *Resource Atlas of Rajasthan*. Eds.: Dr. V.B. Moghe, Dr. K.R. Nahar and Mr. A.K. Parmar). Department of Science & Technology, Govt. of Rajasthan, Jaipur, pp. 231.

87. Tyagi, B.K., 1994. Cannibalism among the larvae of filariasis vector, *Culex quinquefasciatus* Say. *Geobios New Reports* 13: 136-138.

88. Tyagi, B.K., 1994. Effect of the choice of oviposition medium on the breeding behaviour of *Culex quinquefasiatus* Say. *Geobios New Reports* 13: 28-30.

89. Tyagi, B.K., 1994. Thanatosis as an adaptive defense mechanism in *Anopheles Stephensi* larvae breeding in "Tanka" and "Beri" in the desert villages of Rajasthan. *Geobios New Reports* 13: 132-135.

90. Tyagi, B.K., 1994. Impact of extensive canalization on the distribution of malaria vectors in the north-western Rajasthan. *Abstr. Pap. Indian Science congress Association*, Jaipur, pp. 39-40.

91. Tyagi, B.K., 1994. Need for imparting training to personnels undertaking insecticide spraying for vector control, with special reference to xeric conditions in western Rajasthan. *Abstr. Pap. IV International Symposium on Vector & Vector-Borne Diseases*, Bhubaneswar, p. 82.

92. Tyagi, B.K., 1994. A note on the breeding of vector mosquitoes in cement tanks and pit latrines. *Journal of Applied Zoological Researches* 5: 149-151.

93. Tyagi, B.K., 1994. Distribution of malaria transmitting mosquitoes (Culicidae: Anophelinae; Anophelini) in the Thar Desert, western Rajasthan, India. Abstr. Pap. *Meeting of Experts on Gaps in Researches on the Faunal Diversity in the Thar Desert*, Jodhpur, p.37.

94. Tyagi, B.K. and S.P. Yadav, 1994. Impact of agricultural patterns and water management on the distribution of disease transmitting mosquitoes. *ibid.*, p. 36.

95. Tyagi, B.K., K.V. Singh, S.K. Bansal, and S.P. Yadav, 1994. Malaria epidemic in some villages of north-western desert Rajasthan. *Journal of Applied Zoological Researches* 5: 152-155.

96. Tyagi, B.K., 1994. Distribution of arthropod vector-borne communicable diseases and control of their vectors in India. *Indian Review of Life Sciences* 14: 223-243.

1995

97. Tyagi, B.K., R.C. Chaudhary and S.P. Yadav, 1995. Epidemic malaria in Thar Desert, India. *The Lancet* 346: 634-635.

98. Tyagi, B.K., 1995. Malaria in the Thar Desert, *ICMR Bulletin* 25: 85-91.

99. Tyagi, B.K., 1995. A note on termite consumption by *Pantala flavescens* (Fabr.) (Anisoptea: Libellulidae). *Notulae odonatologiceae* (Holland) 4: 92.

100. Tyagi, B.K., 1995. Ecology and control of the primary malaria vector, *Anopheles stephensi* Liston, in the Thar Desert, north-western India. *Pap. Abstr. IX European Meeting of Society for Vector Ecology, Prague* (Czeck Republic), pp. 32-33.

1996

101. Tyagi, B.K. and S.P. Yadav, 1996. Malaria and vector control: Awareness of rural community in the Thar desert following an epidemic. *WHOSEARO Regional Health Forum* 1: 41-45.

102. Tyagi, B.K., 1996. Changing scenario of malaria prevalence in desert Rajasthan. *Geobios* 23: 176-184.

103. Tyagi, B.K. and S.P. Yadav, 1996. Malariological and sociological significance of "Tanka" and "Beri" in the Thar Desert, Western Rajasthan (India). *Journal of Arid Environments* (U.K.) 33: 497-501.

104. Tyagi, B.K., 1996. Chronological and spatial distribution of anopheline vectors of malaria in the Thar Desert, pp. 227-240 (In: *Faunal diversity in the Thar Desert – Gaps in Research.* Eds.: Dr. A.K. Ghosh, Dr. Q.H. Baqri and Dr. I. Prakash) Scientific Publishers, Jodhpur/India, pp.410.

105. Tyagi, B.K., & R.C. Chaudhary, 1996. Geographical diversification of human plasmodia and malaria syndrome in the Thar Desert, pp. 25-36 (In: *Faunal diversity in the Thar Desert – Gaps in Research.* Eds.: Dr. A.K. Ghosh, Dr. Q.H. Baqri and Dr. I. Prakash) Scientific Publishers, Jodhpur/India, pp.410.

106. Tyagi, B.K., 1996. Major vector-borne human diseases in the Thar Desert, India. *Annals of Medical Entomology* 5 (1 & 2): 13-19.

107. Tyagi, B.K., 1996. Mosquito and human diseases: Prospects of their management in the 21st Century, with special reference to India. *Journal of Applied Zoological Researches* 7 (2):

108. Tyagi, B.K., 1996. Thar Marusthal mein malaria: ek vivechanatmaka sameeksha (in Hindi). *Arogya Sandesh*, pp. 915, 22.

109. Tyagi, B.K., 1996. Phenology of Anopheles subpictus Grassi, an unusual vector of malaria in the Thar Desert, India. Abstr. Pap. XX International Congress of Entomology, Firenze, Italy.

110. Tyagi, B.K., 1996. Malaria in the Thar Desert: A bird's eye-view. Abstr. Poster 63rd Annual Meeting of Indian Academy of Sciences, Jodhpur.

111. Shahi, A.K., S. Chandra, **B.K. Tyagi** and B.L. Kaul, 1996. Conservation strategies for Cymbopogon species of Indian Thar Desert. Abstr. Pap. XIV National Symposium on Agronomy, Environment and Food Security for 21st Century. Indian Society for Agronomy and CCS Haryana Agr. Univ., Hisar (Dec. 11-13, 1996), pp. 14-15.

1997

112. Tyagi, B.K., T. Ramnath and A.K. Shahi, 1996. Evaluation of repellent activities of *Tagetus minuta* (Family: Compositae) against the vector mosquito *Anopheles stephensi* Liston, *Culex quinquefasciatus* Say and *Aedes aegypti* (L.). *International Pest Control* (USA) 39: 184-185.

113. Tyagi, B.K., 1996. Emerging and re-emerging vector-borne diseases in the Great Indian Thar desert. *WHOSEARO Regional Health Forum* 2: 9-15.

114. Tyagi, B.K., and S.P. Yadav, 1996. Correlation of irrigation and flood water management with malaria in the Thar desert. *ibid.* 2: 5-11.

115. Tyagi, B.K., 1997. Malaria epidemics in the Great Indian Thar Desert and the validity of the Bauma-van der Kaay's El Nino Southern Oscillation theory as an early warning system for future epidemics. *Annals of Medical Entomology* 6: 19-24.

116. Tyagi, B.K., 1997. Dragonfly wealth of India: extant species – a preliminary systematic list. *Zoos' Print* 12:3-10.

117. Tyagi, B.K., 1998. Major vector-borne human diseases in the Thar desert, with special reference to the changing climate. *Abstr. Pap. Seminary on Research & Development Priorities of Arid Zone Beyond 2000 A.D.*, CAZRI, Jodhpur (Mar. 13-14, 1997), p.11.

118. Tyagi, B.K. and P.K. Sen-Sarma, 1999. Morphology of the spermatozoa of *Microcerotermes beesoni* Snyder with reference to termite phylogeny (Isoptera: Termitidae). *Uttar Pradesh Journal of Zoology* 17: 245-246.

119. Tyagi, B.K. and R.C. Chaudhary, 1997. Outbreak of *falciparum* malaria in the Thar Desert with particular emphasis on physiographic changes brought about by extensive canalization and their impact on vector density and dissemination. *Journal of Arid Environments* (UK) 36: 541-555.

120. Dam, P.K., S.P. Yadav and B.K. Tyagi, 1997. Thar Marukschetra mein pai jane wali aushadhiyukta vanashpatiyan: Ek avalokan [in Hindi] (Medicinal Plants of Thar Desert: An overview). Abstr. Pap. Rashtriya Vaigyanik Sanghoshthi (Natl. Sci. Symp.), Nov. 6-7, 1997; Lucknow, p.54.

1998

121. Tyagi, B.K., A.K. Shahi and B.L. Kaul, 1998. Evaluation of repellent activities of the Cymbopogon essential oils against mosquito vectors of malaria, filariasis and dengue fever in India. *Phytomedicine* 5: 324-329.

122. Tyagi, B.K. 1998. Malaria and its vectors in the Thar Desert, northwestern Rajasthan (India), pp. 1-10 (In: *Advances in Medical Entomology & Human Welfare*, Ed.: Dr. S.C. Goel + *Sectional Experts*: Dr. S.K. Sharma, Dr. P.R. Malhotra, Dr.

Neelam Tandon and Dr. B.K. Tyagi; PREFACE by Prof. N.K. Ganguly, DG (ICMR). The Uttar Pradesh Zoological Society, Muzaffarnagar, India, pp. 123.

123. Tyagi, B.K. and S.P. Yadav, 1998. Impact of paddy cultivation on the prevalence of malaria and vector mosquitoes in the Thar desert – a possible forecasting indicator for future epidemics. *Joint Annual Conference of ISCD and IAE*, Delhi (Oct., 30-Nov.1, 1998) p.17.

124. Yadav, S.P. and B.K. Tyagi, 1998. Socio-epidemiological observations on the experiences of women with history of malaria episodes on management and family support in the rural Thar desert. *ibid.*, p. 18.

125. Yadav, S.P., B.K. Tyagi and P.K. Dam, 1998. Socio-environmental linkages with malaria-induced economy in the rural community of the Thar Desert, India. *XIII Annual Congress of NESA*, Gulberga (Dec. 12-14, 1998), pp.66-67.

126. Dam, P.K., S.P. Yadav, T. Ramnath and B.K. Tyagi, 1998. Constraints in conservation of medicinal plants in the climatologically changing Thar Desert, *ibid.*, p.6.

1999

127. Yadav, S.P., B.K. Tyagi and T. Ramnath, 1999. Knowledge, attitude and practice towards malaria in rural communities of the epidemic-prone. Thar Desert, north-western India. *Journal of Communicable Diseases* 31: 127-136.

128. Dam, P.K., T. Ramnath, S.P. Yadav and B.K. Tyagi, 1999. Ethnomedicinal measures for controlling malaria fevers and vector-biting in the rural Thar Desert, Rajasthan. *Abst. Pap. 4th International Symposium of Vectors & Vector-Borne Diseases,* Gwalior (Feb. 18-20, 1999), pp.36.

129. Tyagi, B.K., S.P. Yadav, P.K. Dam and T. Ramnath, 1999. Enhanced breeding of malaria/dengue fever transmitting vector mosquitoes (*Anopheles stephensi* Liston/*Aedes aegypti* Linnaeus) in multi-storey buildings of Jodhpur township: Prospective problems and remedies. *Abstr. Pap. UGC National Seminar on Multi-storey buildings of growing towns – Problems & remedies,* Jodhpur (Mar. 4-6, 1999), p. 45.

130. Yadav, S.P., B.K. Tyagi, P.K. Dam and T. Ramnath, 1999. On some socio-economic aspects of inhabitants of multi-storey buildings of Jodhpur city. *Ibid.*, p.1.

131. Tyagi, B.K., 1999. Vector-borne diseases of the Thar Desert. *Abstr. Pap. Man in Desert – Medical Problems CME – 99, Military Hospital,* Jodhpur, p.3.

132. Yadav, S.P. and B.K. Tyagi, 1999. A study on some cultural and health seeking behaviour of destitute leprosy patients in the holy place of Ramdeora, Jaisalmer district (north-western India). First Conference of Council of Behavioural Scientists., agra, pp. 41-42.

2000

133. Yadav, S.P. and B.K. Tyagi, 2000. Responses of women with malaria history in relation to family management and support in the rural Thar Desert, India under constant threat of increasing mosquitogenic conditions. *Annals of Medical Entomology* 9: 32-41.

134. Mathur, M.L., S.P. Yadav and B.K. Tyagi, 1999. A study of an epidemic of acute respiratory disease in Jaipur town. *Journal of Postgraduate Medicine* 46: 88-90.

135. Tyagi, B.K. and S.P. Yadav (2000). Paddy cultivation and the prospect of malaria in the Thar Desert, India: A review. *Journal of Applied Zoological Researches* 11: 41-47.

136. Dam, P.K., T. Ramnath, S.P. Yadav and B.K. Tyagi, 2000. Use of ethnomedinal practices in the Thar desert, north western India. *Journal of Economic & Taxonomic Botany* 24: 303-318.

137. Tyagi, B.K., 2000. Water Management and infectious diseases. *Expert lecture in Training Programme for Junior Chemists, Groundwater Department, GOR,* Jodhpur (May 29,2000)

138. Tyagi, B.K., S.P. Yadav, P.K. Dam and T. Ramnath, 2000. Enhanced breeding of malaria/dengue vector mosquitoes, *Anopheles stephensi* Liston/*Aedes aegypti* Linn., in a fast growing Jodhpur township: Future prospects and control methodologies, pp. 373-380 (In: *Multi storey buildings of growing towns – Problems and remedies.* Eds. Dr. M.R. Sethia and Dr. A.K. Gupta), Department of Structural Engineering, MBM Eng. Coll., Fac. of Engineering, JNV Univ., Jodhpur, pp. 513.

139. Yadav, S.P., B.K. Tyagi, P.K. Dam and T. Ramnath, 2000. On some socio-economic aspects of inhabitants of multi-storey buildings of Jodhpur city, pp. 53-58 (In: *Multi storey buildings of growing towns – Problems and remedies.* Eds. Dr. M.R. Sethia and Dr. A.K. Gupta), Department of Structural Engineering, MBM Eng. Coll, Fac. of Engineering, JNV Univ., Jodhpur, pp. 513.

140. Shahi, A.K., B.K. Tyagi, R.C. Chaudhary and B.K. Kaul, 2000. *Cymbopogon schoenanthus* (L.) Spreng. (Poaceae) – a versatile aromatic and medicinal plant of Indian Thar Desert. *Flavours & Fragrances Association of India Journal* 1: 31-33.

141. Dam, P.K., S.P. Yadav and B.K. Tyagi, 2000. Thar Marukschetra mein aushadhiukta vanaspatiyan: ek purvalokana (in Hindi) [The Medicinal Plants of Thar Desert Region: A Review]. *ICMR Patrika* 14 (8): 61-65.

142. Tyagi, B.K., 2000. Significance of genus *Cymbopogon* in the formulation of mosquito repellants. *National Symposium on Propsects and Potentials of Plant Biotechnology in Indian in 21st Century & 23rd Meeting PTCA,* Jodhpur (Oct. 18-21, 2000), p. 146.

2001

143. Tyagi, B.K, K.V. Singh, S.K. Bansal, S.P. Yadav and P.K. Dam, 2001. Thar Marukschetra mein malaria ka prakop wa uska nidaan (in Hindi) [Malaria conflagration in the Thar Desert regional and its control]. *Abstr. Pap. Symposium on Vgyan Pradyogiki ke naye aayaam (New dimensions of science and technology)*, Defense Laboratory, Jodhpur, pp. 66-69.

144. Tyagi, B.K. and S.P. Yadav, 2001. Bionomics of malaria vectors in two physiographically difference areas of the epidemic-prone Thar Desert, north-western Rajasthan (India). *Journal of Arid Environments* 47: 161-172.

145. Tyagi, B.K., 2001. Malaria conflagration under the impact of changing Thar Desert environment (north-western Rajasthan, India). Symposium on the Impact of Human Activities on Thar Desert Environment, CAZRI, Jodhpur (Feb., 15-17,2001), p. 109.

146. Yadav, S.P., B.K. Tyagi, P.K. Dam and R. Sachdev, 2001. A study of association between socio-economic and malaria in rural part of Indian Thar Deser, *ibid.*, pp. 141-142

147. Tyagi, B.K., 2001. Major vector-borne human diseases in the Thar Desert, with special reference to the changing climate, pp. 103-112 (In: Priorities of research for development of arid regions. Eds.: Dr. S.K. Verma, Dr. N.M. Nahar, Dr. R. Raj-Bhansali, Dr. S.K. Jindal and Dr. Satya Vir), Scientific Publishers (India), Jodhpur, pp. 238

148. Tyagi, B.K. and A.K. Shahi, 2001. Evaluation of repellent potential of two *Cymbopogon* species (Poaceae), endemic to the Thar Desert (India), against major vector mosquitoes. *Intern. Confer. Globalization of Indian Essential Oils*, Agra [Sep. 21-23,2001].

149. Tyagi, B.K. and S.P. Yadav, 2001. Evolution of pathways of vectorism and malaria epidemics in the Thar Desert, north western India. *15th Natl. Congress of Parasitology*, Jodhpur. [Oct, 1-3, 2001], pp. 71-72.

150. Tyagi, B.K., S.P. Yadav, R. Sachdev and P.K. Dam 2001. Malaria outbreak in the IGNP command area in Jaisalmer district, Thar Desert, India. *Journal of Communicable Diseases* 33(2): 88-95.

151. Tyagi, B.K., 2001. Dracunculiasis eradication in India, with special reference to the Thar Desert (Rajasthan). (In: *Helminthology in India.* (Ed.) M.L. Sood], International Books Distributors, Dehradun. pp. 543-555. (Invited Chapter).

2002

152. Poopathi, S. and B.K. Tyagi. 2002. Mosquitocidal toxins of bacilli for effective biological control: Recent advancement in mosquito control research. *6th Int. Symp. On vectors & vector-borne diseases*, Bhubaneshvar (Feb 9-11, 2002), pp. 77-78.

153. Poopathi, S. and B.K. Tyagi, 2002. Studies on *Bacillus sphaericus* toxicity related resistance development and biology in the filariasis vector *Culex quinquefasciatus* (Diptera: Culicidae) from south India. *Applied Entomology and Zoology*. 37(3): 365-371.

154. Tyagi, B.K., J.R. Sharma and S.P. Yadav, 2002. Irrigated agriculture and exacerbation of *P. falciparum* dominated malaria in the Thar Desert, North western Rajasthan. *Pap. Abstr. SIMA special Seminar on Malaria in irrigated agriculture at ICID's 18h Intern. Congr. on Irrigation and Drainage,* Montreal, Canada; July 18 – 28, Zool, p.43.

155. Tyagi, B.K., 2002. *Malaria in the Thar Desert: Facts, Figures and Future. Agrobios* (India). Pp. 165.

156. Tyagi, B.K., 2002. Future of phytochemicals against vector mosquitoes. *Pap. Abstr. VI AZRA Congr. on Adv. in Applied Zoological Researches for Food Production and Environmental Safety,* Cuttack, 19-21 Dec, 2002.

157. Tyagi, B.K. and A.K. Shahi, 2002. Genus *Cymbopogon* (Poaceae) – A potent group of mosquito repellent bearing characteristics. *Pap. Abstr. XVI Natl. Congr. Parasitology,* Bareilly (Oct 31, Nov 2, 2002), p. 91.

158. Tyagi, B.K., 2002. Future of phytochemicals against vector mosquitoes. *J. Appl. Zool. Res.* 14(1): 101-106.

2003

159. Tyagi, B.K., 2003. Cutaneous leishmaniasis in the Thar Desert: a critical appraisal of the fast emerging "sand-disfiguring disease". [In: *Advancement in insect biodiversity.* Ed. Prof. Rajiv Gupta, JNV Univ., Jodhpur]. pp. 313-323. (Invited Chapter)

160. Poopathi, S., K. Anup Kumar, N. Arunachalam, B.K. Tyagi and V. Sekar, 2003. Control of *Culex quinquefasciatus* (Diptera:Culicidae) by *Bacillus sphaericus* and *Bacillus thuringiensis* subsp. *israelensis* produced on a new culture medium. *Biocontrol Science and Technology* 13(8): 743-748.

161. Tyagi, B.K. 2003. *Medical Entomology: A handbook of medically important insects and other arthropods.* Scientific Publishers (India). pp. 265.

162. Tyagi, B.K., 2003. Malaria conflagration under the impact of changing Thar Desert environment. [In: *Human Impact on Desert Environment.* Eds. P. Narayanan, S. Kathju, A. Kar, M.P. Singh and Praveen Kumar]. Arid Zone Research Association of India & Scientific Publishers (India), Jodhpur, pp. 60-73.

163. Tyagi, B.K., 2003. *Cymbopogon* essential oils: Their chemistry and mosquito repellent activities. *Pap. Abstr. Natl. Sym. on Biochemical Sci. Health & Environmental Aspects*, Agra, pp. 140-142.

164. Tyagi, B.K., 2003. A review of the emergence of *Plasmodium falciparum* dominated malaria in irrigated areas of the Thar Desert, India. *Acta Tropica* (Special Issue: Malaria & agriculture) 89: 227-239.

165. Tyagi, B.K., J. Hiriyan and S.C. Tewari, 2003. Dengue in India: A review with special emphasis on the climatic and environmental variabilities. *Pap. Abstr. 1st Intern. Seminar of Medical Entomology, Bhopal*, pp. 33.

166. Tyagi, B.K., 2003. Emergence of dengue in India as a major public health problem. *Natl. Sym. on Recent Trends in Zoological Sciences, Patiala*, p. 34.

167. Tyagi, B.K., 2003. A review of plant-based materials active against vector mosquitoes with particular emphasis on repellent potential of *Cymbopogon* (Poaceae). *J. Econ Taxon. Bot.* 28: 489-493.

168. Hiriyan, J., S.C. Tewari and B.K. Tyagi, 2003. Plastic cups inveigling *Aedes albopictus* (Skuse) breeding around tea vendor spots in Ernakulam city, Kerala State. *Dengue Bulletin* 27: 195-196.

169. Tyagi, B.K., 2003. Epidemic dengue/Dengue haemorrhagic fever emergence in India, with special reference to Kerala State. *Pap. Abstr. Natl. Sym. Management of Aquatic Resources for Biodiversity Maintenance & Conservation*, Jodhpur (Feb. 2-4, 2004), pp.56.

170. Poopathi, S., K. Anupkumar, N. Arunachalam, B.K. Tyagi and V. Sekar. 2003. An alternate approach for biospesticide production in mosquito control. *3rd Global Meet of Parasitologic Diseases* pp. 77.

2004

171. Tyagi, B.K., and J. Hiriyan, 2004. Breeding of dengue vector, *Aedes aegypti* (Linnaeus) in the rural Thar Desert, north western Rajasthan, India. *Dengue Bulletin* 28: 220-222.

172. Hiriyan, J. and B.K. Tyagi, 2004. Cocoa pod (*Theobroma cacao*) – a potential breeding habitat of *Aedes albopictus* (Skuse) in dengue sensitive Kerala State, India. *J. American Mosquito Control Association* 20 (3):323-325.

173. Tyagi, B.K., 2004. *The Invincible Deadly Mosquitoes.* Scientific Publishers (India), Jodhpur, pp. 265.

2005

174. Sunish, I.P., R. Rajendan, T.R. Mani, A.P. Dash and B.K. Tyagi, 2006. Community level evidence on the advantage of including albendazole in the global programme for elimination of lymphatic filariasis. *The Lancet Infectious Diseases* 6:125-126.

175. Tyagi, B.K. and Q.H. Baqri, 2005. *Changing faunal ecology in the Thar Desert.* Scientific Publishers (India), Jodhpur, pp. 367.

176. Tyagi, B.K., 2005. Thar Desert's changing climate and emergence/exacerbation of vector-borne diseases: Past, present and future. (In: *Changing faunal ecology in the Thar Desert.* Eds. B.K. Tyagi and Q.H. Baqri) Scientific Publishers (India), Jodhpur, pp. 323-367.

177. Tyagi, B.K., Mohd. Idris and Q.H. Baqri, 2005. Dr. Ishwar Prakash: Polymath and rodentologist – A bio-bibliography (1931-2002). (In: *Changing faunal ecology in the Thar Desert.* Eds. B.K. Tyagi and Q.H. Baqri) Scientific Publishers (India), Jodhpur, pp. 1-25.

2006

178. Thangarartham, P.S., M.K. Jeevan , R. Rajendran and B.K. Tyagi, 2006. Dual infection by dengue virus and Plasmodium vivax in Alappuzha district, Kerala, India. *Japanese Journal of Infectious Diseases* 59: 211-212.

179. Thenmozhi, V., R. Rajendran, K. Ayanar, R. Manavalan and B.K. Tyagi. 2006. A long-term study Japanese encephalitis virus infection in Anopheles subpictus in Cuddalore district. Tamil Nadu, south India. *Tropical Medicine and International Health* 11:288-293.

180. Paramasivan, R., K.J. Dhananjeyan, J. Hiriyan, Tyagi, B.K., A.P. Dash. 2006. Serological and virological studies of dengue outbreak in certain areas of Kanyakumari district. *Indian Journal of Medical Research* 123: 673-677.

181. Philip Samuel, P. and B.K. Tyagi. 2006. Diagnostic methods to detect and isolate dengue viruses from vector mosquitoes: A review. *Indian Journal of Medical Research* 123: 615-628.

182. Poopthi, S. and B.K. Tyagi, 2006. The challenge of mosquito control strategies: from primordial to molecular approaches. *Biotechnology & Molecular Biology Review* 1: 51-65.

183. Tyagi, B.K., J. Hiriyan, P. Philip Samuel, S.C. Tewari and R. Paramasivan. 2006. Dengue in Kerala: A critical review. ICMR Bulletin 36 (4-5): 13-22.

184. Tyagi, B.K., N. Arunacham, R. Rajendran, R. Paramsivan, I.P. Sunish, P. Philip Samuel, V. Thenmozhi, T.R. Mani, S.C. Tewari and M. Muniaraj, 2006. *Abstracts, 8th International Symposium on Vectors & Vector Borne Diseases* (Central Theme: Tropical Diseases and Health), Madurai, Oct. 13-15, 2006. Centre for Research in Medical Entomology, Madurai, pp. 212.

185. Muniaraj, M., D.S. Dinesh, P.K. Sinha, P. Das, R. Paramasivan, R. Rajendran, V. Thenmozhi and B.K. Tyagi. A novel method to determine the interaction of indigenous microbiota of sandfly (phlebotomus agrentipes) gut with promastigotes of leishmania donovani. ibid., pp. 4-5.

186. Thenmozhi, V., R. Selvaraj Pandian and B.K. Tyagi. Comparison of mosquito fauna at the forest fring and far off areas in Natham Taluk, Dindigul District, Tamil Nadu, South India. ibid., pp. 24-25.

187. Ayanar, K., R. Rajendran, R. Manavalan, N. Arunachalam and B.K. Tyagi. Blood feeding patterns of vector mosquitoes of Japanese Encephalitis in Thanjavur District. ibid., pp. 26.

188. Krishnamoorthi, R., V. Ashok kumar, S.C. Tewari, N. Arunachalam, R. Rajendran, R. Manavalan, and B.K. Tyagi. Breeding habitats of the major dengue vector *Aedes Aegypti* (L) in an endemic urban city, Chennai, India. ibid., pp.31.

189. Arunachalam, N., P. Philip Samuel, J. Hiriyan and B.K. Tyagi. Sampling of outdoor resting population of *Culex Tritaeniorhynchus* giles (Diptera: Culicidae), the vector of Japanese Encephalitis in Kerala, South India. ibid., pp.32

190. Arunachalam, N., U.S. N. Murty, D. Narahari, A. Balasubramanian, R. Rajendran, P. Philip Samuel, V. Thenmozhi, R. Paramasivan and B.K. Tyagi. Japanese Encephalitis virus transmission dynamics in a suburban area of Kurnool, Andhra Pradesh, India. ibid., pp.34.

191. Philip Samuel, P., N. Arunachalam and B.K. Tyagi. Host feeding pattern of Japanese Encephalitis vector mosquito *Culex gelidus* from Kerala, India. ibid., pp. 40.

192. Thangaratham, P.S., R. Rajendran, S.C. Tewari, R. Paramasivan, N. Arunachalam, K.J. Dhananjeyan and B.K. Tyagi. Clinical investigations of suspected cases during dengue outbreak in Tiruppur, Coimbatore District, Tamil Nadu, India. ibid., pp. 45.

193. Paramasivan, R., V. Thenmozhi, S. Victor Jerald Leo, R. Rajendran and B.K. Tyagi. Preparation of Japanese Encephalitis virus antigen in cell culture – A preliminary study. ibid., pp. 65.
194. Victor Jerald Leo, S., P. Philip Samuel, R. Paramasivan and B.K. Tyagi. Protein profile of dengue vectors *Aedes (Stegomyia) Aegypti* and *Aedes albopictus.* ibid., pp. 66.
195. Victor Jerald Leo, S., P. Philip Samuel, R. Paramasivan and B.K. Tyagi. Ontogenetic protein variations observed in the predominant dengue vectors, *Aedes Aegypti* and *Aedes Albopictus.* ibid., pp. 67.
196. Dhananjeyan, K.J., R. Paramasivan, S. Victor Jerald Leo, S.C. Tewari, V. Thenmozhi, R. Rajendran and B.K. Tyagi. Use of non-destructive source of genomic DNA in insect studies. ibid., pp. 70.
197. Paramasivan, R., S. Victor Jerald Leo, R. Selvaraj Pandian and B.K. Tyagi. Binding of human IGE and IGG antibodies to the allergens of *Armigeres subalbatus* (Insecta: Culicidae). ibid., pp. 70.
198. Paramasivan, R., V. Thenmozhi, R. Rajendran, A. Thirumalai Arasu, P. Gunasekaran and B.K. Tyagi. RAPD analysis of *Aedes aegypti* populations in Tamil Nadu, India. ibid., pp. 72.
199. Tyagi, B.K., J. Hiriyan, N. Arunachalam, S.C. Tewari and P. Philip Samuel. Monitoring of *Aedes albopictus* (Skuse) populations in epidemic hit Kerala State, India, using an indigenous ovitrap. ibid., pp. 107.
200. Philip Samuel, P., N. Arunachalam, A. Balasubramanian, V. Thenmozh, R. Paramasivan and B.K. Tyagi. Entomological investigations on the Japanese Encephalitis outbreak in Gorakhpur, Uttar Pradesh India during the year 2005. ibid., pp. 109.
201. Philip Samuel, P., N. Arunachalam, P.S. Thangaratham, S. Victor Jerald Leo and B.K. Tyagi. An account of the chikungunya fever outbreak entomological investigation carried out in Chittor District, Andhra Pradesh, India. ibid., pp. 110-111.
202. Rajendran, R., I.P. Sunish, A. Munirathinam, T.R. Mani and B.K. Tyagi. Significance of drug compliance rates in the Lymphatic Filariasis elimination programme in South India. ibid., pp. 115.
203. Sunish, I.P., R. Rajendran, A. Munirathinam, T.R. Mani and B.K. Tyagi. Beyond Filariasis benefits of GPELF programme on soil transmitted helminth infection after four annual mass drug administrations in South India. ibid., pp. 116.
204. Munirathinam, A., R. Rajendran, I.P. Sunish and B.K. Tyagi. Impact of antifilarial drugs on *Pediculus humanus capitis* in rural Filaria endemic villages of Tamil Nadu. ibid., pp. 117.
205. Thenmozhi, V., S.C. Tewari, N. Arunachalam, P. Philip Samuel, R. Paramasivan and B.K. Tyagi. Desiccated vector mosquitoes used for the surveillance of Japanese Encephalitis virus and dengue virus activity in South India – An effective and improved novel surveillance tool. ibid., pp. 136.
206. Thenmozhi, V., J. Hiriyan, S.C. Tewari, P. Philip Samuel, R, Paramasivan, R. Rajendran and B.K. Tyagi. *Aedes albipictus* (Diptera: Culicidae): its role in the maintenance of dengue viruses in nature through vertical transmission in South India. ibid., pp. 141.
207. Thenmozhi, V., R. Rajendran, K. Ayanar, R. Paramasivan, R. Manavalan and B.K. Tyagi. Japanese Encephalitis in Cuddalore District, Tamil Nadu, India: A three – year longitudinal study on vector infection frequency. ibid., pp. 151.

208. Muniaraj, M., R. Paramasivan, R. Rajendran and B.K. Tyagi. Plants showing antileishmanian activitiy – An overview. ibid., pp. 156.

209. Tyagi, B.K., R. Rajendran, I.P. Sunish, P. Philip Samuel, J. Hiriyan, S.C. Tewari and N. Arunachalam. Dengue and climate change: Kerala – An eye opener. ibid., pp. 160.

210. Tyagi, B.K., R. Rajendran and I.P. Sunish. Malaria vector control through integrated approach. ibid., pp. 161.

211. Ramar, M., S. Ignacimuthu, B.K. Tyagi and R. Rajendran. Evaluation of larvicidal and repellency activity of *Myrtus Caryophyllus* against dengue vector, *Aedes Aegypti* (Diptera: Culicidae). ibid., pp. 186.

212. Manavalan, R., V. Ashok kumar, R. Krishnamoorthy, V. Thenmozhi, S.C. Tewari, R. Rajendran, N. Arunachalam and B.K. Tyagi. Community participation is the major source of vector control of chikungunya vector, *Aedes Aegypti*. ibid., pp.198.

213. Ashok kumar, V., S.C. Tewari, N. Arunachalam, R. Rajendran, R. Manavalan, R. Krishnamoorthy and B.K. Tyagi. Role of self help group (women) in control of dengue in Chennai city, India. ibid., pp. 207.

214. Paramasivan, R., V. Thenmozhli, L. Kabilan, S.C. Twewari, N. Arunachalam, K.J. Dhananjeyan and B.K. Tyagi, 2006. Sero-epidemiology of focal outbreak of dengue fever in a rural area of Tamil Nadu. *Indian Journal of Medical Research* 124:718-720.

215. Dash, A.P., B.K. Tyagi, N. Arunacham, R. Rajendran, R. Paramsivan, S.C. Tewari, P. Philip Samuel, K. Ayanar and A. Munirathinam, 2005. Entomological survey carried out in Tsunami affected coastal regions of Tamil Nadu and Kerala. (In: 26 December, 2004 Tsunami: ICMR Response, pp. 1-60. Ed.: M.D. Gupte), National Institute of Epidemiology, Chennai, pp. 37-44.

216. Tyagi, B.K. [Ed.], 2006. Abstracts. 8th International Symposium on Vectors & Vector Borne Diseases. Centre for Research in Medical Entomology, Madurai, pp.212.

217. Tyagi, B.K. (Editor). Vector-borne diseases: Epidemiology and Control. Scientific Publishers. pp. 298.

218. Tyagi and A.P. Dash, 2006. Dengue in India with special reference to the interspecific invasive and virus transmission potential of the Asian tiger mosquito, *Aedes albopictus* (Skuse) in Kerala: An update. (In: Vector Biology. Ed. V.P. Sharma and Jagbir Singh Kirti). The National Academy of Sciences, India, Allahabad, 142-158.

2007

219. Tyagi, B.K., 2007. Emergence of dengue and dengue haemorrhagic fever in Kerala state (South India), impact of climate change on *Aedes albopictus* Skuse as the sole vector. *Tropical Medicine & International Health* 12 (Suppl. 1): 127.

220. Sunish, I.P., R. Rajendan, T.R. Mani, A. Munirathinam, A.P. Dash and B.K. Tyagi, 2007. Vector control complements the effects of mass drug administration against bancroftian filariasis in Trikoilur, south India. *Bulletin of World Health Organization* 85:138-145.

221. Thangarartham, P.S. and B.K. Tyagi, 2007. Severe dengue: the need for new case definitions – the Indian perspective. *The Lancet Infectious Diseases* 59: 81-82.

222. Paramasivan, R., R. Sivaperumal, K.J. Dhananjeyan, V. Thenmozhli and B.K. Tyagi, 2007. Prediction of 3-dimentional structure of salivary odorant-binding protein-2 of the mosquito *Culex quinquefasciatus*, the vector of lymphatic filariasis. *In Silico Biology* 7: 1-6.

223. Tyagi, B.K., 2007 (Ed.). *Odonata: Biology of dragonflies.* Scientific Publishers (India), Jodhpur, pp. 323-366.

224. Tyagi, B.K. and M.A.J.E. Kiauta, 2007. Professor Bastiaan Kiauta: An extraordinary and outstanding odonatologist (In: *Odonata: Biology of dragonflies.* Ed. B.K. Tyagi). Scientific Publishers (India), Jodhpur, pp.1-24.

225. Philip Samuel, P., V. Thenmozhli and B.K. Tyagi, 2007. A focal outbreak of dengue fever in south India: Clinical and entomo-serological observations. *Indian Journal of Medical Research* 125: 179-181.

226. Arunachalam N, U.S.N. Murty, D. Narahari, A. Balasubramanian, P. Philip Samuel, V. Thenmozhi, R. Paramasivan, R. Rajendran & B.K. Tyagi. 2007. Longitudinal studies of JE virus infection in vector mosquitoes in Kurnool district, Andhra Pradesh, South India. *Journal of Medical Entomology* 46: 633-639.

227. Thenmozhi, V., Hiriyan, J., Tewari, S.C., Philip Samuel, P., Paramasivan, R., Rajendran, R., Mani, T.R. and Tyagi, B.K. (2007). Natural Vertical Transmission of Dengue virus in *Aedes albopictus* (Diptera: Culicidae) in southern India, state Kerala. *Japanese journal of Infectious diseases* 60: 245-249.

228. Tyagi, B.K., 2007. Arthropods and their biomedical importance. *Pap. Abstr. Symp. Thrust Areas in Entomological Research.* Entomological Academy of India, Chennai. pp. 16-17.

229. K.J. Dhananjeyan, R. Sivaperumal, R. Paramasivan, S. Victor Jerald Leo, R. Rajendran and B.K. Tyagi, 2007. In-silico homology modeling of three isoforms of insect defensins from the dengue vector mosquito, Aedes aegypti (Linn., 1762). *Journal of Molecular Modeling.* 15: 507-514.

230. Arunachalam N, S.C. Tewari, V. Thenmozhi, R. Rajendran, R. Paramasivan, R. Manavalan, K. Ayanar and B.K. Tyagi. 2007. Natural vertical transmission of dengue viruses by *Aedes aegypti* in Chennai, Tamil Nadu, India. *Indian Journal of Medical Research* 127: 395-397.

231. Tyagi, B.K., Vijay Veer, R. Rajendran, N. Arunachalam and R. Paramasivan, 2007. Professor T.N. Ananthakrishnan – Entomologist and polymath: Bio-bibliography. 1st Conference on Medical Arthropodology, Madurai, pp. 67.

232. Tewari, S.C., J. Hiriyan, K. Ayanar, A. Munirathinam, A. Venkatesh, R. Reuben and B.K. Tyagi, 2007. CRME Mosquito Museum: An Annotated Checklist of Indian Mosquito Species. Contributions of the Centre for Research in Medical Entomology, Madurai, pp.175.

233. Tyagi, B.K. [Ed.], 2007. 1st Conference on Medical Arthropodology: Taxonomy in biodiversity, bioterrorism and biotechnology (Madurai 1-2 September, 2007) – CRME's forthcoming event of great scientific importance. Newsletter 4(2): 2-3.

234. Tyagi, B.K. [Ed.], *CRME Newsletter* 2005, Vol. 2 (1&2); 2006,Vol. 3 (1&2); 2007, Vol. 4 (1&2); 2008, Vol. 5 (1&2); 2009, Vol. 6 (1). Centre for Research in Medical Entomology, Madurai.

235. Tyagi, B.K. et al. 2007. In-Depth Review on Dengue for National Vector Borne Disease Control Programme, Delhi (2006-2007). Centre for Research in Medical Entomology, Madurai, pp.159. [Report].

236. Tyagi, B.K. et al. 2007. In-Depth Review on Japanese Encephalitis for National Vector Borne Disease Control Programme, Delhi (2006-2007). Centre for Research in Medical Entomology, Madurai, pp. 143. [Report].

237. Tyagi, B.K. et al. 2008. Supplementary data on the In-Depth Reviews on Dengue and Japanese Encephalitis for National Vector Borne Disease Control

Programme, Delhi. Centre for Research in Medical Entomology, Madurai, 159. [Report].

238. Tyagi, B.K. [Ed.], 2007. Abstracts. 1st Conference on medical Arthropodology Centre for Research in Medical Entomology, Madurai, pp. 143.

239. Tyagi, B.K., 2007. Chikungunya in India. *Bulletin De La Societe De Pathologie Exotique.* Vol. 100 (5): 321.

240. Philip Samuel P, Thenmozhi V and Tyagi BK (2007). A focal outbreak of dengue fever in a rural area of Tamil Nadu. *Indian Journal of Medical Research* 125: 179 – 181.

241. Tyagi, B.K. [Ed.], 2007. Medical Arhtropodology: Taxonomy and Biodivesity. Proceedings of the First Conference on Medical Arhtropodology. (Sep. 1-2, 2007; Madurai), Centre for Research in Medical Entomology, Madurai; pp. 171.

242. Vennison, S.J., R. Pramasivan, B.K. Tyagi, P. Thirumalai Vasan & I.G. Prabhu, 2007. Application of molecular tools in diagnosing the different dengue serotypes. 1st Conf. on Medical Arhtropodology: Taxonomy in Biodivesity, bioterrorism and biotechnology, Madurai; p. 27.

243. Philip Samuel, P. and Dr. B.K. Tyagi, 2007. Biochemical genetics – An efficient tool for population genetics studies. *ibid.*, pp. 39.40.

244. Munirathinam, A., S.C. Tewari, J. Hiriyan, K. Ayanar, R. Reuben and B.K. Tyagi., 2007. *Anopheles hyrcanus* Group in southern Penninsula, India. . *ibid.*, p.49.

245. Hiriyan, J., R. Krishnamoorthy, R. Rajendran and B.K. Tyagi, 2007. Prevalence of intra-specific variants within *Aedes (Stegomyia) aegypti* populations i south India. . *ibid.*, 65.

246. Muniaraj, M., R. Rajendran, N. Arunachalam, R. Paramasivan and B.K. Tyagi, 2007. Microbial fauna from the breeding habitats of *Aedes aegypti. ibid.*, p. 79.

247. Victor Gerald Leo, S., P. Philip Samuel, R. Param,asivan, K. Ayanar, R. Rajendran and B.K. Tyagi, 2007. Protein profile of Culex vishnui subgroup mosquitoes, the vectors of Japanese encephalitis. . *ibid.*, p. 83.

248. Sharmil, G., R. Paramasivan, R. Rajendran, K.J. Dhananjeyan and B.K. Tyagi, 2007. A comparative analysis of chemical composition of salivary gland extracts of vector mosquitoes. . *ibid.*, p. 85.

249. Arunachalam, A., P. Philip Samuel and B.K. Tyagi, 2007. Biodiversity of mosquito fauna in Kuttanadu, Kerala. . *ibid.*, p. 87.

250. Arunachalam, A., U.S.N. Murty, A. Balasubramanian, R. Rajendran, P. Philip Samuel, V. Thenmozhi, R. Paramasivan and B.K. Tyagi, 2007. Biodiversity of mosquito fauna in Karnool district, Andhra Pradesh. *ibid.*, pp. 89-90.

251. Philip Samuel, P., A. Arunachalam, M. Kannan, R. Rajendran and B.K. Tyagi, 2007. Diverse breeding habitats of the chikungunya vectors investigated in Pattanamthitta district and Idukki district of Kerala, India. *ibid.*, p. 91.

252. Muniaraj, M., R. Rajendran, N. Arunachalam, R. Paramasivan and B.K. Tyagi, 2007. Diversities in microbial relationship with insects. *ibid.*, pp. 93-94.

253. Tewari, S.C., N., Arunachalam, R. Rajendran and B.K. Tyagi, 2007. Species of *Anopheles leucosphyrus* Group in India. *ibid.*, pp. 95-96.

254. Philip Samuel, P., A. Arunachalam, A. Balasubramanian and B.K. Tyagi, 2007. Biodiversity of mosquitoes in Gorakhpur district, Uttar Pradesh, India. *ibid.*, p.97.

255. Manavalan, R, R. Krishnamoorthi, V. Ashok Kumar, S.C. Tewari, R. Rajendran, A. Arunachalam and B.K. Tyagi, 2007. Population dynamics of the principal dengue vector *Aedes aegypti* (L) in Chennai, Tamil Nadu, India. *ibid.*, p.99.

256. Balasubramanian, A., N. Arunachalam, U.S.N. Murty, D. Narahari and B.K. Tyagi, 2007. Population dynamics of *Culex gelidus*, the vector of Japanese encephalitis in a suburban area, Andhra Pradesh, India. *ibid.*, pp.101-102.
257. Krishnamoorthi, R., V. Ashok Kumar, R. Manavalan, S.C. Tewari, R. Rajendran, N. Arunachalam and B.K. Tyagi, 2007. Diversity of breeding habitats of the principal dengue vector *Aedes aegypti* (L) in Chennai, Tamil Nadu, India. *ibid.*, p. 103.
258. Tewari, S.C., J. Hiriyan, K. Ayanar, A. Munirathinam, A. Venkatesh, R. Reuben and B.K. Tyagi, 2007. Biodiversity of tick fauna in western Ghats. *ibid.*, p. 105.
259. Thenmozhi, V., R. Selvaraj Pandian and B.K. Tyagi, 2007. Prevalence of Japanese encephalitis vectors in Alagar hills reserve forest area of Tamil Nadu, India. *ibid.*, p. 107.
260. Thenmozhi, V., R. Selvaraj Pandian and B.K. Tyagi, 2007. Tree hole breeding prevalence of *Aedes mosquitoes (Diptera : Culicidae)* in Alagar hills reserve forest area of Tamil Nadu, India. *ibid.*, p.109.
261. Ayanar, K., R. Rajendran and B.K. Tyagi, 2007. Biodiversity of mosquitoes in the endemic areas of Japanese encephalitis in Cuddalore diustrict, Tamil Nadu. *ibid.*, p.111.
262. Sunish, I.P., R. Rajendran, R. Paramasivan and B.K. Tyagi, 2007. *Wolbachia* and its phylogeny. *ibid.*, p.129.
263. Muniaraj, M., R. Rajendran, N., Arunachalam, R. Paramasivan, P. Philip Samuel and B.K. Tyagi, 2007. *Ascogregarina* spp. From the *Aedes albopictus* of Kerala. *ibid.*, p.131.
264. V. Rajamannar, R. Rajendran, N., Arunachalam and B.K. Tyagi, 2007. E-tools on mosquito taxonomy. *ibid.*, p.133.
265. R. Balasubramanian, R. Rajendran, N., Arunachalam and B.K. Tyagi, 2007. Statistical tools to measure biodiversity. *ibid.*, p. 135-136.

2008

266. Tewari SC, Thenmozhi V, Arunachalam N, Philip Samuel P and Tyagi BK (2008). Desiccated vector mosquitoes used for the surveillance of Japanese encephalitis virus activity in endemic southern India. *Tropical Medicine and International Health* 13 (2): 286-290.
267. Thenmozhi V, Hiriyan J, Tewari SC, Philip Samuel P, Paramasivan R, Rajendran R, Mani TR and Tyagi BK (2008). Natural Vertical Transmission of Dengue virus in *Aedes albopictus* (Diptera: Culicidae) in southern India, state Kerala. *Japanese journal of Infectious diseases* 60: 245-249
268. Tyagi, B.K., I.P. Sunish and R. Rajendran (Eds.), 2008. Professor Dr. Chris F. Curtis: A glimpse onto his life and medico-entomological contributions. Second Conf. on Medoidical Arthropodology (Climate Change & Vector-borne Diseases: In memory of Prof. C.F. Curtis), CRME, Madurai (Dec. 15-16, 2008), pp.1-23.
269. Tyagi, B.K., 2008. Abstracts. 2nd Conf. on Medical Arthropodology (Climate Change & Vector-borne Diseases: Forecasting epidemics, prevention and control), CRME, Madurai (Dec. 15-16, 2008), pp. 1-156.
270. Sharmil, G., S. Victor Jerald Leo, K.J. Dhananjeyan, R. Paramasivan, R. Rajendran and B.K. Tyagi, 2008. Biochemical spectrum of chikungunya patients in Tamil Nadu. Pap. Abstr. 2nd Conf. on Medical Arthropodology (Climate Change & Vector-borne Diseases: Forecasting epidemics, prevention and control), CRME, Madurai (Dec. 15-16, 2008), p. 71.

271. Muniaraj, M., N. Arunachalam, R. Paramasivan, V. Rajamannar and B.K. Tyagi, 2008. Relationship between the incidence of visceral leishmaniasis (Kala-azar) and the annual rainfall in India. *ibid.,* p. 79.

272. S.C. Tewari, K. Ayanar, R. Rajendran and B.K. Tyagi, 2008. Species succession in Japanese encephalitis vectors in southern India – Influence of Climate Change? *Ibid.,* p. 81.

273. Ayanar, K., R. Rajendran, S.C. Tewari and B.K. Tyagi, 2008. Seasonal variations in the abundance of *ulex gelidus,* the vector of Japanese encephalitis in Cuddalore district, Tamil Nadu. *ibid.,* p. 83.

274. Muniaraj, R. Rajendran, M., N. Arunachalam, R. Paramasivan, V. Rajamannar and B.K. Tyagi, 2008. Impact of climate on the trends of major vector-borne diseases of India. *ibid.,* p. 85.

275. Venkatesh, A., Philip Samuel P., S.C. Tewari, R. Rajendran and B.K. Tyagi, 2008. Climatic factors involved in the emergence of vector-borne viral diseases. *ibid.,* p. 87.

276. Victor Gerald Leo, S., P. Philip Samuel, V. Thenmozhi, S.C. Tewari and B.K. Tyagi, 2008. Abiotic factors influencing the immature dengue vector abundance in Trivandrum district, Kerala. *ibid.,* p. 107.

277. Hananjeyan, K.J., R. Paramasivan, R. Sivaperumal, M. Muniaraj, Thiruvarulselvi, V. Thenmozhi, R. Rajendran and B.K. Tyagi, 2008. Molecular modelling of the envelope protein of chikungunya virus. *ibid.,* p. 141.

2009

278. Tyagi, B.K., 2009. Lymphatic filariasis elimination in India, with emphasis on vector control as an adjunct to the mass drug administration programme: the CRME model. Continuing Education Programme "Recent Advances in Disease Vector Management", DRDE, Gwalior (20-24 July, 2009), pp. 54-69.

279. Tyagi, B.K., 2009. Dengue emergence in Kerala under changing impact of climate. *Ibid.*, pp. 167-182.

280. Tyagi, B.K., Hiriyan J., Tewari SC., Ayanar K., Philip Samuel P., Arunachalam N., Paramasivan R., R. Krishnamoorthy, Dhananjeyan K.J., Victor Jerald Leo S., and Rajendran R. (2009): Description of a new species of *Anopheles pseudosundicus* (Diptera : Culicidae) from coastal Kerala. *Zootaxa* 2219, 49-60. (IF 0.740)

281. Tyagi, B.K, R. Rajendran and N. Arunachalam (Eds.). 2008. Dr. M.O.T. Iyengar: Bio-bibliography of an eminent medical entomologist (Authors: R. Iyengar and Tara Gandhi). Third Conf. on Medoidical Arthropodology (Integrated Disease Vector Management; translational research), CRME, Madurai (Dec. 15-16, 2008), pp.1-23.

282. Paramasivan, R., R. Selvaraj Pandian and B.K. Tyagi. 2009. Potential breeding habitats of vector and non-vector mosquitoes in a rural area. Pap. Abstr. Third Conf. on Medical Arthropodology, Madurai; p. 37.

283. Mariappan, T., S.C. Tewari, R. Rajendran, N. Arunachalam and B.K. Tyagi. 2009. Vectors breeding habitats in various geographical locations in India with special reference to dengue and chikungunya. Pap. Abstr. Third Conf. on Medical Arthropodology, Madurai; p. 39.

284. Ramar, M., M. Gabriel Paulraj, S. Ignacimuthu, B.K. Tyagi and R. Rajendran. 2009. Adulticidal activity of essential oil of *Eucalyptus globules* (Eucalyptus) against *Aedes aegypti* (Diptera: Culicidae). Pap. Abstr. Third Conf. on Medical Arthropodology, Madurai; p. 51.

285. Victor Gerald Leo, S., R. Paramasivan, P. Philip Samuel, M. Muniaraj, V. Thenmozhi, K.J. Dhananjeyan, N. Arunachalam, R. Rajendran and B.K. Tyagi, 2009. Laboratory diagnosis of chikungunya virus in fection – An Indian perspective. Pap. Abstr. Third Conf. on Medical Arthropodology, Madurai; p. 61.

286. Philip Samuel, P., M. Muniaraj, P.S. Thangatham, S.Victor Gerald Leo, R. Rajendran and B.K. Tyagi, 2009. Chikungunya fever outbreak investigation carried out in Chinnalampatti village, Dindigal district, Tamil Nadu. Pap. Abstr. Third Conf. on Medical Arthropodology, Madurai; p. 63.

287. Sunish, I.P., R. Rajendran, R., A. Munirathinam, M. Kalimuthu, V. Ashok Kumar and B.K. Tyagi, 2009. Empowering community to implement vector control for the elimination of lymphatic filariasis – an ongoing case study at Trikoilur. Pap. Abstr. Third Conf. on Medical Arthropodology, Madurai; p. 67.

288. Paramasivan, R., R. Selvaraj Pandian and B.K. Tyagi. 2009. Preferential host selection behavior and temporal variation in the feeding behavior of rural mosquitoes. Pap. Abstr. Third Conf. on Medical Arthropodology, Madurai; p. 117.

289. Tewari S.C., N. Arunachalam, R. Rajendran, R. Manavalan, V. Ashok Kumar and Tyagi BK (2008). Entomological investigation during chikungunya fever epidemic in 2006 in Chennai city, southern India. Pap. Abstr. Third Conf. on Medical Arthropodology, Madurai; p. 121.

290. Philip Samuel, P., V. Thenmozhi, S.C. Tewari, S. Victor Gerald Leo, R. Rajendran, N. Arunachalam and B.K. Tyagi, 2009. Dengue vector prevalence in the Thiruvananthapuram district of Kerala, India. Pap. Abstr. Third Conf. on Medical Arthropodology, Madurai; p. 117.

291. R. Krishnamoorthy, S.C. Tewari, A. Munirathinam and B.K. Tyagi, 2009. Biodiversity of mosquito fauna in southern region of Western Ghats. Pap. Abstr. Third Conf. on Medical Arthropodology, Madurai; p. 131.

292. Ayanar, K., R. Rajendsran, S.C. Tewari and B.K. Tyagi, 2009. Seasonal abundance of Japanese encephalitis vectors in endemic areas of Cuddalore district, Tamil Nadu. Pap. Abstr. Third Conf. on Medical Arthropodology, Madurai; p. 133.

293. Venkatesh, A.R. Rajendran and B.K. Tyagi, 2009. Dragonfly – a potential biocontrol agent against vector mosquitoes, a review. India. Pap. Abstr. Third Conf. on Medical Arthropodology, Madurai; p. 139.

294. Kalimuthu, M., R. Selvaraj Pandian, R. Rajendran and B.K. Tyagi, 2009. Impact of synthetic pesticide on the mosquito biocontrol agent, predatory water bug *Diplonychus rusticus* (Fabr.). Pap. Abstr. Third Conf. on Medical Arthropodology, Madurai; p. 143.

295. Paramasivan, R., R. Selvaraj Pandian and B.K. Tyagi, 2009. KAP study: An ideal tool to assess the perception and practices of society in the control of vectors and the management of vectorborne diseases. Abstr. Third Conf. on Medical Arthropodology, Madurai; p. 149.

296. Mariappan, T., R. Rajendran, N. Arunachalam, M. Muniaraj, R. Paramasivan, V. Rajamannar, S. Balavinayagam, P. Udayakumar, S. Umamaheshwari and B.K. Tyagi. 2009. Investigation of chikungunya outbreak in coastal villages of Ramanathapuram district, Tamil Nadu, India. Pap. Abstr. Third Conf. on Medical Arthropodology, Madurai; p. 151.

297. Tewari S.C., N. Arunachalam, R. Rajendran, R. Manavalan, V. Ashok Kumar and Tyagi BK (2008). KAP survey during chikungunya fever epidemic in 2006 in Chennai city, southern India. Pap. Abstr. Third Conf. on Medical Arthropodology, Madurai; p. 153.

298. Philip Samuel, P., S.C. Tewari, R. Rajendran, R. Krishnamoorthy, M. Kannan and B.K. Tyagi, 2009. Entomo-sociological invwstigation of chikungunya outbreak in cherthala area of Kerala. Pap. Abstr. Third Conf. on Medical Arthropodology, Madurai; p. 155.

299. Ashok Kumar, V., T. Karunakaran, R. Rajendran and B.K. Tyagi, 2009. Knowledge, attitude and practices on malaria fever – a sociological study with reference to coastal area of Chennai city. Pap. Abstr. Third Conf. on Medical Arthropodology, Madurai; p. 157.

300. Arunachalam N., Murty US., Narahari D., Balasubramanian A., Philip Samuel P., Thenmozhi V., Paramasivan R., Rajendran R., and **Tyagi, B.K**. (2009): Longitudinal studies of Japanese encephalitis virus infection in vector mosquitoes in Kurnool district, Andhra Pradesh, South India. *Journal of Medical Entomology* 46, 633 - 9. (IF 1.788)

301. Dhananjeyan KJ., Sivaperumal R., Paramasivan R., Thenmozhi V., and **Tyagi, B.K**. (2009): *In-silico* homology modeling of three isoforms of insect defensins from the dengue vector mosquito, *Aedes aegypti* (Linn., 1762). *Journal of Molecular Modeling* 15, 507 - 14. (IF 2.018)

302. Kannan M., Rajendran R., Sunish IP., Balasubramaniam R., Arunachalam N., Paramasivan R., Tewari SC., Philip Samuel P., and **Tyagi, B.K.** (2009): A study on chikungunya outbreak during 2007 in Kerala, south India. *Indian Journal of Medical Research* 129, 311 - 5. (IF 1.883)

303. Munirathinam A., Sunish IP., Rajendran R., and Tyagi, B.K. (2009): Impact of ivermectin drug combination on *Pediculus humanus captitis* infection in the primary school children of south Indian rural villages. *International Journal of Dermatology* 48, 1201-1205. (IF 1.409)

304. Paramasivan R., Philip Samuel P., Thenmozhi V., Rajendran R., Victor Jerald Leo S., Dhananjeyan KJ., Krishnamoorthi R., Arunachalam N., and Tyagi, B.K. (2009): Chikungunya virus isolated in Lakshadweep islands in the Indian Ocean: evidence of the Central/East African genotype. *Japanese J. Infectious Dis.* 62, 67 - 69. (IF 1.07)

305. Philip Samuel P., Ayanar K., Kannan M., Thenmozhi V., Paramasivan R., Balasubramanian A., and Tyagi, B.K. (2009): Sero-entomological investigations on Japanese encephalitis outbreak in Gorakhpur division, Uttar Pradesh, India. *Indian Journal of Medical Research* 129, 329 - 32. (IF 1.883)

306. Poopathi S., Ramesh N., Sundaravadivelu K., Philip Samuel P., and Tyagi BK. (2009): Larvicidal efficacy of various formulations of *Bacillus sphaericus* against the resistant strain of *Culex quinquefasciatus* (Diptera: Culicidae) from southern India. *Tropical Biomedicine* 26, 23 - 9. (0.59)

307. Camilla J. Beech, S.S. Vasan, M. Megan Quinlan, Margareth Lara Capurro, Luke Alphey, Vicente Bayard, Madama Bouaré, Maria Corena McLeod, Pattamaporn Kittayapong, James V. Lavery, Lee Han Lim, Mauro Toledo Marrelli, J. Nagaraju, Kenneth Ombongi, Rofina Yasmin Othman, Vilasini Pillai, Janine Ramsey, Rachel Reuben, Robert I. Rose, Brij Kishore Tyagi, and John Mumford. 2009. Deployment of Innovative Genetic Vector Control Strategies: Progress on Regulatory and Biosafety Aspects, Capacity Building and Development of Best-Practice Guidance. *AsPac. J. Mol. Biol. Biotechnol. 17 (3): 75-85.*

308. Tyagi, B.K. 2009. Scientist. Scientific Publishers (India), Jodhpur; pp. 394.

2010

309. Arunachalam N., Susilowati Tana., Fe Espino., Pattamaporn Kittayapong., Wimal Abeyewickreme., Khin Thet Wai., **Tyagi, B.K**. Axel Kroeger, Johannes

Sommerfeld and Max Petzold (2010): Dengue vectors at the dynamic interface of environmental, biological and social ("eco-bio-social") determinants: a multi-country study in urban and peri-urban Asia. *Bulletin of World Health Organization* 88, 173-184. (IF 4.019)

310. Dhananjeyan KJ., Paramasivan R., Tewari SC., Victor Jerald Leo S., Rajendran R., Venkatesh A., and **Tyagi, B.K**. (2010): Molecular identification of mosquito vectors using genomoc DNA isolated from egg shells, larval and pupal exuvium. *Tropical Biomedicine* 27, 47-53. (IF 0.59)

311. Rajendran R., Sunish IP., Munirathinam A., Ashok Kumar V., and **Tyagi, B.K**. (2010): Role of community empowerment in the elimination of lymphatic filariasis in south India. *Tropical Biomedicine* 27, 68-78. (IF 0.59)

312. Tyagi, B.K. (Ed.) 2010. CRME Commemorative Compendium 2010: Silver Jubilee Publication. Centre for Research in Medical Entomology, Madurai, pp. 243.

313. Tyagi, B.K., 2010. Biocontrol of mosquitoes and success in eliminating human lymphatic filariasis. Abstr. Pap. Natl. Symp. on Recent Advances in Biotechnology. Anna Univ., Triuchirappalli, Tiruchirappalli (23-24 Jan., 2010), p.1. *(Invited Lecture)*.

314. Tyagi, B.K., 2010. Impact of climate change on Aedes albopictus Skuse , the Asian tiger mosquito, and emergence of dengue fever in Kerala State (south India). Abstr. Pap. Natl. Seminar on Climate Change and Vector-borne Diseases. The Asiatic Society, Kolkata (6th Feb., 2010), p. 11-16. *(Invited Lecture)*.

315. Tyagi, B.K., 2010. Faunal wealth of the Thar Desert, India: Need for inventorization of species and conservation of their natural history. UGC-Sponsored Natl. Seminar on Desert: Its Biodiversity and Conservation. JNV University, Jodhpur (9th Feb., 2010), p. 1. *(Invited Lecture)*.

316. Tyagi, B.K., 2010. Functional dynamics of insect fauna in the unique xeric ecosystem of The Thar Desert, northwestern Rajasthan (India): Before and after the advent of Indira Gandhi Nahar Pariyojana. UGC-Sponsored Natl. Seminar on Desert: Its Biodiversity and Conservation. JNV University, Jodhpur (9th Feb., 2010), p. 2. *Invited Lecture)*.

317. Philip Samuel P, Arunachalam N, Rajendran R, Victor Jerald Leo S, Ayanar K, Balasubramainam R and **Tyagi BK** (2011). Temporal variation in the susceptibility of *Culex tritaeniorhynchus* (Diptera: Culicidae) to Japanese encephalitis virus in the Cuddalore District, Tamil Nadu, South India (In press April/May, 2010).

318. Paramasivan, R., Thenmozhi, V., Thangaratham, P.S., Rajendran, R., Tewari, S.C., Dhananjeyan, K.J. and **Tyagi, B.K**. (2011). An outbreak of dengue fever in Tirupur, Coimbatore district, Tamil Nadu. *Indian Journal of Medical Research* (In press April/May 2010 to be printed in July, 2010).

319. Rajendran, R., I.P. Sunish and B.K. Tyagi, 2010. Mass drug administration in the national programme for elimination of lymphatic filariasis. (In: CRME Commemorative Compendium 2010: Silver Jubilee Publication. Ed. B.K. Tyagi). Centre for Research in Medical Entomology, Madurai, pp. 1-17.

320. Sunish, I.P., R. Rajendran and B.K. Tyagi, 2010. Expanded polystyrene beads as an adjunct for the control of bancroftian filariasis vectors in India. (In: CRME Commemorative Compendium 2010: Silver Jubilee Publication. Ed. B.K. Tyagi). Centre for Research in Medical Entomology, Madurai, pp. 18-29.

321. Arunachalam, N., U.S.N. Murthy, D. Narahari, A. Balasubramanian, R. Rajendran, P. Philip Samuel, V. Thenmozhi, R. Paramasivan and B.K. Tyagi, 2010.

Longitudenal studies on the role of *Culex gelidus* in transmission of Japanese encephalitis virus in suburban area, Andhra Pradesh, India. (In: CRME Commemorative Compendium 2010: Silver Jubilee Publication. Ed. B.K. Tyagi). Centre for Research in Medical Entomology, Madurai, pp. 43-52.

322. Philip Samuel, P., R. Rajendran, K. Ayanar, R. Manavalan and B.K. Tyagi, 2010. Control of vectors of Japanese encephalitis. (In: CRME Commemorative Compendium 2010: Silver Jubilee Publication. Ed. B.K. Tyagi). Centre for Research in Medical Entomology, Madurai, pp. 53-68.

323. Philip Samuel, P., S. C. Tewari, N. Arunachalam, V. Thenmozhi, K. Ayanar and B.K. Tyagi, 2010. Host feeding pattern of of vector mosquitoes with emphasis on arboviral disease transmission in southern India (In: CRME Commemorative Compendium 2010: Silver Jubilee Publication. Ed. B.K. Tyagi). Centre for Research in Medical Entomology, Madurai, pp. 83-98.

324. Tyagi, B.K., S.C. Tewari, J. Hiriyan, V. Thenmozhi, R. Paramasivan and P. Philip Samuel, 2010. *Aedes albopictus* and dengue emergence in Kerala state (South India) with special reference to anthropocentric activities and climate changes. (In: CRME Commemorative Compendium 2010: Silver Jubilee Publication. Ed. B.K. Tyagi). Centre for Research in Medical Entomology, Madurai, pp. 99-113.

325. Arunachalam, N., S. C. Tewari, R. Rajendran, R. Manavalan, Mariam Samuel, R. Krishnamoorthi, V. Ashok Kumar, Axel Kroeger, Johannas Sommerfeld, P. Max and B.K. Tyagi, 2010. Eco-bio-socio aspects of dengue. (In: CRME Commemorative Compendium 2010: Silver Jubilee Publication. Ed. B.K. Tyagi). Centre for Research in Medical Entomology, Madurai, pp. 114-130.

326. Paramasivan, R., V. Thenmozhi, S. C. Tewari, P. Philip Samuel, T. Mariappan, R. Rajendran, N. Arunachalam and B.K. Tyagi, 2010. Re-emergence of chikungunya virus in southern India (2006-2010). (In: CRME Commemorative Compendium 2010: Silver Jubilee Publication. Ed. B.K. Tyagi). Centre for Research in Medical Entomology, Madurai, pp. 227-243.

327. Tyagi, B.K. and I.P. Sunish (Eds), 2010. R. Reuben: a renowned medical entomologist. CRME bio-bibliography series No. 4; 4th Conf. Med. Arthropodol., Madurai, pp. 19.

328. Tyagi, B.K. and I.P. Sunish (Eds), 2010. R. Rajendran: an eminent field vector biologist. CRME bio-bibliography series No. 5; 4th Conf. Med. Arthropodol., Madurai, pp. 17.

329. Tyagi, B.K. (Ed.). Proceedings of the second conference of medical arthropodology. 4th Conf. Med. Arthropodol., Madurai, pp. 124.

330. Victor Jerald Leo, P. Philip Samuel, V. Thenmozhi, S. C. Tewari, R. Balasubramaniam and B.K. Tyagi, 2010. Abiotic factors influencing the immature dengue vector abundance in Trivandrum district, Kerala. (In: second conference of medical arthropodology. Ed. B.K. Tyagi). 4th Conf. Med. Arthropodol., Madurai, pp. 77-87.

331. Selvi, S., M.A. Edah, W.A. Nazni, H.L. Lee, B.K. Tyagi, M. Sofia-Azirun and A.H. Azahari, 2010. Insecticide susceptibility and resistance development in malathion selected *Aedes albopictus* (Skuse). (In: Second Conference of Medical Arthropodology. Ed. B.K. Tyagi). 4th Conf. Med. Arthropodol., Madurai, pp. 104-124.

332. Tyagi, B.K., 2010. Abstracts. 4th Conf. Med. Arthropodol., Madurai, pp. 122.

333. Sunish, I.P., R. Rajendran, A. Munirathinam and B.K. Tyagi, 2010. Status of lymphatic filariasis elimination in Tamil Nadu State, south India. 4th Conf. Med. Arthropodol., Madurai, p. 13.

334. Krishnamoorthi, R., G. Baskaran, A. Munirathinam and B.K. Tyagi, 2010. Biodiversity of mosqioto fauna in southern region of Western Ghats. 4th Conf. Med. Arthropodol., Madurai, p. 17.

335. Harun Raseethu, A., J. Sebastian Raj, T. Mariappan and B.K. Tyagi, 2010. Susceptibility status of different strains of *Aedes aegypti* (Linnaeus, 1762) against an organophosphorus compound (Temephos) used in public health programme. 4th Conf. Med. Arthropodol., Madurai, p. 27.

336. Munirathinam, A., D. Ramesh, I.P. Sunish, M. Govindraju and B.K. Tyagi, 2010. Mosquito biodiversity in Sitheri hill range of Eastern Ghats. 4th Conf. Med. Arthropodol., Madurai, p. 31.

337. Thenmozhi, V., R. Paramasivan and B.K. Tyagi, 2010. Forecasting mosquito prevalence and infection to prevent Japanese encephalitis. 4th Conf. Med. Arthropodol., Madurai, p. 45.

338. Thenmozhi, V., R. Paramasivan and B.K. Tyagi, 2010. Role of *Culex gelidus* Theobald in the transmission of Japanese encephalitis virus in nature. 4th Conf. Med. Arthropodol., Madurai, p. 47.

339. Philip Samuel, P., R. Krishnamoorthi, R. Rajendran, N. Arunachalam and B.K. Tyagi, 2010. Comprehensive analysis of dengue awareness scenario in Thiruvananthapuram district, Kerala. 4th Conf. Med. Arthropodol., Madurai, p. 49.

340. Victor Jerald Leo, S., P. Philip Samuel and B.K. Tyagi, 2010. Physico-chemical analysis of water in the mud pot, the potential breeding grounds for the dengue and chikungunya vectors – a longitudinal study in Thiruvananthapuram district, Kerala. 4th Conf. Med. Arthropodol., Madurai, p. 51.

341. Basakar, G., P. Philip Samuel, S. Victor Jerald Leo, R. Krishnamoorthi and B.K. Tyagi, 2010. Implications of rainguard to control *Aedes albopictus* breeding in latex cups in Kerala. 4th Conf. Med. Arthropodol., Madurai, p. 53.

342. Venkatesh, A., I.P. Sunish, K. Pazninathan, G. Baskaran, J. Nagrajan and B.K. Tyagi, 2010. Predatory potential of anisopteran nymphs on *Aedes aegypti* larvae under controlled conditions. 4th Conf. Med. Arthropodol., Madurai, p. 61.

343. Philip Samuel and B.K. Tyagi, 2010. Comprehensive analysis of JE awareness scenario in the Gorakhpur and Basti districts of Uttar Pradesh, India. 4th Conf. Med. Arthropodol., Madurai, pp. 67-68.

344. Ramesh, D., A. Munirathinam, K. Ayanar, R. Rajendran, S.C. Tewari, P. Philip Samuel and B.K. Tyagi, 2010. JE vector prevalence in an endemic and non-endemic areas in Tamil Nadu, south India. 4th Conf. Med. Arthropodol., Madurai, p. 69.

345. Philip Samuel, P., N. Arunachalam, M. Muniaraj, R. Krishnamoorthy, R. Rajendran, S. Victor Gerald Leo and B.K. Tyagi, 2010. Breeding patterns of *Aedes* mosquitoes in the chikungunya epidemic affected districts of Kerala State. 4th Conf. Med. Arthropodol., Madurai, pp. 71-72.

346. Philip Samuel, P., R. Rajendran, N. Arunachalam and B.K. Tyagi, 2010. An in-depth KAP survey on dengue fever in the Kottayam district, Kerala India. 4th Conf. Med. Arthropodol., Madurai, pp. 89-90.

347. Victor Gerald Leo, S., P. Philip Samuel and B.K. Tyagi, 2010. RNA interference RNAi) mediated knock-outs of arboviral infections in vector mosquitoes – a review. 4th Conf. Med. Arthropodol., Madurai, p. 95.

348. Victor Gerald Leo, S. , P. Philip Samuel and B.K. Tyagi, 2010. DNA barcoding of vector mosquitoes – opportunities and challenges. 4th Conf. Med. Arthropodol., Madurai, p. 97.

349. Satish Babu, R., , P. Philip Samuel, S. Victor Gerald Leo, R. Govindrajan and B.K. Tyagi, 2010. Utility of larvivorous fishes for controlling vector mosquitoes. 4th Conf. Med. Arthropodol., Madurai, p. 99.

350. Kalimuthu, M., Selvaraj Pandian and B.K. Tyagi, 2010. Effect of insecticide on the reproductive potential of *Diplonichus rusticus* (Fabr.): a biocontrol agent. 4th Conf. Med. Arthropodol., Madurai, p. 101.

351. Philip Samuel, P., S. Victor Gerald Leo, V. Thenmozhi, S.C. Tewari, R. Paramasivan, N. Arunachalam and B.K. Tyagi, 2010. Dengue vector dynamics in the Thiruvananthapuram district, Kerala India. 4th Conf. Med. Arthropodol., Madurai, p. 111.

352. Thenmozhi, V., R. Selvaraj Pandian and B.K. Tyagi, 2010. Mosquitoes and human health. 4th Conf. Med. Arthropodol., Madurai, p. 115.

353. Tyagi, B.K., 2010. The Indian mosquito taxonomy at crossroads: Glorious past, dwindling present and promising future roadmap with a preliminary catalogue of Indian mosquitoes. (In: *Advances in Entomology*. Eds. Jagbir Singh Kirti and Ashwani Kumar), Punjabi University, Patiala. Pp. 142-160.

2011

354. John Vennison S., S. Gowri Sankar, K.J. Dhananjeyan, R. Paramasivan, V. Thenmozhi, and B.K. Tyagi, 2011. Dengue NS1 1GM detection: a new approach for sensitive acute dengue virus diagnosis and for understanding its role in pathology and pathogenesis. Abstr. Pap. 5th Conf. Med. Arthropodol., Madurai, p.9.

355. Rajamannar V., R. Rajendran, S.C. Tewari, P. Philip Samuel, R. Reuben and B.K. Tyagi, 2011. Synergistic role of water management techniques and natural aquatic mosquito predators in Japanese Encepahlitis vector mosquito control. Abstr. Pap. 5th Conf. Med. Arthropodol., Madurai, p.23.

356. Munirathinam A., R. Krishnamoorthi, A. Veerapathiran, A. Venkatesh, T. Mariappan and B.K. Tyagi, 2011. The past and present scenario of mosquito fauna recorded from different hill ranges in the western ghats region, South India. Abstr. Pap. 5th Conf. Med. Arthropodol., Madurai, p.49.

357. Philip Samuel P., R. Paramasivan and B.K. Tyagi, 2011. Entomological investigations into the fever outbreak in Tirunelveli District, Tamil Nadu during January-March 2010. Abstr. Pap. 5th Conf. Med. Arthropodol., Madurai, p.51.

358. Victor Jerald Leo S., P. Philip Samuel, G. Baskaran, R. Sathish Babu, R. Govindarajan and B.K. Tyagi, 2011. Environmental conditions in discarded tires and influences on the breeding of dengue vectors in Thiruvananthapuram District, Kerala. Abstr. Pap. 5th Conf. Med. Arthropodol., Madurai, p.55.

359. Thenmozhi. V., T. Balaji, K. Venkatasubramani and B.K. Tyagi, 2011. Vector Incriminaiton in Non JE endemic area, Thanjavur District. Abstr. Pap. 5th Conf. Med. Arthropodol., Madurai, p.61.

360. Philip Samuel. P., D. Ramesh and B.K. Tyagi, 2011. Japanese Encephalitis vector abundance in Cuddalore District, Tamil Nadu. Abstr. Pap. 5th Conf. Med. Arthropodol., Madurai, p.65.

361. Philip Samuel, P., S. Victor Jerald Leo, M. Muniyaraj, N. Arunachalam and B.K. Tyagi, 2011. Entomological observations during the Japanese Encephalitis

outbreak investigation carried out in the affected areas of Andhra Pradesh. Abstr. Pap. 5th Conf. Med. Arthropodol., Madurai, p.67.

362. Krishnamoorthy, R., A. Munirathinam, A. Venkatesh, G. Baskaran, T. Mariappan and B.K. Tyagi, 2011. A comparative analysis of mosquito bio-diversity in the western ghats following CRME's first exploration in 1986. Abstr. Pap. 5th Conf. Med. Arthropodol., Madurai, p.93.

363. Thenmozhi, V., R. Paramasivan, P. Philip Samuel, T. Balaji, G. Basaran, K.J. Dhananjeyan, A. Selvam and B.K. Tyagi, 2011. Mosquito Diversity in Tirunelveli District, Tamil Nadu. Abstr. Pap. 5th Conf. Med. Arthropodol., Madurai, p.97.

364. Thenmozhi, V., R. Selvaraj Pandian and B.K. Tyagi, 2011. Type of ecosystem: a determinant factor for the enhancement of diversity and density of mosquito fauna. Abstr. Pap. 5th Conf. Med. Arthropodol., Madurai, p.99.

365. Victor Jerald Leo, S., P. Philip Samuel and B.K. Tyagi, 2011. Molecular modelling of anti-viral drug for chikungunya virus deducted by molecular docking of potential target: a thought. Abstr. Pap. 5th Conf. Med. Arthropodol., Madurai, p.101.

366. Rajamannar, V., P. Philip Samuel, A. Venkatesh and B.K. Tyagi, 2011. Health informatics on mosquitoes and mosquito-borne diseases for India. Abstr. Pap. 5th Conf. Med. Arthropodol., Madurai, p.105.

367. Thenmozhi, V., T. Balaji and B.K. Tyagi, 2011. Feeding behaviour of mosquitoes in Japanese Encephalitis endemic area. Abstr. Pap. 5th Conf. Med. Arthropodol., Madurai, p.111.

368. Victor Jerald Leo., P. Philip Samuel, V. Thenmozhi and B.K. Tyagi, 2011. Applications of remote sensing, GIS and GPS in the control of vector borne diseases: a critical review. Abstr. Pap. 5th Conf. Med. Arthropodol., Madurai, p.115.

369. Philip Samuel, P., V. Thenmozhi, R. Paramasivan, Victor Jerald Leo, R. Satish Babu, R. Govindarajan, Kamaraj and B.K. Tyagi, 2011. Distributional patterns of JE, Dengue and Chikungunya vectors in Alleppey District, Kerala during 2011 epidemic. Abstr. Pap. 5th Conf. Med. Arthropodol., Madurai, p.121.

370. Venkatesh, A., I.P. Sunish, J. Nagaraj, G. Baskaran and B.K. Tyagi, 2011. Bradinopyga Geminata (odonata: libellulidae) a potential biocontrol agent of dengue/chikungunya vector mosquitoes (*Aedes*). Abstr. Pap. 5th Conf. Med. Arthropodol., Madurai, p.125.

371. Baskaran, G., P. Philip Samuel, R. Paramasivan, S. Victor Jerald Leo, K.J. Dhananjeyan, K. Pazhaninathan, A. Venkatesh, T. Kamaraj and B.K. Tyagi, 2011. A simple and cost effective method for the effective removal of rainwater stored in tyres, a potential breeding source of dengue/chikv vectors. Abstr. Pap. 5th Conf. Med. Arthropodol., Madurai, p.127.

372. Govindarajan, R., P. Philip Samuel, S. Victor Jerald Leo, R. Sathish Babu, J. Nagaraj and B.K. Tyagi, 2011. The control of mosquito larvae using fungi as a biological control tool – a critical review. Abstr. Pap. 5th Conf. Med. Arthropodol., Madurai, p.129.

373. Kalimuthu, M., R. Selvaraj Pandian, I.P. Sunish, V. Ashok kumar, A. Venkatesh, E. Raja and B.K. Tyagi, 2011. Impact of combined insecticide application on biochemical constituents of a biocontrol agent *diplonycus rusticus* (Fabr.) (Hemiptera: Belostomatidae). Abstr. Pap. 5th Conf. Med. Arthropodol., Madurai, p.133.

374. Nagaraj, J., T. Mariappan, I.P. Sunish, P. Philip Samuel, E. Raja, R. Krishnamoorthy, R. Sathish Babu, R. Govindarajan and B.K. Tyagi, 2011. Container

survey of chikungunya/dengue virus transmitting *aedes aegypti* in Madurai city, South India. Abstr. Pap. 5th Conf. Med. Arthropodol., Madurai, p.143.

375. Ramesh, D., A. Munirathinam, V. Thenmozhi, P. Philip Samuel and B.K. Tyagi, 2011. Seasonal variation in the abundance of cx. Gelidus, the vector of Japanese Encephalitis in Cuddalore District, Tamil Nadu, India. Abstr. Pap. 5th Conf. Med. Arthropodol., Madurai, p.151.

376. Sathish Babu, R., P. Philip Samuel, S. Victor Jerald Leo, R. Govindarajan, J. Nagaraj and B.K. Tyagi, 2011. Utility of callus induction on somatic embryogenesis in Azadirachta Indica (Neem) for mosquito larval control – a critical review. Abstr. Pap. 5th Conf. Med. Arthropodol., Madurai, p.163.

377. Venkatesh, A., I.P. Sunish, K. Pazhaninathan, G. Baskaran, J. Nagaraj and B.K. Tyagi, 2011. Predatory potential of anisopteran nymphs on *aedes aegypti* larvae under controlled conditions. Proc. 4th Conf. Med. Arthropodol., Madurai, p.40-46.

378. Thenmozhi, V., R. Paramasivan, D. Ramesh, T. Balaji, K. Venkatasubramani and B.K. Tyagi, 2011. Reconfirmation of the role of *culex gelidus* theobald in Japanese Encephalitis virus transmission. Proc. 4th Conf. Med. Arthropodol., Madurai, p.47-56.

379. Thenmozhi, V., R. Selvaraj Pandian and B.K. Tyagi, 2011. Mosquitoes and human health. Proc. 4th Conf. Med. Arthropodol., Madurai, p.57-62.

380. Philip Samuel, P., N. Arunachalam, M. Muniyaraj, R. Krishnamoorthy, R. Rajendran, S. Victor Jerald Leo and B.K. Tyagi, 2011. Proc. 4th Conf. Med. Arthropodol., Madurai, p.73-83.

381. Philip Samuel, P., R. Krishnamoorthi, R. Rajendran, N. Arunachalam and B.K. Tyagi, 2011. Comprehensive analysis of dengue awareness scenario in Thiruvananthapuram District, Kerala. Proc. 4th Conf. Med. Arthropodol., Madurai, p.84-91.

382. Victor Jerald Leo, S., P. Philip Samuel and B.K. Tyagi, 2011. DNA barcoding of vector mosquitoes – opportunities and challenges. Proc. 4th Conf. Med. Arthropodol., Madurai, p.100-111.

383. Philip Samuel, P., S. Victor Jerald Leo, V. Thenmozhi, S.C. Tewari, R. Paramasivan, N. Arunachalam and B.K. Tyagi, 2011. Dengue vector dynamics in the Thiruvananthapuram District, Kerala, India. Proc. 4th Conf. Med. Arthropodol., Madurai, p.112-118.

384. Victor Jerald Leo, S., P. Philip Samuel and B.K. Tyagi, 2011. RNA interference (RNAi) mediated knock-outs of arboviral infections in vector mosquitoes – a review. Proc. 4th Conf. Med. Arthropodol., Madurai, p.119-128.

385. Philip Samuel, P., and B.K. Tyagi, 2011. Comprehensive analysis of JE awareness scenario in Gorakhpur and Basti Districts of Uttar Pradesh, India. Proc. 4th Conf. Med. Arthropodol., Madurai, p.129-134.

386. Victor Jerald Leo, S., P. Philip Samuel and B.K. Tyagi, 2011. Physico-chemical analysis of water in the mud pot the potential breeding grounds for the dengue and chikungunya vectors – a longitudinal study in Thiruvananthapuram District, Kerala. Proc. 4th Conf. Med. Arthropodol., Madurai, p.135-144.

387. Baskaran, G., P. Philip Samuel, S. Victor Jerald Leo, R. Krishnamoorthy and B.K. Tyagi, 2011. Implication of rainguard for the control of *aedes albopictus* breeding mainly in latex cups in Kerala, India. Proc. 4th Conf. Med. Arthropodol., Madurai, p.145-151.

388. Philip Samuel, P., R. Rajendran, N. Arunachalam and B.K. Tyagi, 2011. An in-depth kap survey on dengue fever in the Kottayam District, Kerala, India. Proc. 4th Conf. Med. Arthropodol., Madurai, p.163-171.

389. Ramesh, D., A. Munirathinam, K. Ayanar, R. Rajendran, S.C. Tewari, P. Philip Samuel and B.K. Tyagi, 2011. JE vector prevalence in endemic and non-endemic areas in Tamil Nadu, South India. Proc. 4th Conf. Med. Arthropodol., Madurai, p.172-176.

390. Tyagi, B.K., 2011. Multidimentional behaviour of insect vectors of public health importance. (*In*: Dynamics of Insect Behaviour. Ed. T.N. Ananthakrishnan and K.G. Sivaramakrishnan). Scientific Publishers (India), Jodhpur. Pp. 189-210.

391. Tyagi, B.K., 2011. Dr. Vijay Veer: A renowned defense medico-entomologist. CRME Bio-bibliography Series No. 6; 5th Conf. Med. Arthropodol., Madurai, pp. 30.

392. Gowri Shankar, S., K.J. Dhanenjeyan, R. Paramasivan, V. Thenmozhi, B.K. Tyagi and S. John Vennison, 2011. Evaluation and use NSI IgM antibody detection for acute dengue virus diagnosis: report from an outbreak investigation. *Clinical Microbiology and Infection.*

393. Tyagi, B.K., 2011. Vector surveillance: Current status and scope. Natl. CME on Vector Borne Diseases – Indian Perspectives & Solutions, Pudducherry (Aug. 26, 2011), p. 39.

2012

394. Tyagi, B.K., 2012. Vector borne disease interaction between human and canines, with special reference to cutaneous leishmaniasis prevalent in Western Rajasthan. Int. Cong. Modern Concepts in Canine Health and Diseases of Human Concern and IX Annual Convention of Indian Society for Advancement of Canine Practice, Bikaner (9-11th Feb., 2012).

395. Tyagi, B.K., 2012. Culicide phytotelmata (Pitcher plant dwelling mosquitoes): Evolutionary link to Earth's changing climate – a review. Int. Conf. Entomology, Patiala (Feb. 17-19, 2012), pp. 17-18.

396. Tyagi, B.K., 2012. Transgenic insects: a novel way to control pests and vectors of human diseases (Dengue and malaria). XXIII National Symposium on Chronobiology and Seminar on Diversity and Physiology of Desert Fauna (23rd NSC&SDPDF)", Jodhpur (March 1-3, 2012), pp.

397. Govindraju, I.P. Sunish, B.K. Tyagi, C. Rajina and P. Suganthi, 2011. Development of health information system using spatial technology for Sitheri hills – An integrated approach. 4th Int. Conf. HealthGIS 2011, New Delhi (Aug. 5-6, 2011), pp. .

398. Munirathinam, A., A. Veerapathiran, R. Krishnamoorthy, K. Satish and B. K. Tyagi, 2012. Mosquito fauna recorded in the foot hills of Coimbatore district (Western Ghats) in south India. *Ibid.*, pp. 54-55.

399. Phiplip Samuel, P., S. Victor Gerald Leo, R. Satish Babu, R. Govindrajan and B.K. Tyagi, 2012. Differential distribution pattern of dengue vectors from Kerala State, south India. *Ibid.*, pp. 166.

400. Venkatesh, A., G. Baskaran, J. Nagaraj, P. Philip Samuel and B. K. Tyagi, 2012. Predatory potential of *Bradinopyga geminate* (Odonata: Libellulidae) anisopteran and *Ceraiagrion aurantiacum* (sic! It should be read as *coromandelianum*) (Odonata: Coanagrionidae) zygopteran nymphs on the principal vectors of dengue/ chikungunya *Aedes aegypti* (Diptera: Culicidae). *Ibid.*, pp. 167.

401. Halasa, Yara, A., Vishal Dogra, Narendra Arora, B.K. Tyagi, Deoki Nandan and Donald S. Shepard, 2011. Overcoming data limitations: Design of a multi-component study for estimating the economic burden of dengue in India. Dengue Bulletin (WHO-SEARO) 35: 1-14.

402. Philip Samuel, P., R. Paramasivan and B.K. Tyagi, 2012. Chikungunya in the tribal dominated Lakshdweep ils (India): A review of results of an entomo-epidemiological investigation during 2006. ICMR Tribal Health Research Forum 2012, pp. 96-104.

403. M. Govindaraju, B.K. Tyagi, V. Thenmozhi, I.P. Sunish, P. Suganthi, M. Selvaraj, L. Macklin Rani, P. Kumar, B. Sarojini Devi and Rajiv Das Kangabam. 2012. Development of Spatial Information System to Control Vector Borne Diseases (VBDs) 6th Conf. Med. Arthropod., Madurai, p.5.

404. M. Kalimuthu, V. Ashokkumar, I.P. Sunish, B. Bharanidharan, B. Palanisamy, V.A. Ralph Selvin, A. Subramanian, K.R. Jeyanthi, K. Kamaraj, K. Manivarma, N. Arunachalam and B.K. Tyagi. 2012. Assessment of coverage and compliance of Mass Drug Administration under Elimination of Lymphatic Filariasis Programme in Thiruvarur, Nagapattinam and Thiruvannamalai Districts, Tamil Nadu, South India. 6th Conf. Med. Arthropod., Madurai, p.17-18.

405. T. Balaji, V. Thenmozhi, K. Venkatasubramani, S. Victor Jerald Leo, N. Arunachalam and B.K. Tyagi. 2012. Co-Circulation of Dengue and Chikungunya Virus in Tamil Nadu – A Report of the recent trend. 6th Conf. Med. Arthropod., Madurai, p.35.

406. R. Govindarajan, P. Philip Samuel, G. Baskaran, S. Victor Jerald Leo, V. Rajamannar, R. Sathish Babu, J. Nagaraj and B.K. Tyagi. 2012. Ectoprasitic Infections in Human Life. 6th Conf. Med. Arthropod., Madurai, p.45.

407. V. Rajamannar, P. Philip Samuel, T. Mariappan and B.K. Tyagi. 2012. Biomedical Bibliometrics on Chikungunya Literature for India – A tool for determining the quality of scholarly articles for disease information. 6th Conf. Med. Arthropod., Madurai, p.49-50.

408. J. Nagaraj, P. Philip Samuel, T. Mariappan, S. Karthigai Selvi, V. Rajamannar, A. Venkatesh, J. Victor Jerald Leo, R. Sathish Babu, R. Govindarajan and B.K. Tyagi. 2012. 6th Conf. Med. Arthropod., Madurai, p.51.

409. P. Philip Samuel and B.K. Tyagi. 2012. The succession of the vector borne diseases reported from the smallest union territory of India Lakshadweep Archipelago. 6th Conf. Med. Arthropod., Madurai, p.53-54.

410. P. Philip Samuel, T. Mariappan and B.K. Tyagi. 2012. Emergence of Japanese Encephalitis in Northern West Bengal Region as a multivectoral activity. 6th Conf. Med. Arthropod., Madurai, p.55.

411. V. Thenmozhi and B.K. Tyagi. 2012. Role of dual infections noticed during outbreak studies in Southern India. 6th Conf. Med. Arthropod., Madurai, p.67.

412. K. Venkatasubramani, K.J. Dhananjeyan, T. Balaji, G. Baskaran, V. Thenmozhi, R. Paramasivan and B.K. Tyagi. Chikungunya Fever, an important re-emerging public health problem in Tamil Nadu. 6th Conf. Med. Arthropod., Madurai, p.69.

413. A. Venktaesh, G. Baskaran, A. Munirathinam, P. Philip Samuel and B.K. Tyagi. Odonata vs. Mosquito vectors: An innovative, economic and environmental control methodology – A review. 6th Conf. Med. Arthropod., Madurai, p.71-72.

414. V. Thenmozhi, P. Philip Samuel, D. Ramesh, T. Balaji and B.K. Tyagi. Flavivirus infections in school children of Thanjavur District, Tamil Nadu. 6th Conf. Med. Arthropod., Madurai, p.75.

415. V. Thenmozhi and B.K. Tyagi. 2012. Japanese Encephalitis surveillance and epidemiology in Tamil Nadu. 6th Conf. Med. Arthropod., Madurai, p.77.

416. V. Thenmozhi, T. Balaji, K. Venkatasubramani and B.K. Tyagi. 2012. Subclinical Flavivirus infection in school children of Tirunelveli District, Tamil Nadu. 6th Conf. Med. Arthropod., Madurai, p.79-80.

417. G. Baskaran, R. Krishnamoorthy, R. Krishnamoorthi, T. Mariappan, P. Philip Samuel, R. Govindarajan, A. Venkatesh and B.K. Tyagi. 2012. Elimination of bamboo stump fencing to prevent dengue mosquitoes breeding in the Districts of West Bengal. 6th Conf. Med. Arthropod., Madurai, p.81.

418. R. Krsihnamoorthi, R. Manavalan, V. Ashok kumar, V. Thenmozhi, A. Venkatesh, N. Arunachalam and B.K. Tyagi. 2012. Pupal Index (PI) as a reliable tool for assessing dengue vector adult population and its possible contribution to the community based vector control. 6th Conf. Med. Arthropod., Madurai, p.87-88.

419. P. Philip Samuel, V. Thenmozhi, S. Victor Jerald Leo, R. Satish Babu, R. Govindarajan, K. Venkatasubramani and B.K. Tyagi. 2012. Application of ovitraps for the dengue vectors surveillance in the Thiruvananthapuram District, Kerala. 6th Conf. Med. Arthropod., Madurai, p.91.

420. D. Pradiba, S.L. Hoti, K.P. Paily, V. Vasuki, G. Manikandan, N. Arunachalam, M. Kalimuthu, V. Ashok kumar, V. Thenmozhi and B.K. Tyagi. 2012. Significance of Genotyping of Lymphatic Filarial parasites. 6th Conf. Med. Arthropod., Madurai, p.113.

421. V. Thenmozhi, N. Arunachalam, M. Muniyaraj, T. Balaji, A. Selvam, S. Victor Jerald Leo and B.K. Tyagi. 2012. Transovarial transmission of dengue virus in Aedes Aegypti and Aedes Vittatus in Warrangal District, Andhra Pradesh, India. 6th Conf. Med. Arthropod., Madurai, p.115-116.

422. S. Victor Jerald Leo, P. Philip Samuel, R. Paramasivan and B.K. Tyagi. 2012. Mosquito salivary gland proteins and their role in the transmission of Arbo-viral infections. 6th Conf. Med. Arthropod., Madurai, p.117-118.

423. V. Thenmozhi and B.K. Tyagi. 2012. Dengue Diagnosis, Advances and challenges. 6th Conf. Med. Arthropod., Madurai, p.133.

424. V. Ashokkumar, R. Krishnamoorthi, N. Arunachalam, S.C. Tewari, R. Rajendran, R. Manavalan and B.K. Tyagi. 2012. Communication in behaviour change is an effective tool to control dengue/chikungunya vector Aedes Aegypti breeding control in the urban agglomeration of Chennai Metropolis. 6th Conf. Med. Arthropod., Madurai, p.137-138.

425. A. Karuppasamy, N. Arunachalam, R. Krishnamoorthi, V. Ashok kumar, R. Manavalan and B.K. Tyagi. 2012. A sociological study to understand people's knowledge, attitude and practices on dengue fever among the urban community in Chennai, India. 6th Conf. Med. Arthropod., Madurai, p.139-140.

426. E. Raja, V. Ashokkumar, M. Kalimuthu and B.K. Tyagi. 2012. Biotechnological and sociological "Challenges in vector control". 6th Conf. Med. Arthropod., Madurai, p.147.

427. M. Kalimuthu, V. Ashokkumar, N. Arunachalam and B.K. Tyagi. 2012. Predatory potential of larvivorous fish, Gambusia Affinis on the immature stages of Lymphatic Filariasis and Japanese Encephalitis Vectors. 6th Conf. Med. Arthropod., Madurai, p.157.

428. Kuppusamy Chenniappan, N. Kamaladasan, G. Ramesh kumar, B. Kamala-kannan, B.K. Tyagi and S. Chandrasekaran. 2012. *Eichhornia crassipes* provide conducive environment for breeding mosquitoes. 6th Conf. Med. Arthropod., Madurai, p.183.

429. R. Krishnamoorthy, A. Munirathinam, A. Veerapathiran, R. Krishnamoorthi, G. Baskaran, T. Mariappan and B.K. Tyagi. 2012. An observation of behavioural changes of breeding habitates of Genus Toxorhynchites species in India. 6th Conf. Med. Arthropod., Madurai, p.207.
430. A. Munirathinam, R. Krishnamoorthy, A. Veerapathiran, A. Venkatesh and B.K. Tyagi. 2012. Mosquito biodiversity in Western Ghats after a gap of 25 years. 6th Conf. Med. Arthropod., Madurai, p.211.
431. A. Munirathinam, A. Veerapathiran, R. Krishnamoorthy, A. Venkatesh, T. Mariappn and B.K. Tyagi. 2012. Mosquito fauna survey in Anamalai Tiger Reserve Hill Ranges in Western Ghats, Tamil Nadu, South India. 6th Conf. Med. Arthropod., Madurai, p.213.
432. A. Munirathinam, A. Veerapathiran, T. Mariappan and B.K. Tyagi. 2012. Mosquito fauna survey in Sathyamangalam Hill ranges in Western Ghats, Tamil Nadu, South India. 6th Conf. Med. Arthropod., Madurai, p.215.
433. P. Philip Samuel, R. Paramasivan and B.K. Tyagi. 2012. Entomological surveillance to monitor the Japanese Encephalitis vectors from the Gorakhpur Division of Uttar Pradesh, India. 6th Conf. Med. Arthropod., Madurai, p.217-218.
434. P. Philip Samuel, M. Muniyaraj and B.K. Tyagi. 2012. Biodiversity of mosquitoes from Muzhaffarpur District, Bihar, India. 6th Conf. Med. Arthropod., Madurai, p.219.
435. D. Ramesh, A. Munirathinam, V. Thenmozhi, P. Philip Samuel, M. Govindaraju and B.K. Tyagi. 2012. Mosquito species succession in Sitheri Hills, Dharmapuri District. 6th Conf. Med. Arthropod., Madurai, p.221.
436. V. Thenmozhi, T. Mariappan and B.K. Tyagi. 2012. Japanese Encephalitis outbreak in West Bengal has unusual vectors. 6th Conf. Med. Arthropod., Madurai, p.225.
437. A. Veerapathiran, A. Munirathinam, T. Mariappan and B.K. Tyagi. 2012. Mosquito fauna of Malabar squirrel sanctuary at Srivilliputur Hill ranges in Western Ghats, Tamil Nadu, South India. 6th Conf. Med. Arthropod., Madurai, p.227.
438. S. Victor Jerald Leo, P. Philip Samuel, G. Baskaran, K. Venkatasubramani, R. Sathish Babu, R. Govindarajan, J. Nagaraj, A. Venkatesh, T. Balaji and B.K. Tyagi. 2012. Physico-chemical determinants for the growth of dengue and chikungunya larvae in cement tank – A longitudinal study in Thiruvananthapuram District, Kerala. 6th Conf. Med. Arthropod., Madurai, pp.231-232.
439. Saranya Sugumar, Samantha Katherine Clarke, Tyagi B.K., Amitava Kumar Mukherjee and N. Chandrasekaran, 2012. Eucalyptus oil nanoemulsion: Investigation of Larvicidal activity and its mode of action against Culex quinquefasciatus. BioMed Research International (subm. 220413 Ed.).
440. V. Thenmozhi, T. Balaji, A. Selvam, M. Govindarajan, K. Venkatasubramani and B.K. Tyagi. 2012. Subclinical flavivirus infection in school children of Tirunelveli District, Tamil Nadu. 6th Conf. Medical Arthropodology, Madurai (Oct. 18-19, 2012), p.95-102.
441. V. Thenmozhi, N. Arunachalam, M. Muniaraj, T. Balaji, A. Selvam, S. Victor Jerald Leo and B.K. Tyagi. 2012. Transovarial Transmission of Dengue virus in Aedes Aegypti and Aedes vittatus in Warangal District, Andhra Pradesh, India. 6th Conf. Medical Arthropodology, Madurai (Oct. 18-19, 2012), p.103-107.
442. D. Ramesh, A. Munirathinam, V. Thenmozhi, P. Philip Samuel, M. Govindaraj and B.K. Tyagi. 2012. Mosquito species succession in Sitheri Hills, Dharmapuri District. 6th Conf. Medical Arthropodology, Madurai (Oct. 18-19, 2012), p.108-117.

443. P. Philip Samuel, R. Paramasivan, V. Thenmozhi, A. Venkatesh and B.K. Tyagi. 2012. Entomological surveillance to monitor the Japanese Encephalitis vectors from the Gorakhpur Division of Uttar Pradesh, India. 6th Conf. Medical Arthropodology, Madurai (Oct. 18-19, 2012), p.124-130.

444. T. Balaji, V. Thenmozhi, K. Venkatasubramani, S. Victro Jerald Leo, N. Arunachalam and B.K. Tyagi. 2012. Co-circulation of Dengue and Chikungunya viruses in Tamil Nadu – A report on the recent trend. 6th Conf. Medical Arthropodology, Madurai (Oct. 18-19, 2012), p.140-144.

445. A. Veerapathiran, A. Munirathinam, T. Mariappan and B.K. Tyagi. 2012. Mosquito Fauna of Malabar squirrel sanctuary at Srivilliputur Hill ranges in Western Ghats, Tamil Nadu, South India. 6th Conf. Medical Arthropodology, Madurai (Oct. 18-19, 2012), p.145-151.

2013

446. P. Philip Samuel and B.K. Tyagi. 2013. Dengue and its control in India. Brain Storming Conf. on Dengue Scenario in India. Madurai (July 25-26, 2013), p.43-44.

447. S. Karthigai Selvi and B.K. Tyagi. 2013. India amongst front runners in contributing to Global Burden of Dengue: Time to rethink and revamp the surveillance programme in India. Brain Storming Conf. on Dengue Scenario in India. Madurai (July 25-26, 2013), p.63-64.

448. S. Victor Jerald Leo, P. Philip Samuel and B.K. Tyagi. 2013. Proteomic analysis of salivary glands of the Dengue Vector, Aedes Aegypti. Brain Storming Conf. on Dengue Scenario in India. Madurai (July 25-26, 2013), p.83.

449. T. Mariappan, R. Sathish Babu, V. Kodangialagan, K. Venkatasubramani, K.J. Dhananjeyan, A. Venkatesh and B.K. Tyagi.. Status of Entomological indices during Dengue outbreak (2012) in Dindigul District, Tamil Nadu. Brain Storming Conf. on Dengue Scenario in India. Madurai (July 25-26, 2013), p.93.

450. A. Munirathinam, R. Krishnamoorthi, A. Veerapathiran and B.K. Tyagi. Potential Breeding habitats of Dengue/Chikungunya vectors in Western Ghats, South India. Brain Storming Conf. on Dengue Scenario in India. Madurai (July 25-26, 2013), p.95.

451. P. Philip Samuel, S. Victor Jerald Leo, V. Thenmozhi, K. Venkatasubramani, R. Satish Babu, R. Govindarajan, J. Nagaraj, Dilip kumar and B.K. Tyagi. Dengue vectors prevalence in Thiruvananathapuram District, Kerala. Brain Storming Conf. on Dengue Scenario in India. Madurai (July 25-26, 2013), p.101.

452. V. Thenmozhi and B.K. Tyagi. The pathogenesis of Dengue: A comment. Brain Storming Conf. on Dengue Scenario in India. Madurai (July 25-26, 2013), p.107.

453. S. Prabu, T. Mariappan and B.K. Tyagi. Efficacy of Bacillus Thuringiensis var.israelensis toxicity against dengue vector, Aedes Aegypti. Brain Storming Conf. on Dengue Scenario in India. Madurai (July 25-26, 2013), p.155.

454. N. Arunachalam, B.K. Tyagi, Miriam Samuel, R. Krishnamoorthi, R. Manavalan, S.C. Tewari, V. Ashokkumar, A. Karuppasamy, Axel Kroeger and Johannes Sommerfeld. Studies on Aedes Aegypti Ecology in an urban area for implementing community based integrated Dengue vector management. Brain Storming Conf. on Dengue Scenario in India. Madurai (July 25-26, 2013), p.167-168.

455. V. Ashokkumar, N. Arunachalam, R. Krishnamoorthi, S.C. Tewari, R. Manavalan, A. Karuppasamy, Mirium Samuel, J. Sommerfeld and B.K. Tyagi. Will stakeholders play any significant role for the control of Dengue and its vector mosquitoes in Chennai City, South India. Brain Storming Conf. on Dengue Scenario in India. Madurai (July 25-26, 2013), p.169-170.

456. T. Balaji, V. Thenmozhi, T. Mariappan, J. Nagaraj, G. Manikandan, K. Venkatasubramani and B.K. Tyagi. Biological control of different immature vector mosquitoes using Toxorhynchites splendens (wiedemann) in laboratory condition. Brain Storming Conf. on Dengue Scenario in India. Madurai (July 25-26, 2013), p.171.

457. R. Govindarajan, P. Philip Samuel, V. Thenmozhi, G. Basaran, R. Krishnamoorthy, A. Munirathinam, S. Victor Jerald Leo, R. Sathish Babu, J. Nagaraj and B.K. Tyagi. Control of Dengue/Chikungunya vectors by recycling unused tyres: one of the majaor breeding sources. Brain Storming Conf. on Dengue Scenario in India. Madurai (July 25-26, 2013), p.183.

458. M. Kalimuthu, V. Ashokkumar, I.P. Sunish, N. Arunachalam and B.K. Tyagi. Predatory potential of notonecta species on the immature stages of Aedes Aegypti (Linnaeus), The primary vector of Dengue. Brain Storming Conf. on Dengue Scenario in India. Madurai (July 25-26, 2013), p.189.

459. A. Karuppasamy, N. Arunachalam, V. Ashokkumar, R. Krishnamoorthi, R. Manavalan and B.K. Tyagi. A study to explore young peoples knowledge on Dengue and their role in controlling dengue vector mosquitoes in a cluster of Chennai city. Brain Storming Conf. on Dengue Scenario in India. Madurai (July 25-26, 2013), p.193-194.

460. K.S. Malar, R. Paramasivan, M. Muniaraj, N. Arunachalam and B.K. Tyagi. Wolbachia – A potential vector control agent: Recent advancement and future challenges. Brain Storming Conf. on Dengue Scenario in India. Madurai (July 25-26, 2013), p.195.

461. T. Mariappn and B.K. Tyagi. Human Behaviour towards control of Dengue epidemic in Tamil Nadu – A case study. Brain Storming Conf. on Dengue Scenario in India. Madurai (July 25-26, 2013), p.197.

462. E. Raja, V. Ashokkumar, M. Kalimuthu, J. Nagaraj, A. Venkatesh and B.K. Tyagi. KAP survey on Dengue fever and its control measures among school children in Madurai city following a recent dengue outbreak. Brain Storming Conf. on Dengue Scenario in India. Madurai (July 25-26, 2013), p.203.

463. Tyagi B.K. and A. Venkatesh, 2013. Potential control of dengue vector, Aedes Aegypti (Linnaeus) through "Minimum re-inforcement method" in key breeding containers using certain cohabiting odonates such as bradinopyga geminate rambur and crocothemis servillia drury. Brain Storming Conf. on Dengue Scenario in India. Madurai (July 25-26, 2013), p.217-218.

464. R. Krsihnamoorthy, R. Govindarajan, R. Krishnamoorthi, G. Baskaran, T. Mariappan, P. Philip Samuel, J. Nagaraj and B.K. Tyagi. Distribution of Dengue/ Chikungunya vectors in North-Western Bengal, India. Brain Storming Conf. on Dengue Scenario in India. Madurai (July 25-26, 2013), p.221.

465. A. Munirathinam, R. Krishnamoorthi, G. Baskaran, R. Govindarajan, A. Venkatesh and B.K. Tyagi. A simple and field-based identification key for dengue/chikungunya vectors, *Aedes aegypti* (Linnaeus) and *Aedes albopictus* (skuse) larvae and pupae. Brain Storming Conf. on Dengue Scenario in India. Madurai (July 25-26, 2013), p.227.

466. D. Ramesh, V. Thenmozhi, P. Philip Samuel, A. Munirathinam, S. Vijayakumar, I.P. Sunish and B.K. Tyagi. The denuge vectors of Aedes Aegypti Linnaeus and Aedes Albopictus skuse breeding in the Sitheri Hills, Dharmapuri District, Tamil Nadu. Brain Storming Conf. on Dengue Scenario in India. Madurai (July 25-26, 2013), p.231.

467. N. Arunachalam, B.K. Tyagi, Miram Samuel, R. Krishnamoorthi, R. Manavalan, S.C. Tewari, V. Ashokkumar, A. Karuppasamy, Axel Kroeger and Johannes Sommerfeld. Ecotope identification of Aedes Aegypti, the principal vector of Dengue in Chennai city, India. Brain Storming Conf. on Dengue Scenario in India. Madurai (July 25-26, 2013), p.245.

468. R. Krishnamoorthi, R. Manavalan, V. Ashokkumar, N. Arunachalam, A. Munirathinam, A. Veerapathiran and B.K. Tyagi. Seasonal variation in Aedes (stegomyia) aegypti population in Chennai City (TN, India). Brain Storming Conf. on Dengue Scenario in India. Madurai (July 25-26, 2013), p.249.

469. J. Nagaraj, P. Philip Samuel, T. Mariappan, S. Karthigai selvi, R. Sathish Babu, R. Govindarajan, A. Venkatesh and B.K. Tyagi. Role of biotic and abiotic factors in influencing dengue incidence in Madurai District, Tamil Nadu. Brain Storming Conf. on Dengue Scenario in India. Madurai (July 25-26, 2013), p.251.

470. P. Philip Samuel, S. Victor Jerald Leo and B.K. Tyagi. Effects of minor nutrients on larval growth of Aedes vectors in Thiruvananthapuram District, Kerala. Brain Storming Conf. on Dengue Scenario in India. Madurai (July 25-26, 2013), p.253.

471. V. Rajamannar, T. Mariappan, M. Muniaraj, V. Kodangialagan, A. Selvam and B.K. Tyagi. Dengue surveillance at Madurai City during 2012 outbreak: A literature survey. Brain Storming Conf. on Dengue Scenario in India. Madurai (July 25-26, 2013), p.255.

472. I.P. Sunish, R. Rajendran, M. Kalimuthu, A. Munirathinam, V. Ashok kumar, J. Nagaraj and B.K. Tyagi. Transmission indices over a period of 10 years in an area administered with DEC with and without Albendazole, as part of the LF Elimination Campagin. 2nd Workshop on Lymphatic Filariasis: Highlighting achievements and significance of vector control in LF elimination, Chennai (June 5, 2013). p.33.

473. B.K. Tyagi. Vector control approach – A good adjunct to MDA in Lymphatic Filariasis Elimination. 2nd Workshop on Lymphatic Filariasis: Highlighting achievements and significance of vector control in LF elimination, Chennai (June 5, 2013). p.37.

474. N. Arunachalam, V. Ashokkumar, M. Kalimuthu and B.K. Tyagi. Control of bancroftian Filariasis with common salt medicated with diethylcarbamazine: A review 2nd Workshop on Lymphatic Filariasis: Highlighting achievements and significance of vector control in LF elimination, Chennai (June 5, 2013). p.39.

475. N. Arunachalam, I.P. Sunish, M. Kalimuthu, V. Ashokkumar, J. Nagaraj, A. Munirathinam and B.K. Tyagi. Comparison of vector density and transmission indices of bancroftian filariasis in three group of villages with different control strategies in Tirukoilur, India. 2nd Workshop on Lymphatic Filariasis: Highlighting achievements and significance of vector control in LF elimination, Chennai (June 5, 2013). p.41.

476. M. Kalimuthu, I.P. Sunish, V. Ashokkumar, J. Nagaraj, R. Rajendran, N. Arunachalam, A. Munirathinam and B.K. Tyagi. Integrated vector management through community involvement combined with mass drug administration to augment filariasis control in rural villages of South India. 2nd Workshop on Lymphatic Filariasis: Highlighting achievements and significance of vector control in LF elimination, Chennai (June 5, 2013). p.43.

477. V. Ashokkumar, N. Arunachalam, I.P. Sunish, M. Kalimuthu and B.K. Tyagi. Community acceptance and usage of vector control tools can reduce the transmission of Lymphatic Filariasis in rural areas in Tirukoilur, Tamil Nadu,

India. 2nd Workshop on Lymphatic Filariasis: Highlighting achievements and significance of vector control in LF elimination, Chennai (June 5, 2013). p.45.

478. T. Mariappan and B.K. Tyagi. Vector control (VC) continuation lead success of Lymphatic Filariasis (LF) in globe. 2nd Workshop on Lymphatic Filariasis: Highlighting achievements and significance of vector control in LF elimination, Chennai (June 5, 2013). p.47.

479. A. Munirathinam, I.P. Sunish, M. Kalimuthu, V. Ashokkumar, J. Nagaraj, R. Rajendran, N. Arunachalam, and B.K. Tyagi. 2013. Field based study to assess the efficacy of MDA plus integrated vector control against MDA alone on bancroftian filarial infection in endemic areas in Tamil Nadu, South India. 2nd Workshop on Lymphatic Filariasis: Highlighting achievements and significance of vector control in LF elimination, Chennai (June 5, 2013). p.51.

480. Rajesh, K., Dharumadurai Dhanasekaran and Brij Kishore Tyagi, 2013. Mosquito survey and larvicidal activity of actinobacterial isolates against Culex larvae (Diptera:Culicidae). Journal of the Saudi Society of Agricultural Sciences http://dx.doi.org/10.1016/j.jssas.2013.08.001

481. Tyagi, B.K., 2013. Emergence and evolution of dengue and chikungunya viral infections in India and their control. Natl. Conf. Dengue and Chikungunya Epidemiology and their Management in India. Nagercoil, p. 2.

482. Mariappan, T. And B.K. Tyagi, 2013. Dengue: Can we do source reduction of vectors? Natl. Conf. Dengue and Chikungunya Epidemiology and their Management in India. Nagercoil, p. 3.

483. Philip Samuel, P., Victor Gerald Leo, Venkatasubramani, Satish Babu, Govindan, J. Nagaraj and B.K. Tyagi, 2013. Multi-vector involvement in the dengue transmission in the Thiruvananthapuram district of Kerala. Natl. Conf. Dengue and Chikungunya Epidemiology and their Management in India. Nagercoil, p. 14.

484. Thenmozhi, V. and B.K. Tyagi, 2013. Mosquito-borne dengue and chikungunya fever. Natl. Conf. Dengue and Chikungunya Epidemiology and their Management in India. Nagercoil, p. 13.

485. Balaji, T. V. Thenmozhi, K. Subramani and B.K. Tyagi, 2013. A focal outbreak of dengue in Dindigu district in 2012, Natl. Conf. Dengue and Chikungunya Epidemiology and their Management in India. Nagercoil, p. 16.

486. Venkatesh, A. and B.K. Tyagi, 2013. Capture efficiency of Bradinopyga geminata (Odonata: Libellulidae) against larvae of Aedes aegypti, Culex tritaeniorhynchus and Anopheles stephensi in laboratory condition. Natl. Conf. Dengue and Chikungunya Epidemiology and their Management in India. Nagercoil, p. 15.

487. Karthigai Selvi S. and B.K. Tyagi, 2013. Control vector control strategies with special reference to dengue in India: Beyond conventional strategies – a critical review. Natl. Conf. Dengue and Chikungunya Epidemiology and their Management in India. Nagercoil, p. 17.

488. Venkatesh, A. and B.K. Tyagi, 2013. Capture efficiency of Bradinopyga geminata (Odonata: Libellulidae) against larvae of *Aedes aegypti*, *Culex tritaeniorhynchus* and *Anopheles stephensi* in laboratory condition. Journal of Basic & Applied Biology 7 (4): 21-26.

489. Thenmozhi, V. and B.K. Tyagi, 2013. Mosquito-borne dengue and chikungunya fever. Journal of Basic & Applied Biology 7 (4): 16-20.

490. Balaji, T. V. Thenmozhi, K. Subramani and B.K. Tyagi, 2013. A focal outbreak of dengue in Dindigu district in 2012, Journal of Basic & Applied Biology 7 (4): 6-8.

491. Philip Samuel, P., Victor Gerald Leo, Venkatasubramani, Satish Babu, Govindan, J. Nagaraj and B.K. Tyagi, 2013. Multi-vector involvement in the dengue transmission in the Thiruvananthapuram district of Kerala. Journal of Basic & Applied Biology 7 (4): 11-14.

492. Mariappan, T. And B.K. Tyagi, 2013. Dengue: Can we do source reduction of vectors? Journal of Basic & Applied Biology 7 (4): 92-97.

493. Thenmozhi, V. and B.K. Tyagi, 2013. Role of microbiota in vectorial ability of mosquitoes to disease transmission. Natl. Conf. New Opportunities & Challenges in Microbial Research. Bharathidasan Univ., Tricirappalli, TN (Sep. 5-6, 2013), pp. 9-10.

494. Kalimuthu and B.K. Tyagi, 2013. Microbes in the control of disease transmitting vectors. Natl. Conf. New Opportunities & Challenges in Microbial Research. Bharathidasan Univ., Tricirappalli, TN (Sep. 5-6, 2013), p. 59.

495. Suganya G., J. Asnet Mary, R. Shenbagarathai, R. Paramasivan, V. Thenmozhi, B.K. Tyagi. Genetic profiling of Culex quiquefasciatus in southern India by Single Strand Conformation Polymorphism (SSCP) analysis using mtDNA markers. Subm. to Journal of Vector Borne Diseases. (IF 1.177).

496. Tyagi, B.K. S. Karthigai Selvi, C. Vidya, N.K. Arora, Deoki Nandan, Yara A. Halasa, Jhanshi Charles, C. Mohan Dhanalaxmi, T. Mariappan, P. Philip Samuel, R. Paramasivan, Mukul Gaba, Poovazhagi Varadarajan, T. Porkaipandian, Donald S. Shepard Dengue in India: The Madurai case study and national implications. ASTMH 62nd Ann. Meet., Washington (Nov. 13-17, 2013) p.224.

497. Tyagi B.K., Vijay Veer, S. Chatterjee, 2013. Prof. Dr. Devinder Singh – A forensic entomologist & Academic Statesman of National Significance. 7th Conf. Med. Arthropodol., Tezpur, pp. 24

498. Vijay veer, B.K. Tyagi, Anil Prakash, S. Chatterjee 2013. Dr. Jagadish Mahanta, MD – A biomedical scientist and academic statesman of Global significance in Medico-entomological sciences. 7th Conf. Med. Arthropodol., Tezpur, pp. 42.

499. T. Mariappan and B.K. Tyagi. Integrated vector management (IVM) in vector borne disease control present status in India. 7th Conf. Med. Arthropod., Tezpur, p.21.

500. P. Philip Samuel and B.K. Tyagi. Host feeding pattern of Japanese Encephalitis vector mosquitoes Culex vishnui subgroup with emphasis on disease transmission in India. 7th Conf. Med. Arthropod., Tezpur, p.23-24.

501. B.K. Tyagi. Public Health importance of arthropods. 7th Conf. Med. Arthropod., Tezpur, p.1-3.

502. Thenmozhi, V and B.K. Tyagi. Multi-vector involvement in Japanese Encephalitis virus transmission in Tirunelveli District, Tamil Nadu. 7th Conf. Med. Arthropod., Tezpur, p.22.

2014

503. Sugumar, S., S.K. Clarke, M.J. Nirmala, B.K. Tyagi, A. Mukherjee1 and N. Chandrasekaran, 2014. Nanoemulsion of eucalyptus oil and its larvicidal activity against Culex quinquefasciatus. Bulletin of Entomological Research, Page 1 of 10 doi:10.1017/S0007485313000710.

504. K. Rajesh, D. Dhanasekaran, B. K. Tyagi, 2013. Survey of container breeding mosquito larvae (Dengue vector) in Tiruchirappalli district, Tamil Nadu, India. Journal of Entomology and Zoology Studies 1 (6): 88-91.

505. Kandhasamy Rajesh, Dharumadurai Dhanasekaran and Brij Kishore Tyagi Dhanasekharan, 2014. Mosquito survey and larvicidal activity of actinobacterial

isolates against Culex larvae (Diptera: Culicidae). Journal of the Saudi Society of Agricultural Sciences (In press; Received 25 April 2013; accepted 2 August 2013).

506. T. Balaji, V. Thenmozhi, K. Venkatasubramani, T. Mariappan and B.K. Tyagi. Study on host feeding preferences of Japanese Encephalitis vectors in West Bengal by Agarose Gel Diffusion Technology. Abstr. Pap. Brain Storming Conference on JE, Madurai. P. 55

507. G. Baskaran, A. Munirathinam, R. Govindarajaran, R. Krsihanmoorthi, A. Venkatesh and B.K. Tyagi. Mosquitoes of playing a dual role in transmission of vector-borne pathogens. Abstr. Pap. Brain Storming Conference on JE, Madurai. P. 57.

508. J. Nagaraj, P. Philip Samuel, V. Thenmozhi, D. Ramesh, T. Mariappan, A. Venkatesh, S. Karthigai Selvi, R. Govindarajan and B.K. Tyagi. Survival rate of Culex Tritaeniorhynchus vector mosquito for Japanese Encephalitis in Cuddalore District, Tamil Nadu (2010-2012). Abstr. Pap. Brain Storming Conference on JE, Madurai. P. 63.

509. P. Philip Samuel and B.K. Tyagi. Maintenance mechanisms for the arboviruses in nature. Abstr. Pap. Brain Storming Conference on JE, Madurai. P. 65.

510. V. Thenmozhi, T. Balaji, G. Baskaran, K. Venkatasubramani, K.J. Dhananjeyan, V. Rajamannar, A. Selvam and B.K. Tyagi. Involvement of Non-culcine mosquitoes in transmitting Japanese Encephalitis virus in Tirunelveli District, South India. Abstr. Pap. Brain Storming Conference on JE, Madurai. P. 69.

511. V. Thenmozhi, T. Balaji, K. Venkatasubramani and B.K. Tyagi. Seroprevalence studies on Japanese Encephalitis in Tirunelveli District, Tamil Nadu. Abstr. Pap. Brain Storming Conference on JE, Madurai. P. 71.

512. A. Veerapathiran, N. Arunachalam, U.S.N. Murty, D. Narahari, R. Paramasivan, V. Kodangialagan and B.K. Tyagi. Studies on the role of difference species. Abstr. Pap. Brain Storming Conference on JE, Madurai. P. 73.

513. N. Arunachalam, U.S.No. Murty, D. Narahari, R. Paramasivan, M. Srinivasa Rao, V. Kodangialagan, A. Veerapathiran and B.K. Tyagi. Ecology and Japanese Encephaltis virus infection in vector mosquitoes in peri urban areas in Kurnool District, Andhra Pradesh, South India. Abstr. Pap. Brain Storming Conference on JE, Madurai. P. 75.

514. R. Govindarajan, R. Krishnamoorthy, G. Baskaran, T. Mariappan, P. Philip Samuel, R. Krishnamoorthi and B.K. Tyagi. Japanese Encephalitis vectors survillence in Jalpaiguri and Darjeeling Districts, North West Bengal, India. Abstr. Pap. Brain Storming Conference on JE, Madurai. P. 79.

515. R. Krishnamoorthi, T. Mariappan, R. Paramasivan, P. Philip Samuel, R. Krishnamoorthy, G. Baskaran, R. Govindarajan and B.K. Tyagi. Japanese Encephalitis vector prevalence during outbreak investigation in Northern West Bengal. Abstr. Pap. Brain Storming Conference on JE, Madurai. P. 81.

516. T. Mariappan and B.K. Tyagi. Japanese Encephalitis (JE) Outbreak Integrated vector management (IVM) contrivance is a practical solution in India. Abstr. Pap. Brain Storming Conference on JE, Madurai. P. 83.

517. T. Mariappan, V. Thenmozhi and B.K. Tyagi. New paradigms of Japanese Encephalitis (JE) in West Bengal outbreak with emphasis on changing transmission dynamics with new vectors. Abstr. Pap. Brain Storming Conference on JE, Madurai. P. 85.

518. V. Thenmozhi and B.K. Tyagi. Importance of Japanese Encephalitis surveillance. Abstr. Pap. Brain Storming Conference on JE, Madurai. P. 93.

519. V. Rajamannar, V. Thenmozhi, P. Philip Samuel and B.K. Tyagi. Bibliometrics on Culex Tritaeniorhynchus – The primary vector of Japanese Encephalitis. Abstr. Pap. Brain Storming Conference on JE, Madurai. P. 103.

520. M. Kalimuthu, R. Selvaraj Pandian and B.K. Tyagi. Predatory potential of Diplonychus Rusticus against mosquito Larvae of Culex tritaeriorhynchs. Abstr. Pap. Brain Storming Conference on JE, Madurai. P. 125.

521. T. Mariappan, A. Selvam, V. kodangialagan, M. Muniaraj and B.K. Tyagi. Susceptibility status of Japanese Encephalitis (JE) vector Cx. Tritaeniorhynchus against pesticides in Public health. Abstr. Pap. Brain Storming Conference on JE, Madurai. P. 127.

522. G. Baskaran, A. Munirathinam, R. Govindarajan, R. Krishnamoorthi, A. Venkatesh and B.K. Tyagi. Morphological differences between culex gelidus and culex whitmorei Japanese encephalitis vector mosquitoes. Abstr. Pap. Brain Storming Conference on JE, Madurai. P. 155.

523. A. Munirathinam, R. Krishnamoorthi, R. Govindarajan, A. Veerapathiran and B.K. Tyagi. Japanese encephalitis vector abundance in Western Ghats, Tamil Nadu, South India. Abstr. Pap. Brain Storming Conference on JE, Madurai. P. 163.

524. D. Ramesh, M. Muniaraj, P. Philip Samuel, V. Thenmozhi and B.K. Tyagi. Evaluate Japanese encephalitis vectors diversity during pre and post-monsoon seasons in Thanjavur District. Abstr. Pap. Brain Storming Conference on JE, Madurai. P. 167.

525. A. Veerapathiran, A. Munirathinam, R. Krishnamoorthi and B.K. Tyagi. Japanese Encephalitis vectors recorded in Srivilliputur hill ranges of Western Ghats, Tamil Nadu, South India. Abstr. Pap. Brain Storming Conference on JE, Madurai. P. 171.

526. A. Venkatesh, A. Munirathinam, G. Baskaran, R. Govindarajan, R. Krishnamoorthi, J. Nagaraj and B.K. Tyagi. A simple larval identification key for JE vectors, Mansonia (Mansonioides) Uniformis (Theobald) and Mansonia (Mansoniodies) Annulifera (Theobald) for Kerala State, South India. Abstr. Pap. Brain Storming Conference on JE, Madurai. P. 173.

527. S. Victor Jerald Leo, P. Philip Samuel, R. Selvaraj Pandian and B.K. Tyagi. Species composition of mosquitoes with special reference to Japanese Encephalitis in some selected areas in Madurai, Tamil Nadu, India. Abstr. Pap. Brain Storming Conference on JE, Madurai. P. 175.

528. N. Kamadhasan, T. Sharmila Rani, S. Chandrasekaran and B.K. Tyagi. Impact of physic-chemical characteristics of water on Japanese encephalitis mosquito vectors population at various ecosystem along the vaigai river of Tamil Nadu, South India. Abstr. Pap. Brain Storming Conference on JE, Madurai. P. 179.

529. P. Philip Samuel and B.K. Tyagi. Zooprophylactic role – A natural mechanism to reduce the incidence of JE. Abstr. Pap. Brain Storming Conference on JE, Madurai. P. 181.

530. P. Philip Samuel and B.K. Tyagi. Bionomics of Japanese Encephalitis vectors. Abstr. Pap. Brain Storming Conference on JE, Madurai. P. 183.

531. D. Ramesh, M. Muniaraj, P. Philip Samuel, V. Thenmozhi and B.K. Tyagi. Ascertain the Japanese encephalitis vector abundance and dynamics by agricultural pattern in cuddalore District. Abstr. Pap. Brain Storming Conference on JE, Madurai. P. 185.

532. D. Ramesh, M. Muniaraj, P. Philip Samuel, V. Thenmozhi, A. Venkatesh, S. Vijayakumar and B.K. Tyagi. Blood feeding behavior of mosquitoes in Japanese

Encephalitis endemic and non-endemic areas. Abstr. Pap. Brain Storming Conference on JE, Madurai. P. 187.

533. V. Thenmozhi, T. Mariappan and B.K. Tyagi. Role of culex quinquefasciatus say in Japanese encephalitis virus transmission in West Bengal. Abstr. Pap. Brain Storming Conference on JE, Madurai. P. 195.

534. V. Ashok kumar, M. Kalimuthu, N. Arunachalam and B.K. Tyagi. Studies on social and cultural aspects on Japanese Encephalitis (JE) disease may strengthen JE prevention and control program in JE endemic areas. Abstr. Pap. Brain Storming Conference on JE, Madurai. P. 199.

535. P. Philip Samuel and B.K. Tyagi. Dengue and its control in India. Proc. Brain Storming Conference on Dengue, Madurai, p.41-50.

536. R. Krishnamoorthy, R. Govindarajan, R. Krishnamoorthi, G. Baskaran, T. Mariappan, P. Philip Samuel, J. Nagaraj and B.K. Tyagi. Proc. Brain Storming Conference on Dengue, Madurai, p.51-56.

537. Vijayalakshmi Ghosh, S. Saranya, Brij Kishore Tyagi, K. Gopalrathinam, Amitava Mukherjee and Natarajan Chandrasekaran. Nanoemulsions as potent Aedes aegypti lavicide: A novel strategy for dengue vector control. Proc. Brain Storming Conference on Dengue, Madurai, p.57-63.

538. S. Saranya, Vijayalakshmi Ghosh, N. Chandrasekaran, K. Gopalrathinam, B.K. Tyagi and Amitava Mukherjee. Nanoencapsulation of Eucalypts oil and its potent larvicidal activity against Aedes aegypti. Proc. Brain Storming Conference on Dengue, Madurai, p.64-71.

539. V. Thenmozhi and B.K. Tyai. The pathogenesis of Dengue. Proc. Brain Storming Conference on Dengue, Madurai, p.80-86.

540. T. Balaji, V. Thenmozhi, T. Mariappan, J. Nagaraj, G. Manikandan, K. Venkatasubramani and B.K. Tyagi. The predatory efficacy of Toxorhynchites splendents (wiedemann) for biological control of different immature vector mosquitoes in laboratory condition. Proc. Brain Storming Conference on Dengue, Madurai, p.194-199.

541. T. Mariappan, R. Sathish Babu, V. Kodangialagan, K. Venkatasubramani, K.J. Dhananjeyan, A. Selvam, A. Venkatesh and B.K. Tyagi. Status of Entomological indices during dengue outbreak (2012) in Dindigul District, Tamil Nadu. Proc. Brain Storming Conference on Dengue, Madurai, p.200-208.

542. J. Nagaraj, P. Philip Samuel, T. Mariappan, S. Karthigai Selvi, R. Asthish Babu, R. Govindarajan, A. Venkatesh and B.K. Tyagi. Role of biotic and abiotic factors influencing dengue incidence in Madurai District, Tamil Nadu. Proc. Brain Storming Conference on Dengue, Madurai, p.239-246.

543. Sunish I.P. , M. Kalimuthu, R. Rajendran, A. Munirathinam, V. Ashok Kumar, J. Nagaraj, B.K. Tyagi, 2014. Decline in lymphatic filariasis transmission with annual mass drug administration using DEC with and without albendazole over a 10 year per and priority areas for research on Japanese encephalitis and other acute enciod in India. Parasitology International 64(1): (DOI: 10.1016/j.parint.2014.08.005).

544. Tyagi, B.K., V. Thenmozhi and S. Karthigai Seli, 2014. Transmision dynamics of Japanese encephalitis, with emphasis on gaps in understanding and priority areas on Japanese encephalitis and other acute encephalitis syndromes in India.J. Commun. Dis. 1(1):1-10.

545. Tyagi, B.K., 2014. Chikungunya: a new global challenge, with particular reference to India. (In: Major Tropical Disease: Public Health Perspective. Ashwini Kumar, Savio Rodrigues and Amit Dias (Eds.), Broadway Publishing House, Panjim, Goa, pp. 272-289.

546. Shepard Donald S., Yara A. Halasa, Brij Kishore Tyagi, S. Vivek Adhish, Deoki Nandan, K. S. Karthiga, Vidya Chellaswamy, Mukul Gaba, Narendra K. Arora,* and the INCLEN Study Group†. 2014, Economic and Disease Burden of Dengue Illness in India. Am. J. Trop. Med. & Hyg. 91(6): 1235–1242.

547. Tyagi, B.K., 2014. Extended distribution and variation in morphological features of Disparoneura (Chloroneura) quadrimaculata (Rambur, 1842) (Odonata: Zygoptera; Protoneuridae) in the Mt. Abu ranges of Southern Rajasthan, India. HALTERES 5: 69-72.

548. Munirathinam, A., R. Krishnamoorthi, G. Baskaran, R. Govindarajan, A. Venkatesh and B.K. Tyagi, 2014. Mosquito species biodiversity in Phytotelmata from Western Ghats, south India. HALTERES 5: 56-63.

549. Tyagi, B.K., A.Munirathinam, and A. Venkatesh, 2015. A catalogue of Indian mosuitoes. Intl. J. Mosq. Res. 2(2): 50-97.

550. Chandrasekar, R., B.K. Tyagi, Gui, Z.Z. and Reeck, G.R., 2014. Short views on insect biochemistry and molecular biology. Vol. 1. International Book Mission, South India. pp. 1-370

551. Chandrasekar, R., B.K. Tyagi, Gui, Z.Z. and Reeck, G.R., 2014. Short views on insect biochemistry and molecular biology. Vol. 2. International Book Mission, South India. pp. 371-716.

552. Tyagi, B.K., 2014. Progress on the OECD Consensus document on the biology of Aedes aegypti. 2nd Annual South Asia Biosafety Conference, Colombo, Sri Lanka (Sep 15-16, 2014), p. 26.

2015

553. Mariappan, T., V. Thenmozhi, P. Udayakumar, V. Bhavaniumadevi abd B,K, Tyagi, 2015. An observation on breeding behaviour of three different vector species (Aedes aegypti Linnaeus 1762, Anopheles stephensi Liston 1901 and Culex quinquefasciatus Say 1823) in wells in the coastal region of Ramanathapuram district, Tamil Nadu, India. International Journal of Mosquito Research 2015; 2 (2): 42-44.

554. Sunish, I.P., R. Rajendran , A. Munirathinam , M. Kalimuthu , V. Ashok Kumar , J. Nagaraj & B.K. Tyagi, 2015. Impact on prevalence of intestinal helminth infection in school children administered with seven annual rounds of diethyl carbamazine (DEC) with albendazole. Indian J Med Res 141: 15-24.

555. Tyagi, B.K., A. Munirathinam, R. Krishnamoorthy, G. Baskaran, R. Govindarajan, R. Krishnamoorthi, T. Mariappan, K.J. Dhananjeyan and A. Venkatesh. 2015. A revision of genus Toxorhynchites Theobald, 1901, in the South-East Asian countries, with description of a new species Toxorhynchites (Toxorhynchites) darjeelingensis from West Bengal, India (Diptera, Culicidae). HALTERES, Volume 6, 13-32.

556. Mariappan, T., A. Selvam, V. Kodangialagan and B.K. Tyagi 2015. Efficacy of different insecticides against Culex quinquefasciatus say 1823, vector of bancroftian filariasis in Madurai. Abstr. Pap. 3rd National Conference on Lymphatic Filariasis: Elimination 2015 – Mission possible, Madurai. p. 45.

557. Philip Samuel, P., N. Arunachalam and B.K. Tyagi 2015. Significance of blood meal identification of filariasis vectors in Kerala. Abstr. Pap. 3rd National Conference on Lymphatic Filariasis: Elimination 2015 – Mission possible, Madurai. p. 51.

558. Prabhakar Mishra, A.P.B. Balaji, B.K. Tyagi, Amitava Mukherjee and Natarajan Chandrasekaran 2015. Bio-based oil nanoemulsions: An effective tool for controlling C. quinqufasciatus larval population. Abstr. Pap. 3rd National

Conference on Lymphatic Filariasis: Elimination 2015 – Mission possible, Madurai. p. 55.

559. Thenmozhi, V., T. Mariappan and B.K. Tyagi 2015. Dual role of Culex quinquefasciatus say in the transmission of Lymphatic Filariasis and Japanese Encephalitis virus in India. Abstr. Pap. 3rd National Conference on Lymphatic Filariasis: Elimination 2015 – Mission possible, Madurai. p. 79.

560. Venkatesh, A., A. Munirathinam, G. Baskaran, R. Govindarajan, R. Krishnamoorthi and B.K. Tyagi 2015. A simple larval idenfication key for LF vectors, Culex (Culex) quinquefasciatus say, mansonia (mansonioides) uniformis (Theobald), mansonia (Mansonioides) Annulifera (Theobald) and Aedes Niveus from India. Abstr. Pap. 3rd National Conference on Lymphatic Filariasis: Elimination 2015 – Mission possible, Madurai. p. 81.

561. Kalimuthu, M., I.P. Sunish, N. Arunachalam, V. Ashokkumar, A. Munirathinam and B.K. Tyagi 2015. Impact of MDA with integrated vector control on Lymphatic Filariasis elimination in endemic villages of South India. Abstr. Pap. 3rd National Conference on Lymphatic Filariasis: Elimination 2015 – Mission possible, Madurai. p. 101.

562. Rajamannar, V., N. Arunachalam, T. Mariappan, M. Kalimuth and B.K. Tyagi 2015. Lymphatic Filariasis Research in India – A literature survey. Abstr. Pap. 3rd National Conference on Lymphatic Filariasis: Elimination 2015 – Mission possible, Madurai. p. 107.

563. Krishnamoorthi, R., A. Munirathinam, R. Paramasivan, R. Govindarajan, A. Veerapathiran, A. Venkatesh and B.K. Tyagi 2015. Culex quinquefasicatus say, 1823 vectors prevalence in tribal settlement pockets of Theni & Bodi Hills in Western Ghats. Abstr. Pap. 3rd National Conference on Lymphatic Filariasis: Elimination 2015 – Mission possible, Madurai. p. 115.

564. Balaji, T., V. Thenmozhi, K. Venkatasubramani, V. Rajamannar and B.K. Tyagi 2015. Diagnostic Tools for Lymphatic Filariasis. Abstr. Pap. 3rd National Conference on Lymphatic Filariasis: Elimination 2015 – Mission possible, Madurai. p. 119.

565. Paramaisvan, R., K.J. Dhananjeyan, A. Veerapathiran, R. Krishnamoorthi and B.K. Tyagi 2015. DNA barcoding of Culex quinquefasicatus mosquito, the vector of Lymphatic Filariasis. Abstr. Pap. 3rd National Conference on Lymphatic Filariasis: Elimination 2015 – Mission possible, Madurai. p. 127.

566. Tyagi, B.K. [Ed.], 2015. Abstracts. 3rd National Conference on Lymphatic Filariasis: Elimination 2015 – Mission possible; 5-6 June, 2015, Madurai, pp. 136.

567. Venkatesh A. and B.K. Tyagi, 2015. Bradinopyga geminata (Anisoptera: Libellulidae) as a predator of Aedes aegypti immatures (Diptera: Culicidae). International Journal of Mosquito Research 2 (2): 98-105.

568. World Health Organization [First Draft and Compilation by B.K. Tyagi], 2015. Biosafety for human health and the environment in the context of the potential use of genetically modified mosquitoes (GMMs) - A tool for biosafety training based on courses in Africa, Asia and Latin America, 2008–2011. World Health Organization, pp. 242.

569. Tyagi, B.K., 2015. Chapter 1- Introduction (In: Biosafety for human health and the environment in the context of the potential use of genetically modified mosquitoes (GMMs) - A tool for biosafety training based on courses in Africa, Asia and Latin America, 2008–2011). World Health Organization, pp. 19-34.

570. Tyagi, B.K., 2015. Chapter 2 - Overview of arthropod-borne diseases Introduction (In: Biosafety for human health and the environment in the context of the potential use of genetically modified mosquitoes (GMMs) - A tool for

biosafety training based on courses in Africa, Asia and Latin America, 2008–2011). World Health Organization, pp. 35-62.

571. Tyagi, B.K., 2015. Chapter 3. The GMO project (In: Biosafety for human health and the environment in the context of the potential use of genetically modified mosquitoes (GMMs) - A tool for biosafety training based on courses in Africa, Asia and Latin America, 2008–2011). World Health Organization, pp. 63-80.

572. Kandhasamy Rajesh, Dharumadurai Dhanasekaran and Brij Kishore Tyagi, 2015. Mosquito survey and larvicidal activity of actinobacterial isolates against Culex larvae (Diptera: Culicidae). Journal of the Saudi Society of Agricultural Sciences 14: 116–122.

• • •

3. Dr. Vijay Veer

Dr. Vijay Veer, alongside Dr. B.K. Tyagi, Prof. A.P. Dash, Dr. S. Pattanayak, Dr. R. Reuben and Dr. S. Elango, while prof. T.N. Ananthakrisghnan delivering the inaugural address on the occasion 1st Conference of Medical Arthropodology (Madurai; Sep. 1-2, 2007).

"Do you wish to be great?" asks St. Augustine.
"Then begin by being little.
Do you desire to construct a vast and lofty fabric?
Think first about the foundations of humanity.
The higher your structure is to be,
The deeper must be its foundation.
Modest humility is beauty's crown."

The life of Dr. Vijay Veer, an utterly indefatigable entomological genius *nonpareil*, truly reflects the spirit radiated from this famous saying about the character called Duty (Smiles, 1880*). Dr. Vijay Veer, who epitomizes the principles of *karmayoga* as preached by Srimadbhagwadgita, *"Karmanye vadhikaraste ma phaleshu kadachana..."* , is a glorious example of perseverance, focused attention and application in our own lifetime.

"The primal duties shine aloft like stars,
And charities that soothe, and heal, and bless,
*Are scattered at the feet of men like flowers."**

Honest and upright performance of individual duty constitutes the glory of man. It readily and compulsively inspires the others to emulate, and he who does the self-devotedness – often *incognito* – never pauses to think of himself in performing them. Such is the life of this scientist who keeps "work within heart and God over head." Dr. Vijay Veer personifies utmost simplicity, yet keeping incredibly systematic, in all walks of life, be it professional or public. He never paints upon him any outwardly glitter of fame or recognition and remain firmly grounded to the hilt even when riding high on cloud nine of success! Someone had once said of the world's all time greatest bodybuilding and seven times 'Olympia' champion, Arnold Schwarzeneggar, *"The beauty of this man is, he looks ordinary and yet he is so extraordinary."* Dr. Vijay Veer is a genius of the same genre - a true champion of biological sciences, bestowed with exceptionally diverse and deep knowledge of insect ecology and taxonomy, though thrips (Order Thysanoptera), dermestids (Coleoptera) and tabanids (Diptera) have been his greatest forte.

Born on January 1st, 1956 in a tiny village near Dhampur in north Uttar Pradesh, Dr Vijay Veer's life has been a meandering journey full of ups and downs which gave him a great sense of achieving something high and rich in future on his own. As a village urchin, he grew up under strict watchful solicitude of his well-to-do farmer parents in close association with nature which inherited him a strong sense of resilience and perseverance at a tender age of ten as he had got to walk for 12 km every day to attend his middle school classes. He passed his Intermediate from Raja Ranjit Singh Memorial College, Dhampur and graduated from the Hindu College, Moradabad in Uttar Pradesh. After completing his Masters from the Government College, Kotdwara (Uttarakhand) he moved to Dehradun only to meet his *alma mater*, the DAV Post Graduate College!

During those days, in 1970s, there used to be very few government research fellowship programmes, and, as a result, most research scholars had to undertake their PhD research studies on their own which in every imagination was a daunting task. As in vogue in those times, young Vijay Veer came to Dehradun with a recommendation from his Professor in Kotdwara to try out his luck if he could be accepted as a research scholar in the Department of Zoology of the famous DAV College which was undergoing an internal management turmoil, with one Principal recognized by the Management headquartered in Kanpur and another fielded by the college teaching faculty. The Department of Zoology was a highly recognized seat of commendable research activities commissioned by as

---

* *Smiles, Samuel, 1880. Duty. Wilco Publishing House, Mumbai, pp. 341.*

much celebrated and distinguished professor scientists as Prof. M.B. Lal, Prof. S.K. Sangal and Prof. S.K. Kulshrestha. The young Vijay Veer was taken as a research student by Prof. Kulshrestha. The big question before them, however, was as to what research problem should be chosen for Vijay Veer who was completely novice to research and there was no sign of any funding to back up the programme! This precariousness gave them some time to engage in pondering over the resources and the line of research work to be pursued. Enough time indeed for the young Vijay Veer to befriend with other research scholars working both in Zoology and other departments of the college, and was especially introduced by his guide to one particular research fellow colleague, Brij Kishore Tyagi, already two years in research pursuing his doctorate in odonate cytotaxonomy under the guidance of Prof. S.K. Sangal. This companionship was destined to last forever, even after forty years and having been in different research organizations, and research disciplines!

Days spent in the DAV College founded a solid base for Dr. Vijay Veer. The school of difficulty is the best school of moral discipline. When difficulties have to be encountered, they must be met with courage and cheerfulness. Did not Aristotle, the great Greek teacher and philosopher, say that happiness is not so much in our objects as in our energies? Grappling with difficulties is the surest way of overcoming them. The determination to realize an object is the moral conviction that we can and will accomplish it. Our wits are sharpened by our necessity, and the individual man stands forth to meet and overcome difficulties which stand in his way. Young and novice Vijay Veer, too, faced a string of difficulties but he surmounted them bravely. He mastered the art of articulation and writing good research projects. As if providentially set, at that point of time around mid-1976, Dr. T.N. Ananthakrishnan, a professor from Loyola college, Chennai who was looked with great admiration and perhaps with a sense of amazement for his extraordinary oratory skills, took over the Directorship at the Zoological Survey of India, at Kolkata. Earlier the young Vijay Veer had got a chance to visit him in Loyola College to see his collection of thrips. So, when Prof. Ananthakrishnan visited Dehradun on his maiden visit to the valley town, the first thing he did was to call Vijay Veer and gave him the much needed advice and encouragement to write a research project for funding by the Department of Science & Technology where Prof. Ananthakrishnan apparently had a say in deciding grant of funds to genuine research projects! This opportunity changed the future course of pursuit of research by Vijay Veer, and he for next 20 years was virtually submerged in the thrips bio-ecology, taxonomy biodiversity. From this point onwards he did not look back and persevered hard to make his own path, however meandering and difficult it was. He was awarded PhD degree in 1980 under the newly carved out Garhwal University.

'God helps those,
who help themselves."

Dr. Vijay Veer was mentored and guided by some of the world's most celebrated entomologists like Prof. T.N. Ananthakrishnan, Loyola College, Chennai, Prof. J.S. Bhatti, Delhi University, Delhi, Prof. L.A. Mound, Natural History Museum, London, Prof. T. Lewis, Dr. D.G.H. Halstead, Slough (UK),

Dr G.S. Robinson, London, Prof. R.S. Beal Jr, California, Dr. R. Kzur Strassen, Germany, and Dr. I. Kudo, Japan. He met some of them and other eminent scientists such as Prof. M.F Claridge, Prof. Mark Jervis and Dr. David Pinniger during his visit to Cardiff, UK and had fruitful discussions with all of them pertaining to thrips and other insects' eco-biology. There in UK he befriended several entomologists like Mr. David Edwards, a freelance entomologist, Dr. John Deeming of Wales Museum and Dr. Hassan Ali Dawaha of Cardiff University, with whom he made field trips on insect collection in and around Wales State. Dr. Vijay Veer is an affable personality as he continues to foster intense bonds of camaraderie with all these great old friends by writing about his observations revolving around thrips that he encounters during his frequent peregrinations across the country. Needless to over-emphasize, therefore, that when Dr. Vijay Veer, co-editor with Dr. B.K. Tyagi, invited Dr. Mound to contribute an article for their upcoming book in the commemoration of Prof. Ananthakrishnan, he immediately accepted the offer and wrote a highly thought-provoking taxonomic paper, at a mature age of 80 years!.

Even though Dr. Vijay Veer explored many insect orders in his long career with the Defense Organization, thrips could never be sidetracked simply because these tiny insects offered him a vast opportunity to learn insect ecology and biology in the most weird manner (sexual behaviour including haplo- and diploploidy, parthenogenesis, sex selection, gall formation, plant tissue destruction, flight range, plant disease transmission), all that with only one mandible developed to full scale. Following truly the principles of collecting, preserving and identifying these tiny, yet agriculturally significant pests, Dr. Vijay Veer created, along with his equally dedicated and bold and elegant fiancé, Neena Chauhan, with whom he later tied up the knuptial knot, possibly one of the largest collections of slides of mounted specimens of the thrips from all over India, next only to those of Prof. T.N. Ananthakrishnan and Dr. J.S. Bhatti. Many of these insects still remain to be fully described taxonomically, though these may form a versatile source of knowledge for any young thysanopterist.

Dr. Vijay Veer did not restrict himself to the thrips research alone as he sharpened his explorative inquisitiveness by delving into fungi-insect (dried dragonfly specimens) relationship, white grubs (scarabeids), dermestids, tineid moths, tabanids, mosquitoes etc. After completing his PhD dissertation he joined the world famous Forest Research Institute & Colleges, Dehra Dun (now rechristened as the Forest Research Institute and also the headquarters of the deemed university) where he worked with Dr. P.K. Sen-Sarma, Director, Biological Research, on the toxicity of insecticides against forest insect pests, larval taxonomy of white grubs and on diagnosis of nursery and plantation pests, and took part in several insect pests surveys in the Himalayan and Assam forest expeditions, both as a team leader and sometimes as a member of the team along with Dr. S.C. Mishra. He reported several new pests of poplar tree from WIMCO plantations in West Bengal where the pest caused huge financial loss to the cultivators. His study had shown the unpredictable damage incurred by an exotic plant species when introduced in a new and inaccessible area without proper scientific validation. For his long association with Dr. Sen-Sarma, he, along with Dr. B.K. Tyagi and Dr. Shri Prakash, wrote a commemorative volume titled, "*Pests of Forest Importance and their Management*" (Tyagi, Vijay Veer & Shri Prakash, 2009), as a tribute to this truly great genius of the forest entomology of modern times, possibly one of the best books on forest pests. Dr. Vijay Veer maintained his humility throughout his association with such great leaders and his work on forest insect pests is highly acclaimed and even translated to other languages.

Science and teaching are like two wheels of a bicycle; one is as indispensable as another, and both only can give the bicycle a movement – the life. Both are blended in one inseparable entity. Dr. Vijay Veer is an inborn teacher – quiet, patient and thorough. Although during his post-doctoral period he had been commissioned to teach forest entomology to the Indian Forest Service (IFS) incumbents, State Forest Service (SFS) probationers and the Range officers at the FRI & Colleges, Dehra Dun, yet in fact he had by devine choice put to some teaching courses, possibly against his wish, much earlier in life encountering the most impatient, unruly and inconsolable souls on earth – the geniuses of kindergarten! In the movie *Kindergarten* the cop disguised as a teacher to take classes for the kids, played by the macho Arnold Schwarzeneggar, prepares himself post-breakfast to leave his hotel suite and asked slyly his room mate if she thought he was ready hook, line and sinker? The room-mate gave a casual look at him and remarked half sleepishly, "Don't forget to take your gun!" Dr. Vijay Veer was also going to meet his first students, tiny though, in the kindengarten class at Guru Nanak Academy, Raiwala, Dehra Dun where actually Dr. Tyagi was originally employed on a monthly salary of Rs 225.-. Since, however, on one particular occasion, Dr. Tyagi was to urgently go out of town for a week he made a temporary arrangement of a replacement by requesting Dr. Vijay Veer to teach in his stead for a week. On my return, Dr. Tyagi was surprised to see that he had virtually become an apple of the kids' eyes. He later told me the trick: "*Make one of them –the burlier and naughtier one, the monitor!*" It worked!

Dr. Vijay Veer taking a class during the 3rd Asian Biosafety Training Course held in CRME, Madurai during 2010 (Course Director Dr. B.K. Tyagi).

On tabanids Dr. Vijay Veer is an authority as much as he has a towering impact on mosquito research, as evinced by two of his recent monographs; one on Tabanidae and another on malaria mosquitoes, *Anopheles*. His forte is taxonomy and there his imprint is quite discernible: 20 new species, and many other treatments. His diversity in scientific publications can be gauged simply by having a look on the quantum of papers published (>90 in reputed journals), in addition to 6 patents, 2 Indian Standards, a monograph on tabanid flies and three books. When still in the midst of his doctoral research, he, along with Dr. B.K. Tyagi, ventured to establish at the Zoology Department, DAV college, Dehradun a research association of research scholars, the Young Scientists' Society, of which he was the Secretary with Dr. B.K. Tyagi as its President. Together they even managed to organize three symposia. To great dismay of the organizers none of the science departments faculties attended the first two meeting, possibly it was considered an affront of some sort, but by the time the third symposium was around to be organized feelers came pouring from the faculties of all the science departments, probably viewing the degree of sincerity with which these meetings were being conducted, and to our utmost pleasure half of the audience was only the faculty members. Dr M.K. Sharma Head of the Chemistry Department, distributed the prizes to various speakers and announced Dr. Tyagi the best research scholar of the year. Dr. Vijay Veer sowed the seeds of focused industry early in life and carefully cultivated the sense of organizational skills. He became the champion of dedicated causes and left no stone unturned in arriving his target.

Dr. Vijay Veer moved to Defence Research & Development Organization (DRDO) in 1984 and got posted at Defence R & D Establishment, Gwalior, Madhya Pradesh where over the years he had successfully completed various projects on the pests of defence importance and their management, namely, pests of woollen textiles, semio-chemicals for rodent control, bedbugs, mosquitoes, etc. as well as diagnosis and taxonomy of dermestids, horse flies and the thrips. In the process he brought a lot of advancement in entomological researches under the aegis of his parental department, the Defense Research & Development Establishment at Gwalior, now a pioneer in vector mosquito control through innovation and invention. Dr. Vijay Veer's dedicated and focused researches have taken his entomology division to new heights fostering both basic and applied research programmes. By now almost three decades in the DRDE, Dr. Vijay Veer has left an indelible imprint in researches pertaining to pests of agriculture, forest, storage, veterinary, medical and forensic importance. All along he has advocated a practical utilization of all kinds of researches, keeping in mind the necessity of defense *jawans* in the frontier and/or difficult areas. The result of this well ingrained philosophy by him has paid repeatedly rich tributes to his entomology team's hard work. A few examples can be worth mentioning here: (i) Successful commercialization of the mosquito repellent, DEPA (N' N' diethyl phenyl acetamide), (ii) Invention of a hormonal control of vector mosquitoes, (iii) Discovery of a new sensillum for chemical cues, (iv) Development of an insecticide impregnated mosquito net, (v) Indigenous development of *Wool Care*, for protection of woolen textiles, (vi) Invention of *Ratox*, for rodent control, and (vii) Production of *Roachtox* &

Roachline, for cockroach control etc. His initiatives truly had a bearing on translational research since he did not cease working with the discoveries and inventions of products, but determinedly followed their possible commercialization for the benefit to all, but the *jawans* in the first place. He has also served as an expert to the 'DRDO Think Tank', better known as a Group for Forecasting and Analysis of Systems & Technologies (GFAST) situated in New Delhi during 2003-06. With all these and other great traits it is quite unlikely that his vast knowledge on defense-related entomological science can be surpassed by many of the extant entomologists in the country. Taxonomic feats, emeralded by over a dozen *species nova* and *genera nova*, is an ample proof of his unmatched genius.

New Genus & species discovered by Dr. Vijay Veer

Phylum: Arthropoda
Class: Insecta

1. Order: Thysanoptera

Genus ...1. *Doonthrips* Bhatti, Vijay Veer & Chauhan, 1994

Species

1. *Ramphothrips santokhi* Kulshrestha & Vijay Veer, 1984

   Kulshrestha SK & Vijay Veer 1984. Two new species of Thysanoptera (Insecta) from India. *Bulletin of Entomology*, India. 25 (1): 33–37.

2. *Aeolothrips moundi* Kulshrestha & Vijay Veer, 1984

   Kulshrestha S.K. & Vijay Veer 1984. Two new species of Thysanoptera (Insecta) from India. *Bulletin of Entomology*, India. 25 (1): 33–37.

3. *Thrips sensarmai* Vijay Veer & Srivastava, 1985

   Vijay Veer & Srivastava A.K. 1985. Two new species of Thripidae (Insecta: Thysanoptera) from India. *Entomon.* (1) 10: 75–80.

4. *Liothrips sangali* Kulshrestha & Vijay Veer, 1990

   Vijay Veer & Kulshrestha S.K.1990. A new species of Liothrips Uzel from Dehradun (India) (Insecta: Thysanoptera: Phlaeothripidae). *Annals of Entomology*. (1) 8: 55–60.

5. *Azaleothrips bhatti* Vijay Veer & Chauhan, 1990

   Vijay Veer & Chauhan, N. 1990. New species of genera Azaleothrips Ananthakrishnan and Gigantothrips Zimmermann (Thysanoptera: Tubulifera: Phlaeothripidae) from India. *Annals of Entomology*. 8 : 65–70

6. *Gigantothrips himalayensis* Vijay Veer & Chauhan, 1990

   Vijay Veer & Chauhan, N. 1990. New species of genera Azaleothrips Ananthakrishnan and Gigantothrips Zimmermann (Thysanoptera: Tubulifera: Phlaeothripidae) from India. *Annals of Entomology*. 8 : 65–70.

7. Pseudodendrothrips kulshresthai Chauhan & Vijay Veer, 1992

Vijay Veer & Chauhan N. 1992. A new species of Pseudodendrothrips Schmutz (Insecta: Terebrantia: Thirpidae) from Dehradun (India). *Zoology (Journal of Pure and Applied Zoology).* (2) 3: 160–172.

8. *Anaphothrips doonensis* Vijay Veer & Chauhan, 1992

Vijay Veer & Chauhan, N. 1992. Anaphothrips Doonensis, a new species from the tussock plant, Typha elephantiana (Family Typhaceae) from Dehrandun (India) (Thysanoptera: Terebrantia: Thripidae). *Annals of Entomology.* (1) 10 : 41–44.

9. *Doonthrips setor* Bhatti, Vijay Veer & Chauhan, 1994

Bhatti JS, Veer V & Chauhan N. 1994. Doonthrips setor, a new genus and species of Thripidae (Insecta: Terebrantia) from Dehradun (India) . *Zoology (Journal of Pure and Applied Zoology).* 4: 177–192

10. *Franklionthrips uttarakhandiensis* Vijay Veer, 2011

Vijay Veer. 2011. A new species of *Franklinothrips* (Thysanoptera: Terebrantia: Aeolothripidae) from Uttarakhand, India. *Annals of Entomology*, 28(2): 1-5.

2. Order: Diptera

11. *Tabanus hisarensis* Vijay Veer, Parasahar & Rao, 1999

Vijay Veer, Parashar, B.D. & Rao, K.M. 1999. Notes on Tabanidae (Diptera) that are surra vectors or pestiferous with description of a new species of *Tabanus* from India. *Oriental Insects*, 33: 247- 266.

3. Order: Coleoptera

12. *Trogoderma halsteadi* Vijay Veer & Rao, 1994

Vijay Veer & Rao, K. M. 1994. A new species of *Trogoderma* (Coleoptera: Dermestidae) found damaging stored silkworm cocoons in India. *J. Stored Products Res.*, 30(4): 283-295.

13. *Thaumaglossa indica* Vijay Veer, 2004

Vijay Veer, Chauhan, Neena & Sudhir Singh. 2004. A new species of *Thaumaglossa* (Coleoptera: Dermestidae) from India. *Oriental Insects*, 38: 389-394.

14. *Anthrenus umbellatarum* vijaii Vijay Veer, 2011

Vijay Veer, 2011. Biodiversity of dermestid beetles (Coleoptera: Dermestidae) from India, with identification key to the adults of Indian genera and description of a new subspecies. Chapter. 26, 495-522. *Entomology: Ecology & Biodiversity.* (Edited by Tyagi, BK & Vijay Veer). Scientific Publishers (India), Jodhpur.

Dr. Vijay Veer had even described a bacterium isolated from the filariasis vector mosquito, *Culex quinquefasciatus* Say.

15. *Chrysobacterium culicis* Kämpfer, Chandel, Prasad, Shouche, Vijay Veer, 2009

Kämpfer, P., Chandel, K., Prasad, G. B. K. S., Shouche, Y.S., Vijay Veer. 2009. *Chryseobacterium culicis* sp. nov., isolated from the midgut of *Culex quinquefasciatus* Say, a mosquito from India. *Intl. Journal of Systematic & Evolutionary Microbiology*, 60: 2387-2391, 2010.

Dr. Vijay Veer has widely travelled inland, tracking his *chef d'oeuvre* – the insects everywhere and striking discoveries of a large number of new species and genera. During his sojourn in Oxford, UK he personally interacted with the world renowned authority on thrips, Dr. L.A. Mound. Dr. Mound has recently shifted to Australia for his love of thrips! Later when Dr. Vijay Veer asked him as to why he had had to leave Britain where he was well established at the British Museum (Natural History), Dr. Mound, with a child-like twinkle in his eyes, had replied that he guessed the Australian thrips were not sufficiently explored and he might strike many new and interesting species there. All great scientists loved to move because they knew that Nature provides the largest niche for scientific investigation and not the confines of their comforting laboratories. Dr. Vijay Veer is one of those rare scientists in the modern age who prefer the sweltering heat of the field research since, without an iota of skepticism in his mind, he knows that it is again the material secured from field which provide basis of all laboratory researches. His recent theory about the orthopteran, *Locusta migratoria migratoria*, in the brazen Sindhu river valley in a rather low-oxygenated tough climatic region of Ladakh is an outstanding exhibition of a long, patient and painstaking effort to comprehend a specific point of their being there in the north-frontier area. The moral of this miniscule tale is that it is the sun-loving and not the shade-loving scientist who will always stand in front to unravel mysteries of the Nature, the greatest of all laboratories. So, when a student of archeology who was spending hours in the library asked his professor as to what made him such a greatly honoured scientist in the discipline, the professor quietly advised his protégé, "Go out of the library, to the Nature!"

Dr. Vijay Veer is recipient of several awards including prestigious DRDO Technology Spin-off awards. He is a Fellow and Member of many prestigious national and international scientific societies and advisory council of scientific institutions and universities. He is referee to many journals, and also edits *Annals of Entomology* from Dehra Dun, India. About half a dozen PhD students are working under his stewardship.

The famous maxim, "*Behind every successful man, there is always a woman*" comes true to this highly successful and respected entomologist, Dr. Vijay Veer, as his betterhalf, Dr. Neena, herself a celebrated forest entomologist, lends him a kind of support and environment he gets nowhere. With all these glories behind him, Dr. Vijay Veer remains foremost a thoroughly grounded good human being, continuing great entomological researches, without thinking of any rewards whatsoever:

'Who comprehends his trust, and to the same
Keeps faithful with a singleness of aim;

And therefore does not stoop, nor lie in wait
For wealth, or honour, or for worldly state;
Whom they must follow, on whose head must fall,
Like showers of manna, if they come at all."

Dr. Vijay Veer retired in December 2015 as the Director, Defense Reseach Laboratory (DRDO), Tezpur, Assam after a glorius and much decorated career of nearly three decades. He continues to mentor research from Dehradun.

I. Scientific Publications, Patents and Copyrights of Dr Vijay Veer

2016

1. Meghvansi MK, Khan MH, Gupta R, Chaudhary KK, Siddiqui S and Vijay Veer. 2016. Vermibiotechnology: Relevance, Challenges and Future Prospects for India. *South Asian Journal of Experimental Biology* 5 (6), 222-228
2. Tyagi V, Sharma AK, Dhiman S, Srivastava AR, Yadav R, Sukumaran D, Agrawal OP and Vijay Veer Malaria vector *Anopheles culicifacies* sibling species differentiation using egg morphometry and morphology. *Parasites & Vectors* (in press).
3. Yadav K, Dhiman S, Rabha B, Goswami D, Saikia PK and Vijay Veer. 2016. Disappearance of *An. minimus* and *An. dirus* from Certain Malaria Endemic Areas of Assam, India. *J Arthropod-Borne Dis.* (in press).
4. Datta S, Budhauliya R, Chatterjee S, Vanlalhmuaka, Vijay Veer and Chakravarty R. 2016. Enhancement of PCR Detection Limit by Single-Tube Restriction Endonuclease-PCR (RE-PCR). *Molecular Diagnosis & Therapy* (in press).
5. Chaturvedi R, Banerjee S, Das B, Chattopadhyay P, Bhattacharjee CR and Vijay Veer. 2016. High nitrate content in the surface water of Balipara, North Brahmaputra river basin, Sonitpur district Assam, India: a multivariate approach. *Current Science,* 110 (in press).
6. Dhiman S, Kumar D, Rabha B, Goswami D, Yadav K, Deka M, **Vijay Veer**, Baruah I . 2016. Typing of *Plasmodium falciparum* DNA from 2 years old Giemsa-stained dried blood spots using nested polymerase chain reaction assay. *Indian Journal of Medical Microbiology,* DOI:10.4103/0255-0857.176841
7. Barua S,Raul PK, Gopalakrishnan R.,Das B, Vanlalhmuaka, **Vijay Veer**. 2016. Carbon dot-silver nanohybrid: a strong tool against *Culex quinquefasciatus*, a common disease vector. *ACS Sustainable Chemistry & Engineering.* DOI: 1021/acssuschemeng.6b00015
8. Rabha AJ, Naglot A, Sharma GD, Gogoi HK, Gupta VK, Shreemali DD, **Vijay Veer**. 2016. Morphological and molecular diversity of endophytic *Colletotrichum gloeosporioides* from tea plant, *Camellia sinensis* (L.) O. Kuntze of Assam, India. *Journal of Genetic Engineering and Biotechnology*, (in press).

2015

9. Tyagi V, Sharma AK, Yadav R, Sukumaran D, Agrawal OP, **Vijay Veer**. 2015. Cytotaxonomic Examination for Identification of Anopheles culicifacies Sibling Species in Central India. *Journal of Mosquito Research*, 5 (20): 1-7.

10. Tyagi V, Yadav R, Sukumaran D, **Vijay Veer**. 2015. Larvicidal activity of invasive weed Prosopis juliflora against mosquito species Anopheles subpictus, Culex quinquefasciatus and Aedes aegypti. *International Journal of Applied Research*, 1(13): 285-288
11. Barua, S, Raul P, Gopalakrishnan R, Das B, Vanlal H, Vijay Veer.2015. Carbon dot-silver nanohybrid: a strong tool against *Culex quinquefasciatus*, a common disease vector. ACS Applied Materials & Interfaces. DOI: 10.1021/acssuschemeng.6b00015
12. Yadav KK, Bora A, Datta S, Chandel K, Gogoi HK. Prasad GBKS and Vijay Veer. 2015. Molecular characterization of midgut microbiota of *Aedes albopictus* and *Aedes aegypti* from Arunachal, Pradesh, India. *Parasites & Vectors* (in press).
13. Molecular characterization of midgut microbiota of Aedes albopictus and Aedes aegypti from Arunachal Pradesh, India
14. Tsering A, Hui, PK, Gogoi BJ, Vijay Veer. 2015. An ethnobotanical survey on wild medicinal plants used by the Monpa tribe of Eastern Himalaya, India. Journal Ethnopharmacology
15. Sharma S, Joshi J, Guruprasad KN, Chatterjee S and Vijay Veer. 2015. Insight adaptation of crop plants towards UV-B: Role of antioxidants. *Intl Journal of Agronomy* (in press).
16. Bhuyan SK, Vairale MG, Arya N, Yadav P, Vijay Veer, Singh L, Yadava PK, Kumar P. 2015. Molecular epidemiology of *Vibrio cholerae* associated with flood in Brahmaputra river valley, Assam, India. *Infection, Genetics and Evolution* Doi:10.1016/j.meegid.2015.11.029
17. Raul PK, Mahanta A, Bora U, Thakur AJ, Vijay Veer. 2015. In water homocoupling of arylboronic acids using nano-rod shaped and reusable copper oxide(II) catalyst at room temperature. *Tetrahedron Letters* 11/2015; DOI:10.1016/j.tetlet.2015.11.004.
18. Khan MH, Meghvansi MK, Gupta R and Vijay Veer. 2015. Elemental stoichiometry indicates predominant influence of potassium and phosphorus limitation on arbuscular mycorrhizal symbiosis in acidic soil at high altitude. *Journal of Plant Physiology*, 189:105-112.
19. Tyagi V, Sharma AK, Yadav R, Adak T, Sukumaran D, Agrawal OP, Vijay Veer 2015. Implication of Random Amplified Polymorphic DNA method for differentiating *Anopheles culicifacies* sibling species. *European Journal of Biotechnology and Bioscience*, 3(8), 47-54.
20. Yadav KK, Chandel K, Bora A, **Vijay Veer**. 2015. Isolation and characterization of *Dietzia maris* from mdigut of Aedes albopictus: A suitable candidate for paratransgenesis. *International Journal of Mosquito Research*, 2 (3); 131-135.
21. Mendki MJ, Singh AP, Tikar SN, Parashar BD, **Vijay Veer**, Shukla SV, Shri Prakash. Repellent activity of N, N-diethylphenylacetamide (DEPA) with essential oils against *Aedes aegypti*, vector of dengue and chikungunya. *International Journal Medical Research*, 2015; 2(3): 17-20
22. Naglot A, Goswami S, Rahman I, Shrimali DD, Yadav KK, Gupta VK, Rabha AJ, Gogoi HK and **Vijay Veer**. 2015. Antagonistic potential of native *Trichoderma viride* strain against potent tea (*Camellia sinensis* (L.) O. Kuntze) fungal pathogens in North East India. *Plant Pathol.* J. 31(3):1-13.

23. Datta D, Budhauliya R, Das B, Chatterjee S, Vanlalhmuaka, & **Vijay Veer**. 2015. Next- generation sequencing in clinical virology: Discovery of new viruses. Virology: *World Journal of Virology*, 4(3): 265-276.
24. Datta S, Budhauliya R, Das B, Chatterjee S, Vairale MG, Bora A, Gogoi HK, **Vijay Veer**. 2015. First report of Tomato leaf curl Bangladesh virus (ToLCBV) infecting *Gomphosemma niveum* plants in Assam, India. *New Disease reports*, **31**, 19. htp://dx.doi.org/:10.5197/j.2044-0588.2015.031.019.
25. Deka B, Mandal M, Venkata M, Kumar S, Doley R, Naglot A, Baruah B, **Vijay Veer**, Namsa ND, 2015. Green synthesis, characterization and antibacterial activity of silver nanoparticles using aqueous leaf extract of *Artemisia annua* L. International Journal Nanomedicine, Dove Medical Press
26. Chattopadhayay P, Sunil Dhiman, Borah S, Rabha B, Chaurasia AK, & **Vijay Veer**. 2015. Essential oil based polymeric patch development and evaluating its repellent activity against mosquitoes. *Acta Tropica*, 147:45-53.
27. Das NG, Dhiman S, Talukdar P.K., Goswami D, Rabha B, Baruah I & **Vijay Veer**. 2015. Role of asymptomatic carriers and weather variables in persistent transmission of malaria in an endemic district of Assam, India. *Infection & Epidemiology*, 5: DOI:10.3402/iee.v5.25442 (IF: 1.177).
28. Kshitij C, Parikh R Y, Mendki M J., Shouche Y S. & **Vijay Veer**. 2015. Isolation and characterization of Vagococcus sp from midgut of *Culex quinquefasciatus* (Say) mosquito. *J Vector Borne Dis* 52, pp. 52–57
29. Singh S, Banerjee S, Chattopadhyay P, Borthakur SK, & **Vijay Veer**. 2015. Deoxynivalenol induces cytotoxicity and genotoxicity in animal primary cell culture. *Toxicology Mechanisms and Methods*. DOI:10.3109/15376516.2015. 1006743. (IF: 1.55).
30. Yadav KK, Bora A, Datta SN, Chandel K, Gogoi HK, Prasad G B K S, & **Vijay Veer**. 2015. Molecular characterization of midgut microbiota of *Aedes albopictus* and *Aedes aegypti* from Arunachal Pradesh, India. Plos One.
31. Dykes CL, Kushwah RBS, Das MK, Sharma SN, Bhatt RM, **Vijay Veer**, Agrawal OP, Adak T. & Singh OP. 2015. Knockdown resistance (kdr) mutations in Indian *Anopheles culicifacies* populations. Parasites & Vectors, 8: 333.
32. Das NG, Dhiman S, Talukdar PK, Rabha B, Goswami D, & **Vijay Veer**. 2015. Synergistic mosquito-repellent activity of *Curcuma longa, Pogostemon heyneanus* and *Zanthoxylum limonella* essential oils. Journal of infection and public health. DOI: http://dx.doi.org/10.1016/j.jiph.2015.02.005.
33. Yadav R, Tikar SN, Sharma AK, Tyagi V, Sukumaran D, Jain A K & **Vijay Veer**. 2015. Screening of some weeds for larvicidal activity against *Aedes albopictus*, a vector of dengue and chikungunya. *J Vector Borne Dis,* 52: 88–94.
34. Raul, PK., Mahanta A, Bora U, Thakur AJ, & **Vijay Veer**. 2013. In water homocoupling of arylboronic acid using nano-rod shaped and reusable copper oxide (II) catalyst at room temperature. *Green Chemistry. J.* (in press).
35. Datta S, Gopalakrishnan R, Chatterjee S & **Vijay Veer**. 2015. Phylogenetic Characterization of a Novel Insect-Specific Flavivirus Detected in a Culex Pool, collected from Assam, India. *Intervirology*, 58:149-154.
36. Datta S., Agarwal M, Chatterjeec S, Gogoi HK, & **Vijay Veer**, Singh L. 2015. Detection of human papillomavirus in women attending Pap cervical screening

camp at a peripheral hospital of North-Eastern India. *Medical Journal Armed Forces India*, 71, 2: 105-208.

37. Yadav K, Rabha B, Dhiman S, & **Vijay Veer.** 2015. Multi-insecticide susceptibility evaluation of dengue vectors *Stegomyia albopicta* and *St. aegypti* in Assam, India. *Parasites & Vectors*, 8:143. Doi: 10.1186/s13071-015-0754-0

2014

38. Purkayastha J, Alam SI, Sengupta N, Nath B, Kumar B, Gogoi HK, Singh L, **Vijay Veer**. 2014. World's Hottest Chilli Bhut Jolokia (*Capsicum assamicum*) Proteome Revealed: Comparative Proteomic Analysis of Differentially Expressed Proteins. *J Proteomics Bioinform* ,7(12): 389-402 (2014)
39. Goyary, D., Chattopadhyay P, Giri S, Aher V, Upadhyay A, & **Vijay Veer**. 2014. Ochratoxin A induces cytotoxicity, DNA damage and apoptosis in rat hepatocyte primary cell culture at nanomolar concentration. *World Mycotoxin Journal* 01/2014; 7(3):379-386. (IF: 2.38).
40. Subham Banerjee, Pronobesh Chattopadhyay, Animesh Ghosh, Aseem Bhatnagar & **Vijay Veer**. 2014. Pharmacokinetic and biodistribution study of eserine and pralidoxime chloride in rabbits following a single application of a transdermal patch. *European Journal of Drug Metabolism and Pharmacokinetics* (2014) (IF: 1.312).
41. Das, NG, Dhiman S, Talukdar, PK, Goswami D, Rabha B, Baruah I, **Vijay Veer.** 2014. Role of asymptomatic carriers and weather variables in persistent transmission of malaria in an endemic district of Assam, India. *Infection Ecology & Epidemiology*, 01/2015; 5:25442.
42. Uniyal A, Tikar SN, Singh R, Shukla SV, Agrawal OP, **Vijay Veer** and Sukumaran D. 2014. Repellent effect, knockdown study and electrophysiological responses of essential oils against *Aedes aegypti. Journal of Entomology and Zoology Studies*, 2 (5): 351-357.
43. Kumar D, Dhiman S, Rabha B, Goswami D. Deka M, Singh L, Baruah I, **Vijay Veer**. 2014. Genetic polymorphism and amino acid sequence variation in *Plasmodium falciparum* GLURP R2 repeat region in Assam, India, at an interval of five years. *Malaria Journal* 2014, 13:450 doi: 10.1186/1475-2875-13-450
44. Yadav K, Sarkar PK & **Vijay Veer**. 2014. Utility of male mosquito hypopygium in species identification *International Journal of Mosquito Research*, 1 (4): 50-54
45. Gogoi J, Nakhuru KS, Chattopadhayay P, Rai AK, & **Vijay Veer**. 2014 Hypertrophic scar formation on application of terpenoid fraction of tuberous root of *Mirabilis jalapa* L. on excision wound model in wistar albino rats *International Scholarly Research Notices* Volume 2014, Article ID 583730, 9 pages http://dx.doi.org/10.1155/2014/583730
46. Gogoi J, Nakhuru KS, Goyary D, Rai AK , Chattopadhayay P & **Vijay Veer**. 2014. Effects of tuberous root of *Mirabilis jalapa* L. on antioxidant systems of wound healing in rat *International Journal of Phytopharmacology*
47. Gupta DK, Chatterjee S, Datta SN, **Vijay Veer** & Walther C. 2014. Role of phosphate fertilizers in heavy metal uptake and detoxification to toxic metals. *Chemosphere,* 108: 134-144.
48. Raul PK, Senapathi S, Sahoo, AK, Umlong IM, Devi RR, Thakur AJ, **Vijay Veer**. 2014. CuO Nanorods: a potential and efficient adsorbent in water purification" *RSC Advances*: 40580-40587.

49. Bora BJ, Saha UK, Chatterjee S & **Vijay Veer**. 2014. Effect of compression ratio on performance, combustion and emission characteristics of a dual fuel diesel engine run on raw biogas. *Energy Conversion and Management*, 87: 1000-9.

50. Bora A, Purkayastha J, Gogoi HK, & **Vijay Veer**. 2014. Studies on the effect of different nitrogen source on the photoautotrophic growth of a freshwater microalga *Chlorella ellipsoidea*. *South Asian Journal of Experimental Biology* 4 (1), 8-12

51. Sharma N, **Vijay Veer**, Shrivastava AK and Chauhan RS. 2014. Photo Induced Structural Changes in Insecticide Incorporated Mosquito Net Material. *International Journal of Theoretical & Applied Sciences* 6(2): 31-38

52. Gopalakrishnan R, Baruah I and **Vijay Veer**. 2014. Monitoring of malaria, Japanese encephalitis and filariasis vectors. *Med J Armed Forces India* 70: 129-133.

53. Das M, Gopalakrishnan R, Dharmendra Kumar, Gayan J, Baruah I, **Vijay Veer** & Dutta P. 2014. Spatiotemporal distribution of dengue vectors & identification of high risk zones in district Sonitpur, Assam, India. *Indian J Med Res,* 140: 278-284.

54. Khan MH, Meghvansi MK, Gupta R, **Vijay Veer**, Singh L. & Kalita MC. 2014. Foliar spray with vermiwash modifies the arbuscular mycorrhizal dependency and butirent stoichiometry of Bhut jolokia (*Capsicum assamicum*). *PloS ONE*, 9 (30): e92318 IF: 4.28

55. Chattopadhyay P, Goyary D, Mazumder PM & **Vijay Veer**. 2014. *In vitro* fibroblast growth stimulatory and in vivo wound healing activity of *Cleome viscosa*. *OrientalPharmacy and Experimental Medicine*, 1: 68-73.

56. Upadhyay A, Chattopadhyay P, Goyary D, Mazumder PM & **Vijay Veer**. 2014. *Ixora coccinea* down regulate the inflammation-associated prostaglandins (PGE-2), nitric oxide (NO) and cytokines production in LPS-stimulated RAW 264.7 macrophage. *Journal of Natural Remedies*, 14(1): 76-82.

57. Upadhyay A, Chattopadhyay P, Goyary D, Mazumder PM & **Vijay Veer**.2014. *Euphorbia hirta* accelerates fibroblast proliferation and Smad-mediated collagen production in rat excision wound. *Pharmacognosy Magazine*, 10(39) 534-542.

58. Upadhyay A, Chattopadhyay P, Goyary D, Mazumder PM & **Vijay Veer.** 2014. Anti- inflammatory Effect of *Euphorbia hirta* Leaf Extract in Rats and Modulation of Inflammation-Associated Prostaglandins (PGE-2) and Nitric Oxide (NO) Expression in RAW264.7 Macrophage. *J. Pharmaceutical Sciences and Pharmacology*,1(1): 68-73.

59. Dhiman S, Rabha B, Yadav K, Baruah I**,** & **Vijay Veer**. 2014. Insecticide susceptibility and dengue vector status of wild *Stegomyia albopicta* in a strategically important area of Assam, India. *Parasites & Vectors,* 7(1): 295.

60. Yadav K, Dhiman S, Rabha B, Saikia PK, & **Vijay Veer**. 2014. Socio-economic determinants for malaria transmission risk in an endemic primary health centre in Assam, India. *Infectious Diseases of Poverty*, 2014; 3:19.

61. 116. Yadav Kamlesh, Chandel K, Purkayastha J, **Vijay Veer**. 2014. 16s RNA-RFLP based midgut bacterial diversity of *Aedes (Stegomyia) aegypti* (Diptera: Culicidae) from Gwalior region, India. *International Journal of Biotechnology and Bio Sciences*, 4(1): 70 – 74.

62. Dhiman S, Rabha B, Goswami D, Das NG, Baruah I, Bhola RK, & **Vijay Veer**. 2014. Insecticide resistance and human blood meal preference of *Anopheles*

annularis in Asom- Meghalaya border area, northeast India. *Journal of vector borne diseases*, 51(2): 133-136.

63. Goswami D, Dhiman S, Rabha B, Dinesh Kumar, Baruah I, Sharma DK, **Vijay Veer**. 2014. Pfcrt mutant haplotypes may not correspond with chloroquine resistance. *J Infect Dev Ctries*, 8(6):768-773.

64. Dhiman S, Rabha B, Goswami D, Das NG, Baruah I, Bhola RK,& **Vijay Veer**. 2014. Insecticide resistance and human blood meal preference of *Anopheles annularis* in Asom- Meghalaya border, northeast India. *Journal of vector borne diseases*, 51: 133–136.

65. Banerjee S, Chattopadhyay P, Ghosh A, Bhattacharya SS, Kundu A, & **Vijay Veer**. 2014. Accelerated stability testing of a transdermal patch composed of eserine and pralidoxime chloride for prophylaxis against (±)-anatoxin A poisoning. *Journal of Food and Drug Analysis,* 22(2): 264-270.

66. Banerjee S, Chattopadhyay P, Ghosh A, Pathak MP, Gogoi J and **Vijay Veer**. 2014. Protection by a transdermal patch iontaining eserine and pralidoxime chloride for prophylaxis against (±)-Anatoxin A poisoning in rats. *European Journal of Pharmaceutical Sciences*, 56: 28-36.

67. Chattopadhyay P, Banerjee S, Pathak MP, Agnihotri A, Karmakar S, Goyary D, Dhiman D and **Vijay Veer**. 2014. Acute and tubchronic dermal toxicity of *Vitex negundo* essential oil. *Cutaneous & Occular Toxicology*, 33(1):16-21.

68. Chaturvedi R, Chattopadhyay P, Banerjee S, Bhattacharjee CR, Raul P, Borah K, Singh L, and **Vijay Veer**. 2014. Iron-rich drinking water and ascorbic acid supplementation improved hemolytic anemia in experimental Wistar rats. *Int J Food Sci Nutr*, Early Online: 1–6. DOI: 10.3109/09637486.2014.918589.

69. Chaturvedi R, Chattopadhyay P, Banerjee S, Bhattacharjee CR, Raul P, Borah K, Singh L, & **Vijay Veer.** 2014. Iron-rich drinking water and ascorbic acid supplementation improved hemolytic anemia in experimental Wistar rats

70. Upadhyay, A.,Chattopadhyay,P., Goyary D., Goyary, Mazumder PM and **Vijay Veer**. 2014 Topical Application of *Cleome viscosa* Increases the Expression of Basic Fibroblast Growth Factor and Type III Collagen in Rat Cutaneous Wound. *BioMed Research International*, vol. 2014, Article ID 680879, 7 pages, 2014. doi:10.1155/2014/680879.

71. Bora, A., Purkayastha,J., Gogoi, HK., **Vijay Veer**. 2014. Studies on the effect of different nitrogen source on the photoautotrophic growth of a freshwater microalga *Chlorella ellipsoidea. South Asian Journal of Experimental Biology*, 4(1); 8-12.

72. Rabha, AJ., Naglot, A., Sharma, GD., Gogoi HK & **Vijay Veer**. 2014. In vitro evaluation of antagonism of endophytic *Colletotrichum gloeosporioides* against potent fungal pathogens of *Camellia sinensis. Indian Journal of Microbiology, DOI 10.1007/s12088-014-04*

73. Sharma AK, Tyagi, V, Singh, S, **Vijay Veer**, Agrawal, OP, Sukumaran D. 2014 Distribution of *Anopheles culicifacies* and Detection of its sibling species E from Madhya Pradesh: Central India. *Iranian Journal of Arthropod-Borne Diseases*, 8(2):186-196

74. Banerjee S, Chattopadhyay P., Ghosh, A, Dutta, P, **Vijay Veer**. 2014. Aspect of Adhesives in Transdermal Drug Delivery Systems. *International Journal of Adhesion and Adhesives*, 50:70-84.

75. Goswami, D, Dhiman S, Bipul R, Dinesh Kumar, Baruah I, **Vijay Veer**, Bhola RK, Sharma DK. 2014. High Prevalence of *pfcrt* K76T and *mdr1* N86Y mutations in Sonitpur district of Assam, India. *J. Parasitic Dis*, DOI: 10.1007/s12639-013-0298-1

76. Upadhyay A, Chattopadhyay P, Goyary D, Mazumder PM and **Vijay Veer**. 2014. *Ixora coccinea* enhances cutaneous wound healing by up-regulating the expression of collagen and basic fibroblast growth factor. *ISRN Pharmacology*, 2014, Article ID 751824, 9 pages.

77. Raul PK, Devi RR, Umlong IM, Thakur AJ, Banerjee S, **Vijay Veer**. 2014. Iron Oxide hydroxide nanoflower assisted removal of arsenic from water. *Materials Research Bulletin*, 49: 360-368.

78. Gogoi J, Nakhuru KS, Chattopadhayay P Rai AK, Gogoi HK & **Vijay Veer.** 2013. *In: vivo* evaluation of cutaneous wound healing activity of *Mirabilis jalapa* L radix *Oriental Pharmacy and Experimental Medicine* DOI:10.1007/s13596-013-0122-698. Tyagi Varun, Kumar AK, Yadav R, Agrawal O.P., Sukumaran D., **Vijay Veer**. 2013. Characteristics of the larval breeding sites of *Anopheles culicifacies* sibling species in Madhya Pradesh, India. *International Journal of Malaria Research and Reviews*, Vol. 1(5): 47-53.

80. Aher V, Chattopadhyay P, Goyary D, & **Vijay Veer**.2013.Evaluation of the genotoxic and antigenotoxic potential of the alkaloid punarnavine from *Boerhaavia diffusa. Planta Medica* , 79(11): 939-945.

81. Gopalakrishnan R, Das M, Baruah I, Veer V, Dutta P . 2013. Physicochemical characteristics of habitats in relation to the density of container-breeding mosquitoes in Asom, India. *Journal of Vector Borne Diseases*, 50(3):215-219.

82. Dhiman, S, Rabha, B, Talukdar PK, Das NG, Yadav K, Baruah I, Singh, L, & **Vijay Veer**. 2013. DDT & deltamethrin resistance status of known Japanese encephalitis vectors in Assam, India. *Indian Journal of Medical Research,* 138:114-120.

83. Gopalakrishnan R, Chaurasia AK, Baruah I & **Vijay Veer**. 2013. Evaluation of permethrin impregnated military uniforms for contact toxicity against mosquitoes and persistence in repeated washings. *International Journal of Environmental Science and Technology* DOI 10.1007/s13762-013-0337-6.

84. Tyagi V, Yadav R, Sharma AK, Tyagi V, Yadav S, **Vijay Veer**, Sukumaran D. 2013. Larvicidal activity of leaf extract of some weeds against malaria vector *Anopheles stephensi. International Journal of Malaria Research and Reviews*, 1(3): 35-39.

85. Chattopadhyay P**,** Dhiman S**,** Devi KA**,** Banerjee S**,** Rabha B**,** Chaurasia AK and **Vijay Veer**. 2013**.** Ultra low concentration deltamethrin loaded patch development and evaluation of its repellency against dengue vector *Aedes (S) albopictus. Parasites & Vectors*, 6(1): 284-294.

86. Chandel, K, Mendki, MJ, Parikh RY, Kulkarni G, Tikar SN, Sukumaran D, Shri Prakash, Parashar BD, Shouche YS & **Vijay Veer**.2013. Midgut microbial community of *Culex quinquefasciatus* mosquito populations from India. *PLoS ONE*, 8(11), e80453.

87. Goswami D, Baruah, I., Dhiman, S, Rabha, B, **Vijay Veer**, Singh, L., Sharma DK. 2013. Chemotherapy and Drug Resistance Status of Malaria Parasite in Northeast India. *Asian Pacific Journal of Tropical Medicine*, 6(7): 583-588.

88. Chattopadhyay P, Islam MJ, Goyary D, Agnihotri A, Karmakar S, Banerjee S, Singh L, & **Vijay Veer.** 2013. Subchronic dermal exposure to T-2 toxin produces cardiac toxicity in experimental wistar rats. *Toxicology and Industrial Health*, 1-8, DOI: 10.1177/0748233713503373

89. Banerjee S, Singh S, Policegoudra R, Chattopadhyay P, Ghosh A, & **Vijay Veer**. 2013. Evaluation of the mutagenic potential of a combinational prophylactic transdermal patch by Ames test. *Immunol Biol. Spec.*, http://dx.doi.org/10.1016/j.immbio.2013.07.004.

90. Banerjee S, Chattopadhyay P, Ghosh A, Pathak M, Singh S, & **Vijay Veer.** 2013. Acute Dermal Irritation, Sensitization and Acute Toxicity Studies of a Transdermal Patch for Prophylaxis against (±) Ana toxin-A Poisoning. *International Journal of Toxicology*. 32(4): 308-313.

91. Banerjee S, Chattopadhyay P, Ghosh A, Kaity S, & **Vijay Veer**. 2013. Development and Validation of a Reverse Phase Liquid Chromatography Method for the Simultaneous Quantification of Eserine and Pralidoxime Chloride in Drugs-in-Adhesive Matrix Type Transdermal Patches. *Drug Res (Stuttg)*, 63: 450-456.

92. Dhiman, S, Goswami, D, Dinesh Kumar, Bipul Rabha, Sharma DK, Bhola RK, Baruah I, & **Vijay Veer**. 2013. Nested PCR detection of *Plasmodium malariae* from microscopy confirmed *P. falciparum* samples in endemic area of northeast India. *Folia Parasitologica.* 60 (5): 401–405.

93. Kalita, R, Das AK, Gogoi HK & **Vijay Veer**. 2013. Effect of growth hormones on in vitro mass propagation of orchid species *Rhyncostylis retusa. Journal of cell and Tissue Research*, 13 (2): 3711-3714.

94. Meghvansi MK, Khan HM, Gupta R & **Vijay Veer**. 2013. Identification of new species of *Cercospora* causing leaf spot disease in *Capsicum assamicum* in northeastern India. *Research in Microbiology*, 164: 894-902.

95. Gopalakrishnan R, Chaurasia AK, Baruah I. & **Vijay Veer.** 2013. Evaluation of permethrin- impregnated military uniforms for contact toxicity against mosquitoes and persistence in repeated washings. *International Journal of Environmental Science and Technology*, DOI 10.1007/s13762-013-0337-6.

96. Chattopadhyay, P, Goyary, D, Dhiman,S, Rabha, B, Hazarika S, **Vijay Veer**. 2013. Immunomodulating Effects and hypersensitivity reactions caused by Northeast Indian black fly salivary gland extracts. *Journal of Immunutoxicology*.early online 1-7.

97. **Vijay Veer**, Sharma AK, Sachin S.N, Mendki, M.J, Tyagi, V, Chandel K, & Selvamurthy W. 2013. Molecular characterization of migratory locust, *Locusta migratoria* Linn. (Orthoptera: Acrididae: Oedipodinae) from Ladakh region, India. *International Journal of Veterinary Medicine: Research & Reports*, 2013, Article ID 942894, DOI: 10.5171/2013.942894.

98. Bipul, R, Dhiman S, Yadav K, Hazarika S, Bhola RK, **Vijay Veer**. 2013. Influence of water physico-chemical characteristics on Simuliidae (Diptera) prevalence in some streams of Meghalaya, India. *Journal of Vector Borne Diseases.* 50: 18-23.

99. Chattopadhyay P, Banerjee S, Pathak MP, Agnihothri A, Goyary D, Karmakar S, **Vijay Veer**. 2013. Acute and subchronic dermal toxicity of *Vitex negundo* essential oil. *Cutaneous and Ocular Toxicology*, 33 (1): 16-21.

100. Banerjee S, Chattopadhyay P, Ghosh A, **Vijay Veer.** 2013. Development and Validation of a Reverse Phase Liquid Chromatography Method for the Simultaneous Quantification of Eserine and Pralidoxime Chloride in Drugs-in-Adhesive Matrix Type Transdermal Patches. *Arzneimittelforschung*/Drug Research, 63(9):450-456

101. Chattopadhyay P, Upadhyay A, Agnihotri A, Karmakar S, Ghowary D, & **Vijay Veer.** 2013. Comparative Hematoxicity of Fusirium Mycotoxin in Experimental Sprague Dawley Rats. *Toxicology International*, 20(1):25-29

102. Upadhyay A, Chattopadhyay P, Goyary D, Papiya Mitra Mazumder,& **Vijay Veer**. 2013. *Eleutherine indica* L. accelerates *in vivo* cutaneous wound healing by stimulating S mad- mediated collagen production. *Journal of Ethanopharmacology*, 146 (2): 490-494.

103. Banerjee S, Chattopadhyay P, Ghosh A, Goyary D, Karmakar S, **Vijay Veer**. 2013. Influence of process variables on essential oil microcapsule properties by carbohydrate polymer–protein blends. *Carbohydrate Polymers*, 93(2): 691-697.

104. Datta SN, Agarwal MM, Chatterjee S, Gogoi HK, **Vijay Veer**, & L Singh.2013. Detection of human papillomavirus in women attending Pap cervical screening camp at a peripheral hospital of Q1 North-Eastern India. Medical Journal Armed Forces India (Elsevier) DOI: 10.1016/j.mjafi.2013.04.014; 2013

2012

105. Policegoudra RS, Saikia S, Das J, Chattopadhyay P, Singh L, & **Veer Vijay**. 2012. Phenolic Content, Antioxidant Activity, Antibacterial Activity and Phytochemical Composition of *Garcinia lancifolia. Indian J Pharm Sci.,* 74(3):268-71.

106. Dhiman S, Hazarika, S, Bipul B, Das, NG, Bhola, RK, **Vijay Veer**, & Singh, L.2012. Species composition, daytime biting activity and possible Onchocerciasis infection among *Simulium* in north-east India. *Journal of Environmental Biology*, 01/2014; 35

107. Rajendran C, Kumar D, Baruah I, Gogoi HK, Srivastava RB & **Vijay Veer**. 2012. In vivo antimalarial evaluation of certain medicinal plants against *Plasmodium berghei*, a mice infecting malaria parasite. *J. Parasitic Diseases*. DOl:10.I007/s12639-013-0252-2.

108. Gopalakrishnan, R, Das, M, Baruah, I, **Vijay Veer** & Dutta, P. 2012. Studies on the ovitraps baited with hay and leaf infusions for the surveillance of dengue vector, *Aedes albopictus* in northeastern India. *Tropical Biomedicine*, 29(4): 598-604

109. Dhiman, S., Rabha, B., Chattopadhyay, P, Das, N.G, Hazarika, S, Bhola, R.K, & **Vijay Veer**, Singh, L. 2012. Field evaluation of repellency of a polyherbal essential oil against black flies and its dermal toxicity using model. *Tropical Biomedicine*, 29(3): 391-397.

110. Purkayastha J, Gogoi H.K, Singh, L, Sahoo L, & **Vijay Veer.** 2012. High frequency β-glucuronidase reporter gene expression in shoot tip, leaf and hypocotyl explants of *Jatropha curcas* L. *Journal of Crop Improvement, 26 (5): 711 – 725.*

111. Chattopadhyay P, Pandey A, Goyary D, Chaurasia A, Singh L, **Vijay Veer**. 2012. Technetium-99m-labeled deoxynivalenol from *Fusarium mycotoxin* alters organ toxicity in BALB/c mice by oral and intravenous route .*Journal of Venomous Animals and Toxins including Tropical Diseases*, 2012, 18(3): 258-263.

112. Chattopadhyay P, Upadhyay A, Karmakar S, Goyary S, Goyary D, Policegourda R, Singh L. & **Vijay Veer** 2012. Naga chilli amelioreates ethanol induced liver toxicity in partially hepatectomised Wistar rats. *Asian Pacific J. Trop. Biomed.* 2012: 1-6.

113. Purkayastha, J, Alam SI, Gogoi, H.K., Singh, L & **Vijay Veer.** 2012. Molecular characterization of 'Bhut Jolokia' the hottest chili. *Journal of Biosciences, 37 (4): 757-768.*

114. Gogoi, BJ,Tsering J, Hui Tag, **Vijay Veer**. 2012. Antioxidant potential of *Garcinia* species from Sonitpur district, Assam, North east India. *Int'l J. of Pharmaceutical Sciences and Research*, 3(9): 1000-1003.

115. Gogoi, BJ. Tsering J & **Vijay Veer**. 2012. Antioxidant potential and total phenolic content of *Leucas* of Sonitpur district, Assam. *Int'l. J Res. Pharm. Sci.*, 3(3): 376-378.

116. Datta SN, Chatterjee S, **Vijay Veer**, & Chakravarty R. 2012. Molecular biology of the Hepatitis B virus for clinicians. *Journal of Clinical & Experimental Hepatology*. 2 (4) : 353-365.

2011

117. Tikar SN, Mendki MJ, Sharma AK, Sukumaran D, **Vijay Veer**, Shri Prakash & Parashar BD. 2011. Resistance status of malaria vectors, *Anopheles stephensi* and *Anopheles subpictus* from arid and semi-arid zone of India towards adulticides and larvicides. *Journal of Insect Science,* 11: 85.

118. Mendki MJ, Sharma AK, **Vijay Veer**, Agarwal OP, Shri Prakash & Parashar BD. 2011. Population Genetic structure of *Culex quinquefasciatus* Say (Diptera: Culicidae) the major filariasis vector in India assessed by ISSR marker. *Asian Pacific Journal of Tropical Medicine*. 412-420. *Asian Pacific Journal of Tropical Medicine*, 05/2011; 4(5):357-62. DOI: 10.1016/S1995-7645(11)60103-X

119. Bhatnagar, A. Kaundal P, Kaushal N, Garg ID & **Vijay Veer.** 2011**.** *Thrips palmi* (Thysanoptera : Thripidae) a vector of groundnut bud necrosis virus (GBNV) of early potato crop (*Solanum tuberosum* Linn.) in central India. *Annals of Entomology*, 29 (1): 15-22.

2010

120. **Vijay Veer.** 2010**.** A new species of *Franklinothrips* (Thysanoptera: Terebrantia: Aeolothripidae) from Uttarakhand, India. *Annals of Entomology* 28(2):7–9.

121. Sharma AK, Mendki MJ, Tikar S. N., Kulkarni, G, **Vijay Veer**, Shri Prakash, Shouche YS, and Parashar BD. 2010. Analysis of 16S rRNA mitochondrial gene variations in *Culex quinquefasciatus* mosquito from Different Geographical Regions of India. *Acta Tropica*, 116: 89-94.

2009.

122. Kämpfer P, Chandel K, Prasad GBK.S, Shouche YS, & **Vijay Veer**. 2009. *Chryseobacterium culicis* sp. nov., isolated from the midgut of *Culex quinquefasciatus* Say, a mosquito from India. Intl. *Journal of systematic & Evolutionary Microbiology*, (2010) 60:2387-2391.

123. Sharma, A.K., Mendki, M.J., Tikar, S.N., Chandel, K., Sukumaran, D., Parashar, B.D., & **Vijay Veer**, Agarwal, O.P., & Shri Prakash, 2009. Genetic variability in geographical populations of *Culex quinquefasciatus* Say (Diptera: Culicidae) from India based on Random Amplified Polymorphic DNA analysis. *Acta Tropica,* 112(1): 71-76.

2007

124. Vasudeva, P., Hegde, S. & **Vijay Veer**. 2007. Tabanid flies (Diptera: Tabanidae) of Rajiv Gandhi National Park, Karnataka, India. *J. Ent. Res. Soc.*, 9(1): 53-62.

2006

125. Parashar, B.D., **Vijay Veer**, Shri Prakash & Vinod Kumar.2006. Prevalence of blood sucking flies, vector of Trypanosomiasis in the Nandankanan Zoological Park, Bhubaneswar (Orissa) and their control by integrated pest management. *Indian Zoo Year Book*, Volume, 6:112-125.

2004

126. **Vijay Veer**, Chauhan, Neena & Sudhir Singh. 2004. A new species of *Thaumaglossa* (Coleoptera: Dermestidae) from India. *Oriental Insects*, 38: 389-394.

2003

127. Parashar, B. D., Ganesan, K., Sukumaran, D., Rao, Y.V.S., **Vijay Veer** & Shri Prakash. 2003. Aggregating activity induced by excreta extract in *Cimex hemipterus*. *Entomon*, 28(3): 215-222.
128. Shri Prakash, Kumar S, **Vijay Veer**, Gopalan N, Purnanad, Pandey KS, &. Rao, KM. 2003. Laboratory evaluation of four rodenticide admixed cereal based baits against commennsal rat, *Rattus rattus* Linn. (Rodentia: Muridae: Murinae). *J. Stored Products. Res.*, 39: 141-147.

2002

129. **Vijay Veer**, Gopalan, N., Kumar, S. & Shri Prakash. 2002. Bioassay of some sulphur containing compounds as rat attractant admixed in cereal bait against commennsal rat, *Rattus rattus* Linn. *Indian Journal of Experimental Biology*, 40: 941-944.
130. **Vijay Veer**, B. D. Parashar & Shri Prakash. 2002. Tabanid and muscoid haematophagous flies, vectors of trypanosomiasis or surra disease in wild animals and livestock in Nandankanan Biological Park, Bhubaneswar (Orissa, India). *Current Science*, 82 (5): 500- 503.

2001

131. **Vijay Veer**, Gopalan, N. & Rao, K.M. 2001. Lambda-cyhalothrin sprays as an insect proofing agent for woollen fabrics. *Indian Journal of Textile & Fibre Research*. 26: 308-312.

2000

132. **Vijay Veer** & Rao, K.M. 2000. Descriptions of adult and larval morphology of *Evorinea indica*(Arrow) (Coleoptera: Dermestidae). *Oriental Insects*, 34: 293-300.

1999

133. Bhatti, J.S., **Vijay Veer** & Dahiya, Anu. 1999. The North American species *Kurtomathrips morrilli* Moulton and *Neohydatothrips gracilipes* (Hood) (Terebrantia: Thripidae) have now established large populations from North to South India. *Thrips*, 1: 54-57.

134. **Vijay Veer**, Chauhan, R. S. & Rao, K.M. 1999. Studies on larval feeding of some new insect pests on wool -synthetic blend fabrics in India. *Indian Journal of Textile & Fibre Research*, 24: 139-141.

135. **Vijay Veer**, Parashar, B.D. & Rao, K.M. 1999. Notes on Tabanidae (Diptera) that are surra vectors or pestiferous with description of a new species of *Tabanus* from India. *Oriental Insects*, 33: 247- 266.

1998

136. Bhatti, J.S., **Vijay Veer** & Negi, B.K. 1998. Discovery of the natural habitat of the aquatic Thysanopteran, *Organothrips indicus* (Terebrantia: Thripidae) in India and North America. *Oriental Insects*, (USA) 32: 259-266.

137. Shri Prakash, Kumar, S., **Vijay Veer** & Rao, K.M. 1998. A novel tabletized bait for *Periplaneta americana* (Dictyoptera: Blattidae), *Blattella germanica,* and *Supella longipalpa* (Dictyoptera: Blattellidae). *Zoology (J. of Pure & Applied Zoology)*, 5(2): 133-140, 1998.

1996

138. **Vijay Veer** & Rao, K.M. 1996. Dermestid beetles and some other insect pests associated with stored silkworm cocoons in India including a world list of dermestid species found associated this commodity. *J. Stored Products Res.*, 32(1): 69-89.

1995

139. Kumar, S., Shri Prakash, **Vijay Veer**, Gopalan, N. & Rao, K. M.1995.Laboratory evaluation of a rodenticide difethialone based bait against *Rattus rattus.* Indian J. Toxicology.2:808, 199.

140. **Vijay Veer** & Rao KM. 1995. Annotated checklist of Indian Dermestidae (Coleoptera), *Zoology (J. of Pure & Applied Zoology)*, 5 (1): 1-12.

141. **Vijay Veer** & Rao KM. 1995. Taxonomic and biological notes on three *Attagenus* spp. (Coleoptera: Dermestidae) not previously recorded as pests of stored woollen fabrics in India. *J. Stored Products Res.*, 31(3): 211-219.

142. **Vijay Veer** & Rao KM. 1995. Effectiveness of Azadirachtin against wool insect pests: 387- 389. In: *Pesticides, Crop Protection & Environment*, Oxford & IBH, New Delhi, 1995.

1994

143. **Vijay Veer** & Rao KM. 1994. A new species of *Trogoderma* (Coleoptera: Dermestidae) found damaging stored silkworm cocoons in India. *J. Stored Products Res.*, 30(4): 283-295.

1993

144. **Vijay Veer** & Rao KM. 1993. Deltamethrin sprays as a long term protectant of wool fabrics from insectpests in store. *Indian Journal of Fibre & Textile Research*, 18:156-157.

145. **Vijay Veer**, Prasad R, & Rao, KM. 1993. Evaluation of permethrin in industrial application on wool against *Tinea translucens* and *Anthrenus flavipes. Indian Journal of Fibre & Textile Research*, 18: 25-29.

146. Bhatti JS, **Vijay Veer** & Chauhan N. 1993. *Doonthrips setor,* a new genus and species of Thripidae (Insecta: Terebrantia) from Dehradun (India). In: *Advances in Thysanopterology* (ed. J.S. Bhatti). *Zoology*, 4: 177-192.

1992

147. Prasad R, **Vijay Veer** & Rao KM. 1992.Evaluation of potential insect repellents as insect proofing agents for wool fabrics. *Indian Journal of Textile Research*, 17:56-57.

148. **Vijay Veer**, Gupta, BK. & Sen-Sarma, PK. 1992.Ovicidal action of three synthetic pyrethroids against poplar defoliator, *Pygaerea fulgurita* Walker (Lepidoptera: Notodontidae). *J. Entomol. Res.*, 16(1): 54-56.

149. Chauhan, N & **Vijay Veer**.1992. *Anaphothrips doonensis* new species from the tussock of plant, *Typha elephantiana* (Family- Typhaeceae) from Dehradun (Thysanoptera: Terebrantia: Thripidae). *Annals of Entomology* 10(1): 41-44, 1992.

150. Chauhan, N & **Vijay Veer**. 1992. A new species of *Pseudodenrothrips* Schmutz (Insecta: Terebrantia: Thripidae) from Dehradun (India). *Zoology*, 3(2): 169-172, 1992.

1991

151. **Vijay Veer**, Prasad R & Rao, KM. 1991. Deltamethrin sprays as an insect repellent of woolen fabrics. *Textile Research Journal*, (USA) 61: 655-659.

152. **Vijay Veer**, Prasad R. & Rao, KM. 1991.Taxonomic and biological notes on *Attagenus* and *Anthrenus* spp. (Coleoptera: Dermestidae) found damaging stored woollen fabrics in India. *Journal of Stored Products Research*, 27: 185-198.

1990

153. **Vijay Veer** & Chauhan N. 1990. New species of genera *Azaleothrips* Ananthakrishnan and *Gigantothrips* Zimmermann (Thysanoptera: Tubulifera: Phlaeothripidae) from India. *Annals of Entomology*, 8: 65-70, 1990.

154. Kulshrestha SK. & **Vijay Veer**. 1990. A new species of *Liothrips* Uzel from Dehradun (India) (Insecta: Thysanoptera: Phlaeothripidae). *Annals of Entomology*, 8(1): 55-60, 1990.

1989

155. **Vijay Veer**, Prasad R, & Rao KM. 1989. Studies on feeding of wool-synthetic blend fabrics by larvae of insect pests. *Indian Journal Textile Research*, 14: 169-171.

156. Gupta BK, Ahamad SI. & **Vijay Veer.** 1989. Relative toxicity of some conventional insecticides against adult beetles of *Calopepla leayana* Latr. (Chrysomelidae: Coleoptera). *Indian Forester*, 115(6): 430-434.1989.

1986

157. Gupta BK. & **Vijay Veer**. 1986. Laboratory bioassay of some insecticides as contact poison against third instar larvae of *Glyphodes pyloalis* Walker (Lepidoptera: Pyralidae). *Indian Forester*, 112(6): 528-533.

1985

158. **Vijay Veer**. 1985. Observations on the host preferences and biology of *Thrips flavus* Schrank (Thysanoptera: Thripidae) from Dehradun, India. *Annals of Entomology*, 3(1): 39-48, 1985..

159. Mishra SC, **Vijay Veer** & Chandra, A. 1985. *Aristobia horridula* Hope (Coleoptera: Lamiidae) a new pest of shisham (*Dalbergia sissoo* Roxb.) in west Bengal. *Indian Forester*, 111 (9): 738- 741.

160. **Vijay Veer** & Srivastava, AK. 1985. Two new species of Thripidae (Insecta: Thysanoptera) from India. *Entomon*, 10(1): 75-80.1985.

1984

161. **Vijay Veer** & Chandra A.1984. Insect pests in nurseries and plantations of *Populus deltoids* in Assam and West Bengal. *Indian Forester*, 110 (7): 640-643, 1984.

162. Kulshrestha SK & **Vijay Veer**. 1984. Two new species of Thysanoptera (Insecta) from Dehradun. *Bulletin of Entomology* 25 (1): 33-37, 1984.

163. **Vijay Veer**. 1984. On some new records of predators of *Thrips flavus* Schrank and *Thrips hawaiiensis* (Morgan) (Thysanoptera: Thripidae) from Dehradun (India). *Indian J. of Forestry*, 7(3): 245-246, 1984.

164. Kulshrestha SK, **Vijay Veer** & Srivastava AK. 1984. Observations on the variations in the population of *Haplothrips ganglbaueri* Schmutz (Thysanoptera: Insecta) on *Oryza sativa* L. and *Triticum vulgare* from Doon valley with a note on its phenology. *Indian Journal of Forestry*, 7(2): 124 -30, 1984.

1983

165. **Vijay Veer** & Sen-Sarma PK. 1983. Descriptions of the third instar larvae of some species of genus *Anomola* Samouelle (Coleoptera: Scarabeidae: Rutelinae) from India. *J. Soil Biol. Ecol.*, 3(2): 124-135.

166. Kulshrestha, SK. & **Vijay Veer**. 1983. Systematics of undescribed males of some Thysanoptera from India. *Journal of Bombay natural History Society*, 80 (1): 240-243,

1983.

167. Srivastava AK. & **Vijay Veer**. 1983. Preliminary observations on the life-history of *Lefroyothrips lefroyi* (Bagnall) (Thysanoptera: Thripidae). *Indian Journal of Forestry*, 6(1): 84-85.1983.

1982

168. Gupta BK, **Vijay Veer** & Sen-Sarma PK. 1982. Laboratory evaluation of some insecticides against third instar larvae of *Selpa celtis* Moore (Noctuidae: Lepidoptera). *Indian Forester*, 108(7): 481- 486. 1982.

169. Sen-Sarma PK, Gupta BK. & **Vijay Veer.** 1982. Relative toxicity of some insecticides against third instar larvae of *Dirades adjutaria* Walker (Epiplemidae: Lepidoptera). *Indian J. Forestry*, 5 (1): 42- 46. 1982

170. **Vijay Veer** & Chauhan N. 1982. A study on the genus *Arathrips* Bhatti (Thysanoptera: Thripidae) from Dehradun (India). *Entomon*, 8(2): 131-135.1983.

171. Kulshrestha SK, Srivastava AK. & **Vijay Veer**. 1982. Some observations on the relative abundance and sex-ratio of thrips species (Thysanoptera: Insecta) infesting *Oryza sativa*, *Saccharum officinarum* and *Triticum vulgare* in Doon valley. *Indian Journal of Forestry*, 5(3): 211-215, 1982.

1981

172. **Vijay Veer** & Srivastava AK. 1981.*Ajothrips gara* Bhatti (Thysanoptera) damaging leaves of *Mangifera indica* (mango) in Dehra Dun. *Indian Journal of Forestry*, 4(1): 66, 1981.

1980

173. Tyagi BK. & **Vijay Veer**. 1980. More records of entomogenous fungi attacking preserved dragonfly collections. *J. Bombay Nat. Hist. Soc.*, 77(10): 177-178, 1980.

1979

174. Tyagi BK. & **Vijay Veer.** 1979. A note on some entomogenous fungi attacking preserved dragonfly collections. *J. Bombay Nat. Hist. Soc.*, 75(1): 946-947.

175. **Vijay Veer**. 1979. A new record of internal parasite of *Gynaikothrips flaviantennatus* Moulton from India. *Current Science*, 48(12): 558.

1978

176. **Vijay Veer**. 1978. *Thrips flavus* Schrank as pollinating agent. *Indian Journal of Forestry*, 1 (4): 278.

II. Patents (Process & design) filed/granted & commercialized

1. A method of Defluoridation of Water from Fluoride Contaminated Water using Natural Waste Materials, and to bring down The Fluoride Level below the Permissible Limit Fit for Drinking Purpose. M, Lohborlang, Bodhaditya Das , Vijay Veer, Kochi Ismail, Uttam Thapa, Teiborlangmukhim, Shakira Sultana Islam.
2. An improved process for preparation of a composition of n, n diethyl phenyl acetamide (DEPA) and herbal essential oil for mosquito control. Vijay Veer, Reji Gopalakrishan, Prasant Raul
3. A mall odorant and irritant formulation for personal protective and mob control and an improved process for preparation of the same. Gogoi, BJ, Chattopadhyay, P, Chaurasia, AK, Vijay Veer (3540/DEL/2014).
4. Mosquitocidal trapping Device. Gopalakrishanan R, Vijay Veer & Baruah I (Design patent 264649).
5. Oligonucleotide primer sequences for cercosporin toxin biosynthesis (CTB) genes and method of detection of cercosporin species. Meghavansi MK, Khan MH, Gupta R, & Vijay Veer (2991/DEL/2014).
6. A biotoilet system for complete digestion of human fecal matter with purification of effluent water for reuse. Chatterjee S, Vairale MG, Prakash A, Gogoi HK, Vijay Veer, Singh L. (No. 1654/DEL/2014).
7. Preparation of herbal based wound healing formulation and evaluation thereof Chattopadhyay P, Upadhyay A, Goyary, D., Vijay Veer (No. 660/DEL/2014).
8. Photobioreactor for growth of microalgae. Purkayastha J, Gogoi HK, Bora A, Singh L, Vijay Veer (2559/DEL/2013)
9. Novel Oligonucleotide Primers for Cercosporin Toxin Biosynthesis CTB Genes and a Related Method for the Detection of *Cercospora* Meghvansi MK, Khan MH, Gupta R, Vijay Veer
10. Delivery of eserine and pralidoxime chloride through matrix type transdermal path formulations for prolongs release and a process for the preparation thereof. Chattopadhyay P, Banerjee S, Ghosh A, & Vijay Veer (3033/DEL/2013)
11. Herbal composition for snake repellent spray and a process for the preparation thereof. Chattopadhyay P., Banerjee S., Agnihotri, A, Karmakar S, Goyary D, Rudragoud P, Vairale, MG, Chaurasia Vijay Veer (No.3217/DEL/2013).
12. A Spine Column system and a method of solid phase extraction of target compounds involving the same. Gupta, VK, Naglot, A. Chattopadhyay, P. Chaur-

asia, AK, Shrimali, DD, Kalita MC, Gogoi, H.K., Vijay Veer (No.2560/DEL/ 2013).

13. Synthesis and development of polymeric nano composite for removal of bacteria from contaminated water and process of removal thereof). Devi RR, Raul PK, Policegourda R, Banerjee S, Vijay Veer ((No.1108/DEL/2013).
14. Insect repellent nanogel formulation for clothing impregnation Patanjali PK, Raza SK, Vijay Veer, Shweta, Kala S, Gopalakrishanan R, Baruah I. (No.2053/DEL/ 2013).
15. An improved thermoplastic for development of multi –insect repellent polymer yarn, net, fabric and tarpaulin and development process thereof. Vijay Veer, Selvamurthy W. (No. 1949/DEL/2013).
16. Autocidal ovitrap for monitoring dengue and chikungunya vectors.. Gopalakrishnan R, Vijay Veer, Baruah I, Das M. ((No 1016/DEL/2013).
17. Insect repellent spray formulation for treating cotton fabrics. Gopalakrishnan R, Vijay Veer, Baruah I, Chaurasia AK, Kumar D, (No. 512/DEL/2013).
18. Natural mosquito repellent lotion formulation and process for the preparation thereof. NG Das NG, Talukdar PK, Rabha B, Baruah I & Vijay Veer. (No. 1085/DEL/2013).
19. Bio-Tank: A subsurface process for the human waste disposal and treatment system. Singh L, Chatterjee S, Vairale M G, Kamboj DV, Pandey A, Gogoi HK, Vijay Veer. (Aplication No. 2249/DEL/2012).
20. Bio-toilet. Singh L, Chatterjee S, Vijay Veer, Kamboj DV. Design patent No. 247200.
21. Wool care composition. United States Patent No.US 8,524,770 B2, Date of Patent: Sep. 3, 2001. Vijay Veer, Chandel K, Mendki MG, Shri Prakash, Vijayaraghavan R, Sekhar K (PCT). (Granted. In UK, France, The Netherlands, USA).
22. Attractant composition for bed bugs. Parashar BD, Ganesan K, Rao YVS, Sukumaran D, Mendki MJ, Suryanarayana MVS, Vijay Veer, Shri Paraksh, Vijayaraghavan R, Sekhar K. (545/DEL/2009).
23. An Improved formulation N, N diethyl phenyl acetamide (DEPA) - a multi insect repellent for protection of humans from flying, biting and blood sucking insects. Mendki MJ, Tikar SN, Gautam A, Pravin Kumar, Parasahar BD, Vijay Veer, Shri Prakash, Vijayaraghavan R, Sekhar K. (1676/DEL/2010).
24. An improved process for preparation of insecticide spray for protection of woollen items from insect pests. Vijay Veer, Shri Prakash & Rao KM. (Grant No. 189729).
25. Process for preparation of rodenticide improved bait. Rao KM, Shri Prakash, Vijay Veer, Purnanand, Kumar S & Gopalan N. (Grant No. 189767).
26. A process for preparation of an improved rodenticide bait Rao KM, Shri Prakash, Vijay Veer, Purnanand, Kumar S, Gopalan N, Malhotra RC, Pandey KS. (Grant No. 196997).
27. A process for preparation of bio-degradable controlled release insecticide matrix composition for protection of woollen fabrics. Chauhan RS, Vijay Veer, Ramana KV, Gopalan N & Rao KM. (Grant No. 190832).

III. Trademarks Registration (7 nos)

DrlPure™ (Grant No.2574253)
Defender Net™, (2391126)
Capsispray™, 2676308
Capsigrenade™ (2831116)
Wormipower™ (2826675)
Toxmos™ (2826677)
Dentrap™ (2826676)

IV. BIS Standards/ specifications prepared (2 nos)

1. Method of Test for Evaluation of insectproofness of woollen textiles. First Revision IS: 7250. 1995. The Bureau of Indian Standards, New Delhi,
2. Textiles- Estimation of common preservatives -part -2, IS: 3522, 1989, amendment no.1, 1992.The Bureau of Indian Standards, New Delhi.

V. Books:

1. Chatterjee S, Meghvansi MK, and Vijay Veer (eds). Special Issue on Recent Advances in Biodegradation, Sanitation, and Bioremediation. *South Asian Journal of Experimental Biology*, 5, No 6, 304 pp.
2. Vijay Veer & Reji Gopalakrishnan. 2016. Herbal insecticides, repellents and biomedicines: effectiveness & commercialization. Springer, India, Private Ltd., 270 pages, ISBN-10: 8132227026, ISBN-13: 978-813222702
3. Tyagi BK, Vijay Veer *et al.* 2015.Biosafety for human health and the environment in the context of the potential use of genetically modified mosquitoes (GMMs), WHO/TDR, 242 pp
4. Meghvansi, MK & Vijay Veer. 2014. Vermibiotechnology for Solid waste management. Published by Defence Research Laboratory, Tezpur, 109 pp
5. Vijay Veer & Purkayastha, J (editors). 2012. *Dr W Selvamurthy – a journey dedicated to the Armed Forces.* Xiii+114 pp. Published by DESIDOC, Delhi.
6. Tyagi, B.K. & Vijay Veer (editors). 2011. *Entomology: Ecology and Biodiversity.* Scientific Publishers (India), Jodhpur, 523 pp.
7. Sarkar, P.K. & Vijay Veer. 2010. *The Field Guide on Malariology.* Bishen Singh Mahendra Pal Singh, Dehra Dun. Vii+249 pp.
8. Tyagi, B.K., Vijay Veer, & Shri Prakash, (edited). 2007. *Pests of forest importance and their management in India.* Scientific Publication, Jodhpur. xiv+299 pp.

VI. Technical reports:

1. **Vijay Veer**. 2011. Invasion of migratory locust in Ladakh. Technical report submitted to DRDO/ MoD, 34 pp. **Vijay Veer** (Principal Investigator).
2. **Vijay Veer**. 2004. Biodefence and Biowarfare agents in Indian Perspective- An analysis, 40 pp, 2004. Report submitted to Scientific Adviser to Defence Minister.
3. Sen-Sarma PK, Sujan Singh & **Vijay Veer**. 1982. Strengthening of plant protection in India and South-east Asian countries. FAO Publication, 1982.

VII. Technical Bulletins:

1. Sarma MK. & **Vijay Veer**. 2013. Medicinal mushrooms of North-east India. *Technical Bulletin*, No. DRL/TB/2013/1. 16 pp.
2. Sarma MK. & **Vijay Veer.** 2013. Package of practices for Mushroom spawn production. *Technical Bulletin*, No. DRL/TB/2013/2. 27pp.

VIII. Monographs

1. Meghvansi MK, Khan HM & **Vijay Veer**. 2013. Biology of *Cercospora* leaf spot disease. Printed by Createspace, USA. 158 pp.

2. **Vijay Veer**. 2004. Tabanidae flies (Diptera) from the Indian subregion. *Annals of Forestry*, 12(2): 301- 447. **A monograph on Horse flies- vector of animal trypanosomiais.**

IX. Bio-bibliographies

1. Chatterjee, S, Rudragoud P, Megahvansi MK, Sharma S, & **Vijay Veer**. 2014. Dr Dilip R Ranade: A scientist of global significance in the field of microbiology**.** Sanicon-2014-Recent Advances in Biodegradation of Human Wastes at held DRL, Tezpur on 16-17 December, 2014., 20 pp
2. Purkayastha, J, Sharma, J, Chatterjee S & **Vijay Veer**. 2014. Bio-bibliography of Dr Lokendra Singh: A renowned Defence Life Science Scientist. Sanicon-2014-Recent Advances in Biodegradation of Human Wastes at held DRL, Tezpur on 16-17 December, 2014, 48 pp

3. Tyagi BK, **Vijay Veer** & Chatterjee S. 2013. Bio-bibliography of Prof (Dr) Devinder Singh: A forensic Entomologist & Academic statesman of national significance. 7th Medical Arthropodology held at DRL, Tezpur, 12-13 Dec., 24 p

4. **Vijay Veer**, Tyagi BK, Anil Prakash & Chatterjee S. 2013. Bio-bibliography of Dr J. Mahanta: a biomedical scientist and statesman of global significance in Medico –entomological sciences. 7th Medical Arthropodology held at DRL, Tezpur, 12-13 Dec., 2013. 42 p

5. **Vijay Veer**. 2012. Bio-bibliography of Dr B K. Tyagi: Arthropodologist and polymath. CRME Bio-bibliography Series No. 8; Year 2012. 6th Conference of Medical Arthropodology held at Madurai; Oct. 18-19, 2012)
6. Tyagi BK & **Vijay Veer**. 2012. Bio-bibliography of Prof (Dr) Jagbir Singh: Entomologist and conservationist. CRME Bio-bibliography Ser. No. 7, 24 pp.
7. Tyagi BK & **Vijay Veer**, Rajendran R, Arunachalam,N. & Paramasivan R. 2007. Professor Dr T.N. Ananthakrishnan- Entomologist and polymath: Bio-bibliography. Ist conf. on Medical Arthropodology- Taxonomy in biodiversity, bioterrorism and biotechnology, Sept., 1-2, 2007, CRME, Madurai., 67 pp.
8. **Vijay Veer** & Tyagi BK. 2007. Prof. P. K. Sen-Sarma Ph.D., FNASC. FASC, FESI, FIAWAS, FNAAS - A Termitologist and Forest Entomologist: Biography and Contributions (1929 - 2002). Commemorative Volume Prof. P.K. Sen-Sarma (edited by Tyagi & Vijay Veer).

X. Book Chapters

1. Khan HK, Meghvansi MK, Gupta R, Chaudhary KK, Prasad K, Siddiqui S, Vijay Veer, Varma A. 2015. Combining Application of Vermiwash and Arbuscular Mycorrhizal Fungi for Effective Plant Disease Suppression: 479-493. In: *Organic*

Amendments and Soil Suppressiveness in Plant Disease Management. Springer International Publishing.

2. Meghvansi MK, Khan MH, Gupta R, Chaudhary KK, Siddiqui S, and Vijay Veer. 2015. Vermibiotechnology: Relevance, Challenges and Future Prospects for India: 222-228. In: Recent Advances in Biodegradation, Sanitation, and Bioremediation (Chatterjee S, Meghvansi MK, and Vijay Veer (eds). *South Asian Journal of Experimental Biology*, 5, No 6, 304 pp.
3. Chatterjee S, Sharma S, Prasad RK, Datta S, Dubey D, Meghvansi MK, Vairale MG and Vijay Veer. 2015. Cellulase Enzyme based Biodegradation of Cellulosic Materials: An Overview. 271-282. In: Recent Advances in Biodegradation, Sanitation, and Bioremediation (Chatterjee S, Meghvansi MK, and Vijay Veer (eds). *South Asian Journal of Experimental Biology*, 5, No 6, 304 pp.
4. Sharma S, Chatterjee S, Datta S, Prasad RK, Sharma A, Dubey D, Vairale MG and Vijay Veer. 2015. Isolation and Characterization of Cellulose Degrading Bacteria of Termite Gut from North Eastern Region of India: 283-290. In: Recent Advances in Biodegradation, Sanitation, and Bioremediation (Chatterjee S, Meghvansi MK, and Vijay Veer (eds). *South Asian Journal of Experimental Biology*, 5, No 6, 304 pp.
5. Barua S, Gopalakrishnan R , and Vijay Veer. 2016. .Chapter 6. Extraction, Purification and Characterisation of Insecticidal Compounds from Plants: 101-116. In: *Herbal Insecticides, Repellents and Biomedicines: Effectiveness and Commercialization.* Springer, xv+258 pp.
6. Tsering J, Tag H, Gogoi BJ and Vijay Veer. 2016. Chapter 10. Traditional Anti-poison Plants Used by the Monpa Tribe of Arunachal Pradesh. . 189- 204. In: *Herbal Insecticides, Repellents and Biomedicines: Effectiveness and Commercialization.* Springer, xv+258 pp.
7. Bora A, Gogoi HK, and Vijay Veer. 2016. Chapter 11. Molecular Farming for Production of Biopharmaceuticals and Edible Vaccines in Plants. 205-216. In: *Herbal Insecticides, Repellents and Biomedicines: Effectiveness and Commercialization.* Springer, xv+258 pp.
8. Chattopadhyay P and **Vijay Veer.** 2016. Chapter 13. Regulatory Approval of Botanical Products Including Herbal Drugs, Food, and Insecticides for Commercialization: 243-258. In: *Herbal Insecticides, Repellents and Biomedicines: Effectiveness and Commercialization.* Springer, xv+258 pp.
9. Mitra, A. Chatterjee S, Datta SN, Sharma S, **Vijay Veer**, Razafindrabe BHN, Walther C, Gupta DK. 2014. Mechanism of metal transporters in plants.1-28. Book: *Heavy Metal Remediation: Transport and Accumulation in Plants*; (Editors: DK Gupta, S Chatterjee) Nova Science Publishers, Inc. USA.
10. Sharma S, Chatterjee S, Datta SN, Mitra A, Vairale MG, **Vijay Veer,** Chourasia A, Gupta, DK. 2014. In vitro selection of plants for the removal of toxic metals from contaminated soil: Role of genetic variation in phytoremediation. 155-178. Book: *Heavy Metal Remediation: Transport and Accumulation in Plants*; (Editors: DK Gupta, S Chatterjee); Nova Science Publishers, Inc., USA.
11. Chatterjee S, Mitra A, Datta SN & Vijay Veer. 2014. Chapter 1. Phytoremediation Protocols: An Overview. 1-18. In: *Plant-based remediation processes.* (Ed. DK Gupta). xii, 299 p. 62 illus., 14 in colour. Springer.
12. Chatterjee S, Datta SN, Mallick PH, Mitra A, Vijay Veer & Mukhopadhyay SK. 2013. Chapter 7. Use of Wetland Plants in Bioaccumulation of Heavy Metals. 117-158. In: *Plant-based remediation processes.* (Ed. DK Gupta). xii, 299 p. 62 illus., 14 in colour. Springer.

13. 6. Purkayastha, J, H.K. Gogoi, Singh L & Vijay Veer. 2012. Biofuel from Microalgae: Myth vs. Reality. *In: Microbes as Energy Source*, CABI, Oxfordshire, USA. Pp. 240-249.

14. Vijay Veer. 2011. Biodiversity of Dermestid Beetles (Coleoptera: Dermestidae) from India, with Identification Key to Adults of Indian Genera and Description of A New subspecies: 495-522. In: *Entomology: Ecology & Biodiversity* (Edited By Tyagi, B.K. & Vijay Veer), Scientific Publishers (India), Xvii+ 522 Pp.

15. Shri Prakash, **Vijay Veer**, Parashar, BD, Mendki, MJ, Tikar, SN. & Vijayaraghavan R. 2008. Vectors of diseases and their management: 169-184. In: Life Sciences Perspective in service of the soldier. (Eds. Mandal, M.K., Ramachandran, K & Gupta, A), Macmillan, India Ltd. New Delhi.

16. **Vijay Veer**, Mendki, MJ, Chauhan N. & Shri Prakash. 2007. Potential of pheromone and kairomone (semiochemicals) based eco-friendly insect control technologies for the suppression of forest insect pests in India. Chapter 6: 121-134. In: Pests of forest importance and their management (Edited Tyagi, B.K, Vijay Veer & Shri Prakash). Scientific Publishers (India), Jodhpur, xiv + 299 pp.

17. Chauhan, N. & **Vijay Veer**. 2007. Notes on diagnosis and bio ecology of thrips pests of floral, fruit, vegetable, agronomic crops and forestry plants in the Dehra Dun valley (Uttaranchal), India: Chapter 10: 195-228. In: Pests of forest importance and their management (Edited Tyagi, B.K, Vijay Veer & Shri Prakash). Scientific Publishers (India), Jodhpur, xiv + 299 pp.

18. Shri Prakash, **Vijay Veer**, Mendki, MJ., Gupta, GP. & Verma, K C. 2005. Some new records of mosquitoes (Culicidae) and haematophagous flies of Tabanidae and Muscidae from the Thar Desert (Rajasthan, India):135-142. Chapter 7. In: *Changing Faunal Ecology in Thar Desert* (Ed. B.K.Tyagi & Q.H., Baqri).

XI. Conference Proceedings:

1. BJ Bora, UK Saha, Chatterjee S, & **Vijay Veer**. 2014. Effect of load level on performance and emission characteristics of a biogas run dual fuel diesel engine. 2014. 5th Intl conf Fluid Mechanics & Fluid Power, 12-14 Dec., 14/IITG Springer-Verlag Berlin Heidelberg, 201, 8 pp.

2. **Vijay Veer** & Shri Prakash, 2008. Tabanid flies (Diptera: Tabanidae) of India and their management with a note on disease vectors. Chapter 35. 280-298. In: *Vector Borne diseases: Epidemiology and control.* CRME, Madurai publication.

3. 6. **Vijay Veer**, Prasad R, & Rao KM.1991.Studies on insect proofing of woollen fabrics with Eulan WA New and Permethrin: 244-253. In: *Proc. Symp. Entomology for Defence Services.* (ed. Ramachandran *et al.*).

4. Kulshrestha SK, **Vijay Veer** & Srivastava AK. 1983. Some observations on the host preferences and population structure of *Amblythrips lefroyi* (Bagnall) (Thysanoptera: Thripidae) on *Camellia thea* Linn. From Dehra Dun: 195-202. In: *Insect interrelations in Forest and Agroecosystem* (ed. Sen-Sarma, P.K. Kulshrestha, S.K. & Sangal, S.K.), 1983.

5. **Vijay Veer**, Srivastava AK. & Kulshrestha SK. 1983. Observations on seasonal fluctuations in population of *Thrips flavus* Schrank (Thysanoptera: Thripidae) on *Tegetes erecta* L. from Dehra Dun: 189-194, ibid, 1983.

6. Srivastava AK, **Vijay Veer** & Kulshrestha SK. 1983. Observations on the population ecology of *Sorghothrips jonnaphilus* (Ramakrishna) (Thysanoptera: Thripidae) on *Saccharum officinarum* from Doon valley: 203-208, ibid, 1983.

7. Sen-Sarma PK. & **Vijay Veer** .1985. Vertebrate pests of forest trees in India: 224-230. In: *Non-insect pests and predators*. All India Scientific Writers' Society of India, New Delhi, 279 pp.

---

4. Dr. R.K. Jauhari

'tis not in mortals to command success,
'll do more, deserve it."
- Samuel Smiles

Dr. Rakesh Kumar Jauhari (b. 28th December, 1955), Head of the Department of Zoology, D.A.V. (PG) College, Dehradun (Uttarakhand), is a self-made man with admirable human qualities. A highly celebrated scientist-culicidologist (expert on the dipteran family of mosquitoes), a much extolled academician and an astute administrator, Dr. Jauhari is a role model to the younger generation of budding scientists for his unsatiable panache to dig out wealth of insect research-subjects from seemingly abysmal resources just as the great Nobel Laureates Prof. C.V. Raman and Dr. J.C. Bose encountered throughout much of their meandering early research life!

He graduated (1973) and postgratduated (1975) from Agra University and completed Ph.D. (1979) from Banaras Hindu University in Zoology/ Parasitology. He won a couple of fellowships like JRF/UGC; SRF/CSIR and PDF/CSIR between July 1975 and January 1980 to support his doctoral research work on protozoan and helminth parasites of fishes of eastern Uttar Pradesh at the Parasitology Laboratory, Deptt. of Zoology, Banaras Hindu University, Varanasi. He served briefly as a lecturer at the Deptt. of Zoology, Janta (P.G.) College, Bakewar (Etawah) from Sept. 1982 to Dec 1985 and continued his research interests in parasitologist. On 21st Sept. 1982 he was selected to the post of Lecturer to his *Karmaschetra,* the Departmetment of Zoology, DAV (PG) College, Dehradun and intensified his lifelong interests in teaching and research mainly in the field of (i) parasites of hill stream fishes and Physiological, ecological and bio-chemical aspects of helminth parasites of vertebrate hosts, (ii) biological control of malaria vectors in peri-urban areas of Dehra Dun Valley, (iii) ecology and bionomics of plant Nematodes under tea plantation and Tomato crop, and (iv) mosquito systematics and identification with special reference to "Mosquito fauna of Doon Valley". He is currenly collaborating with (a) ISRO-IIRS, Dehra Dun on a RESPOND project, and (b) Malaco-faunal diversity of Doon Valley under U-COST Project. In course of time of his long and chequered career he continued to evolve both professionally and scientifically; (i) as Senior Lecturer from Jan. 1986 to Dec. 1993, (ii) as Reader from Jan. 1994 Dec. 2005, (iii) as Associate Professor from Jan. 2006 to present, and finally (iv) as Head of the Zoology Dept. from 16th July 2009 onward to date.

Dr. Jauhari has by now more than 36 years' experience of Undergraduate and Postgraduate teaching, besides, of course, research spanning over three decades with guiding writing of dissertations for M.Sc. Students and PhD Scholars on various aspects of human parasites and their vectors.

He has guided 28 candidates for doctoral research of which 25 have been awarded PhD degree. These scholars were greatly benefitted by his extensive resourcefulness as he was continuously winning funds for his research projects from major national organizations, e.g., the ICMR (2 major research projects), CSIR (3), UGC (2), RESPOND-ISRO (2), DST (1), and COST (2).

Dr. Jauhari is a recipient of many a laurel, such as the SERC visiting fellowship of DST (1994-95) availed at VCRC Pondicherry; UGC visiting fellowship (2001-2003), and SERC visiting fellowship of DST (2003-04). Famous for his agility and promptness in settling scientific and administrative issues, he also associated as (i) Joint Secretary, Indian Soc. for Parasitology of Varanasi region (1978-81), (ii) Asstt. Editor of Indian Journal of Physical and Natural Sciences (1983-86), (iii) Asstt. Co-ordinator, IGNOU study Centre (2705) since Jan. 2010, looking after M.P. Programme/Assignments. He has a great event organization skill and has so far executed many national/ international conferences prominent of which include, (a) Third National Congress of Parasitology held at Banaras Hindu University, Varanasi, 1979 and acted as the member of organising committee, (b) Fourth All India Seminar on Ichthyology at D.A.V. (P.G.) College, Dehradun 1983 and acted as Jt. Organising Secretary, (c) Seminar on Odonatology at D.A.V. (P.G.) College, Dehradun in 1985 and acted as member of organising committee, (d) Workshop on Biostatistics at D.A.V. (P.G.) College, Dehradun in 1988 and acted as member of organising committee, (e) Tenth National Congress of Parasitology at Dept. of Zoology, D.A.V. (P.G.) College, Dehradun, March 1990 and acted as Organising Secretary, (f) National Convention on Environmental Management for vector borne disease control at Dept. of Zoology in Collaboration with Chemistry Dept., D.A.V. (P.G.) College, Dehradun in 1995 and acted as Organising Secretary, (g) National Seminar on Perspectives in Environmental Heterogeneity in the Management of Parasitic Diseases at D.A.V. (P.G.) College, Dehradun in 1995 and acted as Convener, (h) National Convention Water Quality and Bio-Resources Environment at D.A.V. (P.G.) College, Dehradun in 1997 and acted as Convener, (i) National Seminar on Current Perspectives in Environmental Health Concerns at D.A.V. (P.G.) College, Dehradun in 2005 and acted as Convener, (j) National Seminar on Recent advances in Vector Biology & Control at D.A.V. (P.G.) College, Dehradun in 2008 and acted as Convener, and (k) National Seminar on Recent advances in Parasitological research at D.A.V. (P.G.) College, Dehradun in 2011 and acted as Chairman.

A prolific writer, Dr. Jauhari has authored a book on "Basic and Comparative Endocrinology", besides a manual on "Application of Remote Sensing in Vector-borne diseases", both in preparation. He has to his credit about 140 research titles generated from a great range of research disciplines on the Doon Valley's insects particularly Culicidae.

Scientific Publications of Dr. R.K. Jauhari

1. Agarwal, G.P. and Jauhari, R.K. (1977): *Faustula varanasiensis* n.sp. from a clupeid fish. *Indian J. Parasitol.*, 1(1): 67-68.
2. Agarwal, G.P. and Jauhari, R.K. (1977): On a new monogenetic trematode, *Bifurcohaptor vishwanathai* n.sp. from a fresh water fish, *Bagarius bagarius* (Ham.). *J. Zool. Soc. India*, 29(1&2): 73-79
3. Jauhari, R.K. and Agarwal, G.P. (1978): On a new species of the genus *Lobotrema* Tripathi, 1959 (*Monogenea: Dactylogyridae*) from a fresh water fish, *Sciaena coitre* (Ham.) at Varanasi, India. *Indian J. Helminth.*, 30(2): 90-95.
4. Agarwal, G.P. and Jauhari, R.K. (1979): On a new species, *Eucreadium thapari* from the intestine of fresh water fish, *Chela bacaila* (Ham.) at Gorakhpur, India. *Riv. di Parassit.*, 40(3): 323-326.
5. Jauhari, R.K. and Agarwal, G.P. (1980): *Oudhia hardayali* n.sp. from the intestine of a fresh water fish, *Mystus vittatus. Proc. Indian Acad. Parasitol.*, 1(2): 41-44.
6. Jauhari, R.K. and Agarwal, G.P. (1981): On a new monogenetic trematode, *Bychowskyella tripathii* n.sp. from the gills of a fresh water fish, *Wallago attu* (Bl. & Schn.). *Jap. J. Parasitol.*, 30(1): 1-8.
7. Agarwal, G.P. and Jauhari, R.K. (1981): *Gangatrema chauhani* n.g., n.sp. (Trematoda: Opecoelidae, Plagioporinae) from the intestine of a fresh water eel, *Mastacembelus armatus. Jap. J. Parasitol.*, 30(1): 9-13.
8. Jauhari, R.K. and Agarwal, G.P. (1981): On a new monogenetic trematode, *Mazocreas chauhani* n.sp. from the gills filaments of a clupeid fish, *Gudusia chapra. Riv. di Parassit.*, 42(3): 467-473.
9. Jauhari, R.K. and Agarwal, G.P. (1982): On a new *Bifurcohaptor* species (Monogenea) from *Bagarius bagarius* (Pisces) in India. *Angew. Parasitol.*, 23: 135-137.
10. Srivastava, J.K., Saxena, R.M. and Jauhari, R.K. (1983): Investigations on the trematode fauna of Doon Valley. Part 1. Parasites of fishes (A): *Eucreadium pandeyi* n.sp. from a fresh water eel, *Mastacembelus armatus* (Lac.) *Indian J. Phy. Nat. Sciences*, 3(A):33-37.
11. Srivastava, J.K. and Jauhari, R.K. (1983): On a new monogenetic trematode, *Lobotrema rajendrai* n.sp. from the gills of a fresh water fish, *Garra gotyla* (Grey). *Indian J. Forestry*, 6(4): 330-331.
12. Agarwal, G.P. and Jauhari, R.K. (1983): On *Pleurogenoides anabasi* n.sp. (*Trematoda: Lecithodendriidae*) from the intestine of a fresh water fish, *Anabas testudinaeus* (Bl.). *Riv. di Parassit.*, 44(1): 61-64.
13. Jauhari, R.K. and Agarwal, G.P. (1983): On a new monogenetic trematode, *Yogendrotrema rajghatai* n.g., n.sp. from the gills of a fresh water fish, *Rhinomugil corsula* (Ham.). *Riv. di Parassit.*, 44(2): 187-192.
14. Agarwal, G.P. and Jauhari, R.K. (1983): On a new digenetic trematode *Pseudoparamacroderoides raychaudhurii* n.sp. from the intestine of a fresh water fish, *Mystus vittatus* (Bl.) at Varanasi, India. *Riv. di Parassit.*, 44(3): 313-316.
15. Srivastava, J.K., Saxena, R.M. and Jauhari, R.K. (1984): Phosphate activity in a digenetic trematode, *Gastrothylax indicus* Dutt, 1979. *Indian J. Forestry*, 7(2): 168-173.

16. Saxena, R.M., Srivastava, J.K. and Jauhari, R.K. (1984): Histopathological studies on the intestine of *Xenentodon cancila* (Ham.) infected with *Bucephalopsis belonea*. *Int. J. Acad. Ichthyol.*, 5(1&2): 191-193.

17. Jauhari, R.K. and Agarwal, G.P. (1984): On two new species of the genus *Faustula* Poche, 1926 (Trematoda: Fellodistomidae) from the intestine of a clupeid fish, *Hilsa ilisha* (Ham.). *Indian J. Helminth.*, 36(1): 45-50.

18. Jauhari, R.K. and Agarwal, G.P. (1984): On two new species of the genus *Bucephalus* (Trematoda: Bucephalidae) from the intestine of *Bagarius bagarius* (Ham.). *Riv. di Parassit.*, 46(2): 361-365.

19. Jauhari, R.K. and Sharma, V. (1986): Seasonal population fluctuations of *Hoplolaimus indicus* (Nematoda: Hoplolaimidae) associated with tomato crops of Doon Valley. *Indian J. Phy. Nat. Sciences*, 6(A): 27-31.

20. Sharma, V., Jauhari, R.K. and Agarwal, G.P. (1986): Seasonal variations in the population of *Xiphinema* sp. (Nematoda: Longidoridae) in tomato crops of Doon Valley. *Indian J. Forestry*, 9(1): 82-84.

21. Srivastava, J.K., Saxena, R.M. and Jauhari, R.K. (1986): Nuclear extrusion in the oocytes of *Eucreadium pandeyi*. *Riv. di Parassit.*, 47: 283-287.

22. Agarwal, G.P. and Jauhari, R.K. (1986): On a new digenetic trematode *Neopodocotyle gorakhpurensis* n.sp. from the intestine of a fresh water eel, *Amphipnous cuchia* (Ham.). *J. Sci. Res., B.H. U.*, 36(1&2): 145-149.

23. Sharma, V. and Jauhari, R.K. (1986): The effect of soil temperature on the population of *Hoplolaimus indicus* (Nematoda: Hoplolaimidae) under tomato crops in Doon Valley. *Riv. di Parassit.*, 47: 215-217.

24. Jauhari, R.K. and Sharma, V. (1987): Studies on the minimum threshold and tolerance limit of tomato crops to *Hoplolaimus indicus* (Nematoda: Hoplolaimidae). *J. Sci. Res. B.H.U.*, 37(1): 65-68.

25. Jauhari, R.K., Grover, S.P. and Agarwal, G.P. (1987): Studies on the effect of *Orientocreadium batrachoides* (Digenea: Allocreadiidae) on the pH of infected organs in *Clarias batrachus* (Linn.). *Indian J. Phy. Nat. Sciences*, 7(A): 25-27.

26. Agarwal, G.P. and Jauhari, R.K. (1987): On a new digenetic trematode *Opisthorchis dayali* n.sp. from the intestine of a siluroid fish *Rita rita* (Ham.). *Riv. di Parassit.*, 48: 85-88.

27. Jauhari, R.K. (1988): Phosphate activity in *Ommatobrephus lobatum* (Trematoda: Digenea). *Angew Parasitol.*, 29: 169-171.

28. Jauhari, R.K. and Sharma, V. (1988): The influence of soil types at 25 degree centigrade on the population of *Hoplolaimus indicus* (Nematoda: Hoplolaimidae) on tomato crops in Doon Valley. *Indian J. Forestry*, 11(3):214-216.

29. Jauhari, R.K. (1988): Studies on the effect of *Gangatrema dehradoonense* (Digenea: Opecoelidae) on the pH of small. *Angew Parasitol.*, 29: 89-90.

30. Sharma, V. and Jauhari, R.K. (1989): Field observations on the control of a lance nematode, *Hoplolaimus indicus* (Nematoda: Hoplolaimudae) by rotation of crops. *J. Sci. Res., B.H.U*, 39: 5-10.

31. Agawal, G.P. and Jauhari, R.K. (1989): On a new monogenetic trematode *Diplozoon dasashwamedhai* n.sp. from a fresh water fish, *Barilius bola* (Ham.). *Fodedar Comm.* Vol., (*Indian J. Helminth* special issue), 1989: 148-151.

32. Srivastava, N. and Jauhari, R.K. (1990): Bio-chemical variations in the tissues of infected and uninfected helminthic vector snail, *Lymnaea auricularia*. *Angew. Parasitol.*, 31: 155-157.

33. Jauhari, R.K. (1990): Lipid composition of *Gastrothylax indicus* (Trematoda: Digenea). *Angew. Parasitol.*, 31: 151-153.

34. Jauhari, R.K. and Srivastava, N. (1990): Predatory action of a fresh water fish, *Channa striatus* towards vector snail *Lymnaea acuminata. Uttar Pradesh J. Zool.*, 10(1): 91-93.

35. Jauhari, R.K. and Srivastava, N. (1990): Effect of *Clinostomum* (Trematoda) metacercariae on the oxygen composition of *Colisa fasciatus* (BL.& Schn.). *Indian J. Parasitol.*, 14(2): 135-136.

36. Agarwal, G.P. and Jauhari, R.K. (1990): On a new monogenetic trematode, *Paramazocreas kasiensis* n.sp. from the gills of a fresh water fish, *Setipinna phasa. Indian J. Parasitol.*, 14(1): 67-70.

37. Srivastava, N. and Jauhari, R.K. (1990): Influence of certain abiotic factors on the occurrence of larval trematodes in *Lymnaea acuminata* (Lamarck) in Doon Valley. *Riv. di Parassit.*, 51(3): 335-341.

38. Jauhari, R.K. and Agarwal, G.P. (1990): Studies on the effect of *Oudhia horai* Gupta, 1955 (Digenea: Opecoelidae) on the pH of small intestine in *Heteropneustes fossilis* . *Riv. di Parassit.*, 51(3): 342-345.

39. Srivastava, N. and Jauhari, R.K. (1990): Population fluctuation of fresh water snail, *Thiara (Melanoides) tuberculata* in Doon Valley in relation to certain abiotic factors. *J. Zool. Soc. India*, 41(1&2):63-67.

40. Tripathi, L.K. and Jauhari, R.K. (1990): On a new digenetic trematode, *Gangatrema hanumanthai* n. sp. from the intestine of a fresh water fish, *Mystus seenghala. Riv. di Parassit.*, 52(1):89-92.

41. Tripathi, L.K. and Jauhari, R.K. (1991): *Dactylostomus gayaprashadi* n.sp. (Trematoda: Digenea) from the intestine of a fresh water fish, *Xenentodon cancila* (Ham.). *Uttar Pradesh J. Zool.*, 11(1): 70-72.

42. Sharma, V. and Jauhari, R.K. (1991): Suitability of cultivated crops as hosts for *Hoplolaimus indicus* (nematode: Hoplolaimidae). Ecosegregation in Parasitocoenosis, [Ed. S.K. Malhotra], 131-133.

43. Srivastava, N. and Jauhari, R.K. (1992): A preliminary report on the Anopheline fauna of Doon Valley. *Indian J. Parasitol.*, 16(1):45-46.

44. Jauhari, R.K., Srivastava, N., Singh, R.P. and Singh, S. (1992): A note on the collection of mosquitoes from different habitations in peri-urban areas of Doon Valley. *Bioved,* 3(2): 237-238.

45. Srivastava, N. and Jauhari, R.K. (1992): Seasonal density of *Anopheles culicifacies* and *An. subpictus* (Diptera: Culicidae) in relation to certain abiotic factors. *Uttar Pradesh J. Zool.*, 12(2): 123-127.

46. Singh, R.P., Singh, S. and Jauhari, R.K. (1994): Incrimination of *Anopheles stephensi* Liston, 1901 as a vector of Malaria in Southeastern Doon Valley. *J. Parasit. Appl. Anim. Biol.*, 3(2): 103-105.

47. Singh, S., Singh, R.P. and Jauhari, R.K. (1995): Biting activity of the malaria vector, *Anopheles culicifacies* on man and cattle in Doon Valley, India. *Appl. Parasitol.*, 36: 185-191.

48. Singh, S., Singh, R.P. and Jauhari, R.K. (1995): Studies on the age composition of *Anopheles culicifacies* Giles, 1901 (Diptera: Culicidae) in rural areas of Doon Valley. *J. Parasit. Appl. Anim. Biol.*, 4(1): 43-47.

49. Singh, R.P., Singh, S. and Jauhari, R.K. (1995): Natural parasitism of Anopheline larvae in rural areas of South eastern Doon valley. *Riv. di Parassit.*, XII (LVI) N-2: 233-236.

50. Singh, S., Singh, R.P. and Jauhari, R.K. (1995): Studies on the longevity of a malaria vector *Anopheles culicifacies* Giles, 1901 in Doon valley. *Riv. di Parassit.*, XII (LVI) N-2: 237-241.

51. Jauhari, R.K., Singh, R.P. and Singh, S. (1995): Breeding and resting of mosquitoes in the forested areas of Doon Valley. *Indian J. Forestry*, 18(3): 249-250.

52. Mahesh, R.K., Singh, R.P., Singh, S. and Jauhari, R.K. (1995): Breeding of mosquitoes in urban areas of city Dehra Dun. *Uttar Pradesh J. Zool.*, 15(2): 100-102.

53. Singh, R.P., Singh, S., Mahesh, R.K. and Jauhari, R.K. (1995): Predation by *Anisops bouveri* Kirkaldy (Hemiptera: Notonectidae) on immature forms of Anopheline mosquitoes in the laboratory. *Hexapoda*, 7(2): 63-70.

54. Jauhari, R.K., Saxena, R.M., Srivastava, J.K. and Mahesh, R.K. (1996): Larvivorous potential of some indigenous fishes of Doon Valley with special reference to their efficacy in control of Anophelines. *J. Parasitol. Appl. Anim. Biol.*, 5(1): 49-55.

55. Jauhari, R.K., Saxena, A.K. and Agarwal, N.K. (1996): Mosquito problem in irrigation systems - A case study of Doon Valley. *Recent Researches in Aquatic Environment* (Eds. A. Gautam and N.K. Agarwal): 119-123.

56. Pokhriyal, B.P., Mahesh, R.K., Singh, C.P. and Jauhari, R.K. (1997): Prevalence of liver fluke metacercariae in different larval stages of *Anopheles stephensi. Uttar Pradesh J. Zool.,* 17(2): 215-217.

57. Mahesh, R.K., Pokhriyal, B.P. and Jauhari, R.K. (1997): A report on the Anopheline mosquito fauna in the forested areas at different altitudes in Doon Valley. *Proc. Nat. Conv. Rec. Adv. Vector Borne Diseases* 1994 (Published in 1997) pp.115-118.

58. Singh, R.P., Singh, S., Mahesh, R.K., Pokhriyal, B.P. and Jauhari, R.K. (1997): A quantitative survey of Anophelines in rural areas of Doon Valley. *Proc. Nat. Conv. Rec. Adv. Vector Borne Diseases* 1994 (Published in 1997) pp.108-114.

59. Mahesh, R.K. and Jauhari, R.K. (1997): Observations on the survival and development of *Anopheles culicifacies* (Diptera: Culicidae) larvae in the rice fields of Dehra Dun valley. *Riv. di Parassit.*, XIV (LVIII) N-2: 257-262.

60. Singh, R.P., Mahesh, R.K. and Jauhari, R.K. (1997): Preferential resting habits of a recognized malaria vector *Anopheles stephensi (Diptera: Culicidae)* in Doon Valley. *Riv. di Parassit.*, XIV (LVIII) N-2: 263-270.

61. Pokhriyal, B.P., Mahesh, R.K. and Jauhari, R.K. (1997): Prevalence of trematode cercarial infection in the snail *Lymnaea (Pseudosuccinea) acuminata* Lamarck, 1822 in different localities of Dehra Dun Valley. *Bioved,* 8(1&2): 15-19.

62. Dua, V.K., Kumari, R., Jauhari, R.K., Ojha, V.P., Shukla, R.P. and Sharma, V.P. (1998): Organochlorine Insecticide residues in water from five lakes of Nainital (U.P.), India. *Bull. Environ. Contam. Toxicol.*, 60: 209-215.

63. Pokhriyal, B.P., Jauhari, R.K. and Sudarshana, R. (1998): Trematode cercarial infection in the snail *Thiara (Melanoides) tuberculata* (Muller, 1774) in different localities of Doon Valley, India. *J. Exp. Zool. India*, 1(2): 107-110.

64. Pokhriyal, B.P., Mahesh, R.K. and Jauhari, R.K. (1998): Occurrence of recognized helminthic vector snails in different habitats in Dehra Dun Valley. *Him. J. Env. Zool.*, 12: 167-172.

65. Pokhriyal, B.P., Mahesh, R.K. and Jauhari, R.K. (1998): Studies on the molluscan diversity in Asan river system in Doon valley with special reference to vectors of trematode parasites. *Indian J. Forestry*, 21(2): 167-170.

66. Lal, Madan and Jauhari, R.K. (1998): The influence of soil types on the population dynamics of migratory nematode *Pratylenchus penetrans* (Cobb, 1917) (*Nematoda: Hoplolaimidae*) on the tea plantations in Doon Valley. *Current Nematology*, 9(1-2): 1-6.

67. Khanna, K., Jauhari, R.K. and Sudarshana, R. (1999): Risk of emergence of epidemics vis-à-vis in Gulf of Cambay. *Proc. Subtle Issues in Coastal Manag. IIRS*, Dehra Dun, pp. 143-147.

68. Mahesh R.K. and Jauhari, R.K. (2000): Resurgence of sylvatic mosquitoes in Doon Valley. *J. Parasit. Appl. Anim. Biol.*, 9(1): 9-16.

69. Jauhari, R.K., Mahesh, R.K., Singh, C.P., Pemola Devi, N., Teotia, R. and Baqer, D. (2000): Morphological variations in the natural population of *Anopheles fluviatilis* James, 1902 (Diptera: Culicidae) collected from forests of Doon Valley. *The Annals Med. Entomol.*, 9(2): 25-31.

70. Mahesh, R.K. and Jauhari, R.K. (2000): Interspecific association and index of association among different aquatic forms of Anophelines collected from Sahaspur block of Doon Valley. *J. Parasitic Dis.*, 24(2): 147-150.

71. Pokhriyal, B.P., Mahesh, R.K. and Jauhari, R.K. (2001): Assessment of adaptation of liver fluke cercariae in the immature stages of *Anopheles fluviatilis* under different ecological conditions. *Uttar Pradesh J. Zool.*, 21(1): 35-39.

72. Jauhari, R.K., Mahesh, R.K., Singh, C.P., Teotia, R. and Pemola Devi, N. (2001): On certain morphological variations in *Anopheles annularis* Van der Wulp, 1884 and *An. subpictus* Grassi, 1889 from Tehri Garhwal region. *J. Exp. Zool. India*, 4(1): 175-180.

73. Pemola Devi, N., Mahesh, R.K. and Jauhari, R.K. (2001): On certain morphological variations in relation to seasonal prevalence of natural population of *Anopheles maculatus* Theobald (1901) collected from Doon Valley. *Uttar Pradesh J. Zool.*, 21(3): 211-220.

74. Mahesh, R.K. and Jauhari, R.K. (2001): Incrimination of *Anopheles fluviatilis* James, 1902 as a vector of malaria in forested areas of Doon Valley. *Indian J. Forestry*, 24(1): 72-76.

75. Mahesh, R.K., Pemola Devi, N. and Jauhari, R.K. (2001): Morphological variations in the natural population of *Anopheles vagus* Donitz (1902) collected from two different localities of Manipur State. *J. Parasit. Appl. Anim. Biol.*, 10(1&2): 33-39.

76. Mahesh, R.K. and Jauhari, R.K. (2001): Dispersal of *Anopheles culicifacies* (Diptera: Culicidae) larvae by water currents in the rice fields of Doon Valley, UA. *Flora & Fauna*, 7(2): 95-97.

77. Jauhari, R.K. and Lal, M. (2001): Effect of soil moisture on the population of migratory nematode *Pratylenchus penetrans* (Cobb, 1917) (Nematoda: Hoplolaimidae) on tae plantation in Doon Valley. *J. Parasit. Appl. Anim. Biol.*, 10(1&2): 1-8.

78. Pemola Devi, N., Mahesh, R.K. and Jauhari, R.K. (2001): Morphological variations in the natural population of *Anopheles subpictus* Grassi, 1899 collected from a forest and an urban area of Doon Valley. *J. Adv. Zool.*, 22(2): 109-113.

79. Jauhari, R.K. and Lal, M. (2002): Effect of soil organisms on the population of migratory nematode, *Pratylenchus penetrans* (Cobb, 1917) (Nematoda: Hoplolaimidae) on tea plantations in Doon Valley. *J. Exp. Zool., India,* 5(1): 45-48.

80. Baqer, D., Jauhari, R.K., Mitra, D. and Sudarshana, R. (2002): Remote sensing as an epidemiological tool in forecasting and prevention of Malaria. *J. Exp. Zool., India*, 5(1): 1-8.

81. Mahesh, R.K. and Jauhari, R.K. (2002): Resting preferences of Anophelines in different parts of Doon Valley, Uttaranchal. *J. Natcon.*, 14(1): 7-14.

82. Baqer, D., Jauhari, R.K., Sudarshana, R. and Mitra, D. (2002): Integration of GIS and Remote sensing in predicting the abundance of malaria vector *Anopheles culicifacies* in Doon Valley, Uttaranchal. Resource & Env. monitoring, *IAPRS & SIS's "Orange Book"*, 34(7): 49-52.

83. Pemola Devi, N., Mahesh, R.K. and Jauhari, R.K. (2002): Distribution patterns of *Anopheles nigerrimus* Giles, 1900 (Diptera: Culicidae) in district Dehra Dun with a note on morphological variations. *Proceedings of the XVI National Congress of Parasitology* 31st Oct.-2nd. Nov. 2002, MJP Rohilkhand University, Bareilly (U.P) pp. 56-67.

84. Pemola Devi, N. and Jauhari, R.K. (2003): Morphological abnormalities among the specimens of *Anopheles annularis* Van der Wulp, 1984 from two hilly regions with diverse climatic conditions. *Proceedings of the National Symposium on Biochemical Sciences health and Environmental Aspects (BSHEA, 2003)* Oct. 2-4, 2003, Dayalbagh Agra (U.P) pp. 295-307.

85. Mahesh, R.K. and Jauhari, R.K. (2003): Seasonal abundance of vector Anophelines in Sahaspur area of Doon Valley, Uttaranchal. *Proceedings of the National Symposium on Biochemical Sciences health and Environmental Aspects (BSHEA, 2003)* Oct. 2-4, 2003, Dayalbagh Agra (U.P). pp. 308-312.

86. Mahesh, R.K. and Jauhari, R.K. (2003): Mosquito fauna of the forested areas of Doon Valley (U.P.) India. *Entomon,* 28(2): 185-190.

87. Pemola Devi, N., Mahesh, R.K. and Jauhari, R.K. (2003): Morphological variations in the natural population of *Anopheles vagus* Donitz (1902) collected from Doon Valley. *Transc. Zool. Soc. India*, 7(1): 11-16.

88. Pemola Devi, N. and Jauhari, R.K. (2004): Altitudinal distribution of mosquitoes in mountainous areas of Garhwal region - Part I. *J. Vect. Borne Dis.*, 41:17-26.

89. Pemola Devi, N. and Jauhari, R.K. (2004): Mosquito records from Garhwal region (Uttaranchal). *J. Expt. Zool. India,* 7 (2) : 237-244.

90. Pemola Devi, N. and Jauhari, R.K. (2004): Mark-Recapture studies for evidence of memorized site-fidelity in *Anopheles culicifacies* Giles, 1901 in Garhwal region. *Entomon,* 29(1): 31-38.

91. Mahesh, R.K. and Jauhari, R.K. (2004): Relationship of some climatological factors on the mosquitoes of Mohobawala and Malsi deer park (Dehradun : Uttaranchal). *Geobios,* 31: 228-230.

92. Pemola Devi, N. and Jauhari, R.K. (2004): A study of habitat biodiversity hotspots towards mosquito occurrence in some parts of Garhwal (Uttaranchal). New Horizon of Animal Sciences, Eds. B.N. Pandey *et al.*, *Publ. Zool. Soc. Ind., Bodh Gaya*, 2004, pp. 244-251.

93. Mahesh, R.K. and Jauhari, R.K. (2004): Seasonal abundance of vector Anophelines in Doiwala area of Doon Valley, Uttaranchal. *J. Parasit. Appl. Anim. Biol.*, 13(1&2): 65-70.

94. Pokhriyal, B.P. and Jauhari, R.K. (2004): Studies on the adaptation of trematodes cercariae in the immature stages of the tiger mosquito, *Aedes aegypti* (Diptera: Culicidae) under different ecological features. *Indian J. Helminth.*, 22: 15-22

95. Pemola Devi, N. and Jauhari, R.K. (2005) : Mosquito records from Manipur State. *J. Expt. Zool. India,* 8(1) : 61-68.

96. Pemola Devi, N. and Jauhari, R.K. (2005) : A study on habitat biodiversity of mosquito richness in certain parts of Garhwal (Uttaranchal). *South East Asian J. Trop. Med. & Pub. Hlth.,* 36(3): 616-622.

97. Sinha, M., Sharma, R.C. and Jauhari, R.K. (2005): Impact of environmental factors on the activity of pathogenic bacterial strains in mosquito control. *Biochem. Cell. Arch.*, 5(1): 73-79.

98. Sinha, M., Sharma, R.C. and Jauhari, R.K. (2005): Distribution and toxicity of larvicidal Bacilli (*Bacillus thuringiensis* and *B. sphaericus*) in water samples collected from different mosquito habitats of Garhwal Himalaya. *J. Parasit. Appl. Anim. Biol.*, 14(1&2): 183-188.

99. Pemola Devi, N. and Jauhari, R.K. (2005): Species diversity patterns among mosquitoes (*Diptera: Culicidae*) from certain parts in Garhwal Himalayas, India. *J. Appl. Biosciences,* 31(2): 105-115.

100. Jauhari, R.K. and Pemola Devi, N. (2005): Satellite derived information in delineating mosquitogenic conditions in Jaisalmer, Rajasthan. *Proceedings of the Appl. Remote Sensing & GIS in Environmental Planning & Management.* 14-15 Mar. 2005, Bhagalpur.

101. Jauhari, R.K. and Pemola Devi, N. (2006): A note on the occurrence of fleas (*Siphonaptera: Pulicidae*) associated with small mammals in western Doon Valley. *J. Expt. Zool. India*, 9(1): 229-232.

102. Pemola Devi, N. and Jauhari, R.K. (2006): A relationship between *Anopheles fluviatilis* and *An. stephensi* (*Diptera: Culicidae*) catches and the prevalence of malaria cases at Kalsi area in Dehra Dun district (Uttaranchal). *Indian J. Med. Res.*, 123:151-158.

103. Pemola Devi, N. and Jauhari, R.K. (2006): Climatic variables and malaria incidence in Dehradun, Uttaranchal. *J. Vector Borne Dis.,* 43: 21-28.

104. Mahesh, R.K. and Jauhari, R.K. (2006): Role of Aquatic vegetation in the breeding of Anophelines in Doon Valley (Uttarakhand) India. *J. Parasit. Appl. Anim. Bio.*, 15(1&2): 57-62.

105. Pemola Devi, N. and Jauhari, R.K. (2007) : Mosquito species associates along some parts of phytogeographic zones in Garhwal region. *J. Insect Sciences,* 7(32):10pp. (available online: insectscience.org / 7.33)

106. Lal, M. and Jauhari, R.K. (2007): Effect of soil temperature on the population of migratory Nematode *Pratylenchus penetrans* (Cobb, 1917) (Nematoda: Hoplolamidae) on tea plantations. *UP J. Zool.,* 27(2): 183-188.

107. Lal, M. and Jauhari, R.K. (2007): Effect of soil pH on the population of migratory nematode *Pratylenchus penetrans* (Cobb, 1917) (Nematoda: Hoplolamidae) on tea plantations in Doon Valley. *J. Exp. Zool. India,* 10(2): 469-471.

108. Baqer, D., Rizwan Ahmed and Jauhari, R.K. (2008). Use of Remote Sensing Technology in quantifying and mapping mosquitogenic conditions in the Gulf of Cambay, Gujarat. Journal of Experimental Zoology. 11:1-4.

109. N. Pemola Devi and Jauhari, R.K. (2008): Diversity and cercarial shedding of malaco fauna collected from water bodies of Ratnagiri district, Maharashtra. *Acta Tropica*, 108: 249-252.

110. N. Pemola Devi and Jauhari, R.K. (2008): Reappraisal on Anopheline mosquitoes of Garhwal region, Uttarakhand. *J. Vector Borne Dis.,* 45: 112-123.

111. N. Pemola Devi and Jauhari, R.K. (2009): Study on digestive tract contents of fish: Preliminary step for identification of indigenous species in mosquito larval control. *J. Applied & Natural Sci.,* 1(1):18-23.

112. N. Pemola Devi and Jauhari, R.K (2009): Survey application on haematophagous insects occurring in different zones of district Dehradun (Uttarakhand). *J. Entomol. Res.,* 33(2): 149-156.

113. Rai, R and Jauhari, R.K. (2009): New record of a Pulmonate snail, *Physa acuta* Draparnaud, 1805 (Gastropoda: Physidae) from Doon valley (Uttarakhand). *Annals of Forestry*, 17(2):293-297.

114. Upadhyay, J. and Jauhari, R.K. (2010): Gills infestation in a clupeid fish *Gudusia chapra* (Ham.) collected from river Khoh at Kotdwar (Pauri-Garhwal). *J. Expt. Zool. India*, 13(1): 229-231.

115. Saxena, R.M., Jauhari, R.K. and Mamta Pandey (2010): Seasonal demography of trematode parasites of fishes in relation to sex, size and physicochemical parameters of water. *J. Expt. Zool. India*, 13(2): 613-617.

116. Rai, R. and Jauhari, R.K. (2010): Malaco faunal diversity in water bodies of Doon valley. *Uttar Pradesh J. Zool.*, 30(1):101-106.

117. N. Pemola Devi, Jauhari, R.K. and Hasan, S.F. (2010): Water quality and larvivorous activity of a killifish, *Aplocheilus panchax* (Ham.) against *Anopheles annularis* larvae in fragments of Loktak lake (Manipur). *J. Expt. Zool. India*, 13(2): 509-512.

118. Upadhyay, J., N. Pemola Devi and Jauhari, R.K. (2010): Investigation on food composition and parasitic infestation of an omnivore fish *Clarius batrachus* Linnaeus, 1758 collected from river Yamuna at Saharanpur (U.P.) India. *J. of Advanced Zool.*, 31(1):55-61.

119. Pandey, M., Saxena, R.M. and Jauhari, R.K. (2010): Statistical analysis of trematode infection in relation to ecological factors of some fishes of Dehradun. *Ann. Forestry*, 18(2): 317-322.

120. N. Pemola Devi, and Jauhari, R.K.(2011): Food preference of *Aplocheilus panchax* (Cyprinidontiformes : Aplocheilidae) with special reference towards mosquito larvae. *Researcher*, 3(6):55-59.

121. Upadhyay, J., Jauhari, R.K and N. Pemola Devi (2011): Abundance of a Monogenoidea parasite *Lobotrema rajendrai* in different host species (*Garra gotyla* and *Tor putitora*) at Kotdwar (Uttarakhand). *Deccan Current Science*, India 5: 284-287.

122. Pemola Devi, N. and Jauhari, R.K. (2011): Surveillance of Culicine mosquitoes in different zones of district Dehradun (Uttarakhand). *Annals of Entomology*. 29(2) : 99-103.

123. Upadhyay, J., Jauhari, R.K and N. Pemola Devi (2012): Parasitic infestation in a Cyprinid fish *Labeo rohita* (Ham.) of river Song in Doon Valley (Uttarakhand). *J. Parasitic Diseases*, 36(1):56-60.

124. N. Pemola Devi and Jauhari, R.K. (2012): Records of potential dengue vectors mosquitoes in Doon Valley, Uttarakhand, India. *Flora and Fauna*, 18(2): 228-232.

125. N. Pemola Devi and Jauhari, R.K. (2013): Meterological variables and malaria cases based on 12 years data analysis in Dehradun (Uttarakhand), India. *European J. Exp. Biol.,* 3(1): 28-37.

126. N. Pemola Devi and Jauhari, R.K. (2013): Dipterous haematophagous insects occurring in Dehradun district and its adjoining areas in Uttarakhand, India. *J. Trop. BioMed.*, 30(4): 608-614.

127. N. Pemola Devi, Jauhari, R.K. and J.P. Bahuguna (2013): Mosquitoes and mosquito borne diseases in Dehradun district, Uttarakhand. *J. Env. & Bio-sciences*, 27(2): 177-182.

128. Jauhari, R.K. and N. Pemola Devi (2013): Occurrence of snail borne disease, Cercarial dermatitis (swimmer itch) in Doon Valley (Uttarakhand), India. *Iranian J. Pub. Hlth.,* 43(2): 162-167.

129. N. Pemola Devi, Jauhari, R.K. and Ritwik Mondal (2013): Ovitrap surveillance of *Aedes* mosquitoes (Diptera: Culicidae) in selected areas of Dehradun district, Uttarakhand, India. *Global J. Med. Res.*, XIII(V) Version I: 53-57.

130. Jauhari, R.K. and N. Pemola Devi (2014): Satellite derived information in delineating malacologic conditions responsible for cercarial dermatitis in Ratnagiri, Maharashtra. A Compendium of Research Papers, SAC, Ahmedabad, pp.157-163.

131. Ritwik Mondal, N. Pemola Devi and Jauhari, R.K. (2014): Occurrence of *Aedes* mosquitoes (Diptera: Culicidae) in urban areas of Doon Valley (Uttarakhand). *Modern Parasitology,* 255-262, Eds. N. Gupta and D.K. Gupta, Publ., Narendra Pub. House, Delhi, India.

132. N. Pemola Devi, Ritwik Mondal and Jauhari, R.K. (2014): Physico-chemical assessment of natural breeding habitats of mosquito larvae in outskirts of Dehradun city, Uttarakhand. *J. Commun. Dis.,* 46(3): 29- 39.

133. Shweta Singh, N. Pemola Devi and Jauhari, R.K. (2015): Seasonal fluctuation on the Abundance of vector snails under the impact of certain meteorological factors in district Dehradun (Uttarakhand). *Deccan Current Science, India,* 11(1): 101 - 109.

134. Ritwik Mondal, N. Pemola Devi and Jauhari, R.K. (2015): Characterization of bacterial isolates found in breeding habitats of mosquitoes in Dehradun city, Uttarakhand. *J. Adv. Res. MicroBiol.* 2(1): 1-7.

135. Ritwik Mondal, N. Pemola Devi and Jauhari, R.K. (2015): Landmark-based geometric morphometric analysis of wing shape among certain species of *Aedes* mosquitoes in district Dehradun (Uttarakhand), India. *J Vector Borne Dis .,* 52: 122–128.

136. Ritwik Mondal, N. Pemola Devi and Jauhari, R.K. (2015): Bacterial characterization in natural breeding habitats of *Aedes* mosquitoes and their role on ovipositional response. *Inter. J. Mosq. Research*, 2(3):175-181.

137. N. Pemola Devi, Shweta, S. and Jauhari, R.K. (2015): Biotic and abiotic components associated with freshwater snails at Gularghati in Doon Valley, Uttarakhand. *Flora & Fauna*, 21(2): 219-224.

138. Sangeeta, M., Shweta S. and Jauhari, R.K. (2015): Morphometric analysis of freshwater snails along with associated planktonic community at Dudhlee in Doon Valley (Uttarakhand). *J. Global Bioscience* 4(6): 2604-2610.

139. Ritwik Mondal, N. Pemola Devi and Jauhari, R.K. (2015): Association between dengue incidence and meteorological parameters in Dehradun city, Uttarakhand, India. *Travel Medicine and Infectious Diseases* (Communicated).

140. Jayti Upadhyay, Jauhari, R.K. and N. Pemola Devi (2015): Zoo-geographical distribution of Monogenoidea parasites in fish community in water bodies of Garhwal – Uttar Pradesh belt. *J. Parasitic Diseases* (Communicated).

141. Ritwik Mondal, N. Pemola Devi and Jauhari, R.K. (2016): Studies on epidemiology of dengue from district Dehradun, Uttarakhand. *J. Commun. Dis.* (Communicated).

142. Ritwik Mondal, N. Pemola Devi and Jauhari, R.K. (2016): Mark-release recapture (MRR) study to determine the dispersal of *Aedes aegypti* and *Ae. albopictus* in Dehradun city, Uttarakhand, India. *Dengue Bulletin* (Communicated).

• • •